Decolonising Geography? Disciplinary Histories and the End of the British Empire in Africa, 1948–1998

## RGS-IBG Book Series

For further information about the series and a full list of published and forthcoming titles please visit www.rgsbookseries.com

**Published**

*Decolonising Geography? Disciplinary Histories and the End of the British Empire in Africa, 1948–1998*
Ruth Craggs and Hannah Neate

*Theory and Explanation in Geography*
Henry Wai-chung Yeung

*How Cities Learn: Tracing Bus Rapid Transit in South Africa*
Astrid Wood

*Defensible Space on the Move: Mobilisation in English Housing Policy and Practice*
Loretta Lees and Elanor Warwick

*Geomorphology and the Carbon Cycle*
Martin Evans

*The Unsettling Outdoors: Environmental Estrangement in Everyday Life*
Russell Hitchings

*Respatialising Finance: Power, Politics and Offshore Renminbi Market Making in London*
Sarah Hall

*Bodies, Affects, Politics: The Clash of Bodily Regimes*
Steve Pile

*Home SOS: Gender, Violence, and Survival in Crisis Ordinary Cambodia*
Katherine Brickell

*Geographies of Anticolonialism: Political Networks Across and Beyond South India, c. 1900–1930*
Andrew Davies

*Geopolitics and the Event: Rethinking Britain's Iraq War through Art*
Alan Ingram

*On Shifting Foundations: State Rescaling, Policy Experimentation And Economic Restructuring In Post-1949 China*
Kean Fan Lim

*Global Asian City: Migration, Desire and the Politics of Encounter in 21st Century Seoul*
Francis L. Collins

*Transnational Geographies Of The Heart: Intimate Subjectivities In A Globalizing City*
Katie Walsh

*Cryptic Concrete: A Subterranean Journey Into Cold War Germany*
Ian Klinke

*Work-Life Advantage: Sustaining Regional Learning and Innovation*
Al James

*Pathological Lives: Disease, Space and Biopolitics*
Steve Hinchliffe, Nick Bingham, John Allen and Simon Carter

*Smoking Geographies: Space, Place and Tobacco*
Ross Barnett, Graham Moon, Jamie Pearce, Lee Thompson and Liz Twigg

*Rehearsing the State: The Political Practices of the Tibetan Government-in-Exile*
Fiona McConnell

*Nothing Personal? Geographies of Governing and Activism in the British Asylum System*
Nick Gill

*Articulations of Capital: Global Production Networks and Regional Transformations*
John Pickles and Adrian Smith, with Robert Begg, Milan Buček, Poli Roukova and Rudolf Pástor

*Metropolitan Preoccupations: The Spatial Politics of Squatting in Berlin*
Alexander Vasudevan

*Everyday Peace? Politics, Citizenship and Muslim Lives in India*
Philippa Williams

*Assembling Export Markets: The Making and Unmaking of Global Food Connections in West Africa*
Stefan Ouma

*Africa's Information Revolution: Technical Regimes and Production Networks in South Africa and Tanzania*
James T. Murphy and Pádraig Carmody

*Origination: The Geographies of Brands and Branding*
Andy Pike

*In the Nature of Landscape: Cultural Geography on the Norfolk Broads*
David Matless

*Geopolitics and Expertise: Knowledge and Authority in European Diplomacy*
Merje Kuus

*Everyday Moral Economies: Food, Politics and Scale in Cuba*
Marisa Wilson

*Material Politics: Disputes Along the Pipeline*
Andrew Barry

*Fashioning Globalisation: New Zealand Design, Working Women and the Cultural Economy*
Maureen Molloy and Wendy Larner

*Working Lives - Gender, Migration and Employment in Britain, 1945–2007*
Linda McDowell

*Dunes: Dynamics, Morphology and Geological History*
Andrew Warren

*Spatial Politics: Essays for Doreen Massey*
Edited by David Featherstone and Joe Painter

*The Improvised State: Sovereignty, Performance and Agency in Dayton Bosnia*
Alex Jeffrey

*Learning the City: Knowledge and Translocal Assemblage*
Colin McFarlane

*Globalizing Responsibility: The Political Rationalities of Ethical Consumption*
Clive Barnett, Paul Cloke, Nick Clarke & Alice Malpass

*Domesticating Neo-Liberalism: Spaces of Economic Practice and Social Reproduction in Post-Socialist Cities*
Alison Stenning, Adrian Smith, Alena Rochovská and Dariusz Świątek

*Swept Up Lives? Re-envisioning the Homeless City*
Paul Cloke, Jon May and Sarah Johnsen

*Aerial Life: Spaces, Mobilities, Affects*
Peter Adey

*Millionaire Migrants: Trans-Pacific Life Lines*
David Ley

*State, Science and the Skies: Governmentalities of the British Atmosphere*
Mark Whitehead

*Complex Locations: Women's geographical work in the UK 1850–1970*
Avril Maddrell

*Value Chain Struggles: Institutions and Governance in the Plantation Districts of South India*
Jeff Neilson and Bill Pritchard

*Queer Visibilities: Space, Identity and Interaction in Cape Town*
Andrew Tucker

*Arsenic Pollution: A Global Synthesis*
Peter Ravenscroft, Hugh Brammer and Keith Richards

*Resistance, Space and Political Identities: The Making of Counter-Global Networks*
David Featherstone

*Mental Health and Social Space: Towards Inclusionary Geographies?*
Hester Parr

*Climate and Society in Colonial Mexico: A Study in Vulnerability*
Georgina H. Endfield

*Geochemical Sediments and Landscapes*
Edited by David J. Nash and Sue J. McLaren

*Driving Spaces: A Cultural-Historical Geography of England's M1 Motorway*
Peter Merriman

*Badlands of the Republic: Space, Politics and Urban Policy*
Mustafa Dikeç

*Geomorphology of Upland Peat: Erosion, Form and Landscape Change*
Martin Evans and Jeff Warburton

*Spaces of Colonialism: Delhi's Urban Governmentalities*
Stephen Legg

*People/States/Territories*
Rhys Jones

*Publics and the City*
Kurt Iveson

*After the Three Italies: Wealth, Inequality and Industrial Change*
Mick Dunford and Lidia Greco

*Putting Workfare in Place*
Peter Sunley, Ron Martin and Corinne Nativel

*Domicile and Diaspora*
Alison Blunt

*Geographies and Moralities*
Edited by Roger Lee and David M. Smith

*Military Geographies*
Rachel Woodward

*A New Deal for Transport?*
Edited by Iain Docherty and Jon Shaw

*Geographies of British Modernity*
Edited by David Gilbert, David Matless and Brian Short

*Lost Geographies of Power*
John Allen

*Globalizing South China*
Carolyn L. Cartier

*Geomorphological Processes and Landscape Change: Britain in the Last 1000 Years*
Edited by David L. Higgitt and E. Mark Lee

# Decolonising Geography? Disciplinary Histories and the End of the British Empire in Africa, 1948–1998

Ruth Craggs
King's College London, UK

Hannah Neate
Manchester Metropolitan University, UK

WILEY

This edition first published 2024


*Registered Offices*
John Wiley & Sons, Inc., 111 River Street, Hoboken, NJ 07030, USA
John Wiley & Sons Ltd, The Atrium, Southern Gate, Chichester, West Sussex, PO19 8SQ, UK

For details of our global editorial offices, customer services, and more information about Wiley products visit us at www.wiley.com.

Wiley also publishes its books in a variety of electronic formats and by print-on-demand. Some content that appears in standard print versions of this book may not be available in other formats.



A catalogue record for this book is available from the Library of Congress

Paperback ISBN: 9781119549307; Hardback ISBN: 9781119549284; ePDF ISBN: 9781119549338; ePub ISBN: 9781119549369; oBook ISBN: 9781119549352

Cover Design: Wiley
Cover Image: The Nigerian Geographical Association Silver Jubilee Conference, 1982, University of Ibadan. Courtesy of the Department of Geography, University of Ibadan.

Set in 10/12pt Plantin Std by Integra Software Services Pvt. Ltd, Pondicherry, India

# Contents

# List of Figures and Table

# Acknowledgements

This book would have been nothing without the contributions of the geographers who shared their experiences and shaped our project in oral history interviews: Tony Allen, Elizabeth Ardayfio-Schandorf, Olusegun Areola, Albert Aweto, 'Bola Ayeni, Johnson Ayoade, Randall Baker, George Benneh, Leonard Berry, Michael Chisholm, Hugh Clout, Jonathan Crush, Patricia Cunnan, Ronnie Donaldson, Michael Dyssel, Adetoye Faniran, Calvyn Gilfellan, Keith Hoggart, Anthony Lemon, Akin Mabogunje, Alan Mabin, Brij Maharaj, Adolfo Mascarenhas, Jeff McCarthy, Ellsworth McPherson, Ngaka Mosiane, Linda Newson, Sue Parnell, Gordon Pirie, Gina Porter, Debbie Potts, Christian Rogerson, Stephen Rule, Victor Savage, Dianne Scott, Cecil Seethal, David Smith, Dhiru Soni, Manfred Spocter, Michael Sutcliffe, and Reuban Udo. Brian Berry, John Iliffe and Anthony O'Connor offered their memories via letter and email, and Stanley Okafor, David Narracott and Mike Pugh offered additional contributions. It was a privilege to speak with and learn from them all. We are extremely grateful to Joanne Sharp for allowing us to use additional oral history interviews conducted as part of research about the University of Dar es Salaam.

In addition to those who we interviewed, many others contributed by facilitating research trips and making us welcome. We would like to thank especially James Esson, Chris Ikporukpo, Tolulope Osayomi, Ibidun Adelekan, Trevor Hill, Oliver Mtapuri, Michelle Hatch, Sithembiso Myeni, Gustav Visser, Sarah Bracking, Mark Pelling, George Owusu, Eben Amankwaa, and Samuel Agyei-Mensah. Without the invaluable support and expertise of archivists in Germany, Ghana, Nigeria, South Africa, the United States and the United Kingdom, this book would also have been impossible. Thanks especially to Thirunagaren Munsamy, Sandra Brits, Karlien Breedt, Clive Kirkwood, Lucy McCann, Bethany Antos, staff at the Archiv für Geographie and Abraham Olayemi, whose generosity and knowledge have improved this book immensely.

We were grateful for the feedback from audiences for papers we gave at the University of Ibadan, National University of Singapore, University of the Western Cape, University of KwaZulu Natal, University of Oxford, University of Sussex, University of Cambridge, University of Nottingham and the London Group of Historical Geographers, as well as to audiences at International Geographical Union meetings in Dublin and Paris, the International Conference of Historical Geographers in Warsaw, and at the Royal Geographical Society with Institute of British Geographers in London.

At Wiley, we are grateful to Jacqueline Scott and Grace Ong, and at the Royal Geographical Society to Phil Emmerson, Ed Armston-Sheret and Catherine Souch who supported the project and the wider teaching focused activities that were associated with it. James Esson, Iram Sammar and Jo Norcup have been a brilliant team in delivering these activities and inspiring with their work on teaching African geographies. Dave Featherstone, as Book Series Editor, has been encouraging as well as challenging throughout, and as a result the book is much improved. The reviewers of the manuscript and proposal provided very useful comments that shaped the argument. Colleagues at King's College London, Manchester Metropolitan and elsewhere have provided critical engagement and encouragement in equal measure. Thank you to Innes Keighren, Felix Driver, Miles Ogborn and the rest of the London Group of Historical Geographers for their historical geography wisdom. Thanks especially the Contested Development Research Group at KCL, Steve Legg, James Sidaway and Paul Ashmore for providing detailed feedback on drafts, and to Katherine Brickell for reading many versions of proposals, grant applications and chapters and being an enthusiastic cheerleader and coffee shop companion throughout.

Travel for research was funded by Manchester Metropolitan University, King's College London and the British Academy, who through a British Academy Mid-Career Fellowship also supported research leave to write most of the book. The KCL SSPP Publication Subvention Fund supported the production of the index. Some of the research has previously been published in different forms in the *Journal of Historical Geography* (Post-colonial careering and the discipline of geography: British geographers in Nigeria and the UK, 1945-1990. Ruth Craggs and Hannah Neate, Journal of Historical Geography, copyright: © 2018 The Authors) and *Annals of the American Geographers Association* (reprinted by permission of Taylor & Francis Ltd, http://www.tandfonline.com on behalf of © 2019 by American Association of Geographers).

A big thank you to Dave Weatherall for being supportive throughout, to Vic Craggs, for his company and enthusiasm in South Africa and to Jean Craggs for her support for the trip and always. Tim has provided academic ideas, archival contacts, and most of all love and encouragement to keep me going, and Daisy has provided a (mainly) delightful distraction since her arrival in the middle of the project and the start of a pandemic.

This book is dedicated to the first generations of African geographers who worked so hard to create a decolonised discipline of geography.

# Chapter One
# Decolonisation and Geography in Africa

## Introduction

*Decolonising Geography?* focuses on the experiences and contributions of academic geographers to decolonisation in the former British empire in Africa. Whilst geography's engagements with the imperial project have been well documented, accounts have tended to end in the early twentieth century rather than continuing to examine the period of constitutional decolonisation itself (Craggs 2014). However, geographers, and geography, were entangled with the end of empire in a number of ways, even if, as Power and Sidaway argue, 'this connection was not always acknowledged' (2004, p. 588). The book focuses on both how the professional lives of academic geographers in this era were shaped through decolonisation *and* how their work shaped that same process. It sheds new light on the influence of late colonial development, decolonisation, and post-colonial geopolitics on geography, and demonstrates how the discipline contributed – positively and negatively – to the broader politics of decolonisation and national development.

Examining the period 1948–1998, and with a final chapter that addresses debates about the decolonisation of the discipline (and university) today, the book explores the careers of geographers working in colonial and post-colonial universities in Africa during and after constitutional decolonisation. It examines the practice of geography within the universities of the (former) British empire, as well as considering the impact of decolonisation on geography in the UK as British geographers returning from posts in the colonies joined UK departments. It draws on case studies from Ghana, Nigeria, South Africa,

*Decolonising Geography? Disciplinary Histories and the End of the British Empire in Africa, 1948–1998*, First Edition.
Ruth Craggs and Hannah Neate.

Tanzania, and Uganda, as well as the UK. The first universities in these African countries were almost all university colleges associated with the University of London, set up in the mid-twentieth century. They made up part of an expanded British academic world which connected the UK and British colonies in Africa (Pietsch 2013).

Whilst Ghana became independent in 1957, followed shortly after by Nigeria, Tanzania, and Uganda, the 1980s and 1990s saw liberation for Zimbabwe, and the end of apartheid in South Africa. Much of the historiography of British decolonisation focuses on the 1940s to 1960s, yet the formal constitutional process was ongoing into the 1980s. The book charts geography's engagements with decolonisation in the fifty years following the opening of the first University Colleges in Ghana, Nigeria, and Uganda in the late 1940s. It extends the chronology of these histories into the late twentieth century as anti-colonial struggles continued across the continent and academics, students, politicians, and publics continued to ask questions about the extent to which decolonisation had meant a complete break from colonialism. The struggle for democratic rule in South Africa was understood by many as part of the broader process of decolonisation and the demand for racial justice across Africa, so whilst the issues in South Africa were somewhat different to those in other decolonising states, it is revealing to explore these alongside one another.

Questions about colonialism and decolonisation reverberated around universities in Africa in the mid twentieth century and continue to shape the discipline of geography to the present day. Our book makes three key contributions to debates about decolonisation. First, whilst *Decolonising geography?* is a historical study, it contributes to the rich and diverse debates which see decolonisation as an incomplete and increasingly urgent project in the twenty-first century. It does so by bringing the meanings, strategies, and lived experiences of a group of primarily African academics working to decolonise the discipline of geography and the university in the mid twentieth century, into conversation with those attempting the same thing today. Second, the book contributes to the decolonisation of geography's histories, by highlighting the contributions of scholars from Africa, as well as the ongoing legacies of colonialism in the discipline. Third, we contribute to the interdisciplinary study of decolonisation by demonstrating the value of biographical approaches for understanding the end of empire as a transformation experienced and produced through individuals' careers worked across continents, linking postcolonial states and former colonial powers, colleagues and institutions, in complex ways.

Following the professional lives of a cohort of geographers over many years the book utilises biographical methods as well as institutional histories to explore a wide range of academic labour. Whilst the published research of geographers forms one element of the analysis, the book also explores teaching, curriculum design, and student work; academic exchanges, grants, scholarships, and funding; and the often-hidden academic labour of departmental administration including aspects such as reference writing, mentoring, and promotion decisions. We examine a set of individual professional lives, from undergraduate studies through often-lengthy academic careers, alongside the institutions, networks, and disciplinary knowledges through which they were worked. Through this approach we are able to explore not only how geographers conceptualised decolonisation in their research, but also how they practised it through their broader academic labour. Disciplinary histories tend to focus on publications, and, in the context of empire and decolonisation, on contributions to either the pursuit, or the radical critique, of these processes (Livingstone 1992; Clayton 2013). Yet, decolonisation was (and is) not only about publications, but also about everyday academic practices. How departments were managed and people treated were as much part of the practice of decolonisation as geographical work like border commissions or anti-colonial critique. The book demonstrates that we must take seriously the everyday work of geographers, as well as their publications, in assessing the interconnections between geography and decolonisation. The biographical approach taken allows the book to bridge a critical gap in historical understanding, by uncovering the *praxis* of geography in all its messiness.

The rest of this chapter sets out the contours of our arguments, contexts, methods, and case studies. The next section first describes the periodisation of the book before moving on to contextualise the place of African universities within debates from this period about decolonisation and post-colonial state-building. The third section sets out our three substantive contributions to the lively and wide-ranging debate about decolonisation unfolding today. Following this we set out our methods and then establish our case study institutions. Finally we introduce the chapters and set out the structure of the rest of the book.

## African Decolonisation

### *Periodisation*

In this book we engage with decolonisation in multiple registers, but our empirical discussion is anchored in the historical period leading up to and following formal processes of political decolonisation. Clayton (2020, p.2) has provided a useful description that captures the different overlapping phases of this process which are explored in the book:

> Postwar decolonisation encompasses three phenomena: first, attempts by Western powers to defend and reform their colonial empires and deal with a rising tide of anticolonial sentiment (dubbed late colonialism); second, the sometimes peaceful and quick but often violent and protracted means by which independence was obtained (and with nationalist and independence struggles often stretching much further back in time); and, third, the ensuing affairs of postcolonial nations and question of whether independence heralded a complete break with the colonial past. (Clayton 2020, p. 2)

Ngũgĩ wa Thiong'o (1993, p. 60) was more succinct in his account, personalising this history into the experiences of African writers like him, who, in the 30 years from the middle of the twentieth century, had gone through 'the age of anti-colonial struggle, the age of independence, and the age of neo-colonialism'. He grew up in colonial Kenya, becoming one of the first generation of African students at Makerere University College, Uganda, around the time of independence. Ngũgĩ is not only a key theorist of decolonisation, but a counterpart, colleague and contemporary of many of the geographers whose contributions are explored in this book. His academic life reflects the broad trajectories of higher education in East Africa, as well as in Ghana and Nigeria. He studied as an undergraduate in a University College in Africa (affiliated to the University of London) that was a British late colonial development project. Ngũgĩ provides an evocative description of the campus and status of the institution in the colonial and early post-colonial period, which captures the optimism and excitement of decolonisation:

> Kampala is a city of high hills. Makerere, after which the college was named, is one of the nine hills on which the city stands. But the name Makerere had come to symbolise higher learning in East Africa for those who ascended the hill it meant a passage into the membership of the band of the very elect. But the college was more than that. In the fifties and early sixties Makerere was the intellectual capital of East and Central Africa ... What a time it was those days at Makerere, in East Africa! It was a replica of the Wordsworthian bliss at being alive at the birth of a revolution and the possibilities of a new future. Africa, Our Africa was coming back. (Ngũgĩ 1993, p. 164; p. 166)

Ngũgĩ's powerful account captures the affective experience of decolonisation at an African university, where political hope combined with new educational infrastructures, generous funding, and intellectual innovation to produce a heady mix for the tiny elite who were part of it all.

However, from as early as the mid-1960s, and increasingly into the late 1970s and 1980s, a combination of dependency, authoritarian governments, shifting aid priorities, spiralling oil prices, and structural adjustment combined to leave many African universities hugely underfunded, subject to government attack and even physical violence (Ajayi et al. 1996). After post-graduate study in the UK, Ngũgĩ worked as a lecturer at the University of Nairobi from 1967. A well-known radical figure he was jailed in 1977 for authoring a play which was critical of the increasingly authoritarian Kenyan government. Ngũgĩ then worked in exile in the United States for many years. The excitement and then increasing challenges of working in African universities impacted academic careers and these material and affective experiences changed substantially over the fifty years explored in this book. These shifting experiences also fed into the theorisations of decolonisation produced by African scholars, politicians and activists – like Ngũgĩ – many of whom studied and worked at these very African universities.

In the following account we explore the contributions of African scholars negotiating, theorising, and attempting to do things differently in the post-colonial university. In doing so we follow Sabelo Ndlovu-Gatsheni (2018) in arguing that African scholars (and scholars writing in Africa) have made valuable contributions to the theorisation and contestation of coloniality, and to the elaboration of decolonial visions for the future university. Constitutional decolonisation brought only 'independence with a question mark', but that question mark provided space for possible futures to be imagined, debated, and researched by African scholars (Ngũgĩ 1993, p. 65). In returning to this period, we hope to recover what Mbembe (2021, p. 43 emphasis in original) reminds us was the '*eventfulness*, singularity, and intensity' of decolonisation as it played out in the lives and work of geographers and geography in Africa, as well as the creativity of the period.

## *Decolonisation, Education, and the Place of African Universities*

The book places African geographers (and other geographers in Africa) within the context of debates in the mid-twentieth century about the place of the university – and education more broadly – in a decolonising continent. Fundamentally at issue was the role that education had played in imperialism, and the extent to which these legacies continued after independence, including through the institutions of higher education. Ngũgĩ argued forcefully that education was central in colonialism:

> Berlin of 1884 [the conference seen as the highpoint of European colonial expansion in Africa] was effected through the sword and the bullet. But the night of the sword and the bullet was followed by the morning of the chalk and the blackboard. The physical violence of the battlefield was followed by the psychological violence of the classroom. But where the former was visibly brutal, the latter was visibly gentle …
>
> Imperialism is total: it has economic, political, military, cultural and psychological consequences for people of the world today… The biggest weapon wielded and actually daily unleashed by imperialism… is the cultural bomb. The effect of the cultural bomb is to annihilate a people's belief in their names, in their languages, in their environment, in their heritage of struggle, in their unity, in their capacities and ultimately in themselves. (Ngũgĩ 1986, p. 9; pp. 2–3)

Articulated through the ideology of 'Black Consciousness', Steve Biko made similar arguments about the impact of colonialism and apartheid in South Africa. Biko was a student in the 'Non-European' section of the largely white University of Natal, a student leader, and political activist who was murdered by the apartheid government in 1977. Biko argued that through colonialism and apartheid, 'the black man has become a shell, a shadow of a man, completely defeated' (1978 [1969–1972], p. 29). Drawing on Frantz Fanon (2019 [1961]), Biko argued this impact was produced in part through (mis)education:

> In an effort to destroy completely the structures that had been built up in the African Society and to impose their imperialism with an unnerving totality the colonialists were not satisfied merely with holding a people in their grip and emptying the Native's brain of all form and content, they turned to the past of the oppressed people and distorted, disfigured and destroyed it. (Biko 1978 [1969-72], p. 29)

Education had supported imperialism, reinforcing Eurocentricity and undermining African cultures and knowledges. Universities globally embodied colonial power – Western, white and capitalist – in everything from their architecture to their disciplinary knowledges. As Walter Rodney put it:

> As I come onto one of these university campuses and I stand up and stare upwards and I see this tremendous structure which I know doesn't belong to us.
>
> Black people are here in these institutions as a part of the development of black struggle, but only as a concession designed to incorporate us into the structure .... Going beyond the symbolism of the building, I'm thinking also of the books, the references, the theoretical assumptions, and the entire ideological underpinnings of what we have to learn in every single discipline. (Rodney 1990 [1975], p. 111)

Rodney, like Ngũgĩ, was entangled with these structures, building a career through (post)colonial universities in the UK, Africa and the Caribbean, studying first at University College of the West Indies (at that time a college of the University of London), then at the School of Oriental and African Studies in London, before lecturing at University College Dar es Salaam (and being appointed to this post by authorities in London) (Rodney 1990 [1975]). These experiences fed into his theorisation of underdevelopment and of the role of the 'guerrilla intellectual' working from the inside to transform the colonial university (Rodney 1990 [1975], pp. 111–113).

Ngũgĩ wa Thiong'o (1993, p. 3) has argued that decolonisation in twentieth century Africa was 'the political struggle to move the centre' from Europe and to Africa. To what extent could the university – as a key site for education shaped through colonialism – contribute to (or hinder) that process? These questions were often central in the thinking not only of those within the university but also in society more broadly. Indeed, they preoccupied African nationalist politicians who lead their countries to independence. Kwame Nkrumah, first Prime Minister and President of independent Ghana was one of many of the new wave of African leaders to explicitly concern themselves with the role of the university in Africa. For Nkrumah (2009 [1964], p. 3), studying abroad, or at a colonial university college in Africa, could strip African students of their local connections and critical awareness of their own position: 'The colonial student can be so seduced by [the western philosophies] ... that he surrenders his whole personality to them. When he does this, he loses sight of the fundamental social fact that he is a colonial subject', he argued. Nkrumah himself had studied in the US and reflected on this experience in his most well-known political book *Consciencism* (2009 [1964]). When Ghana became independent, Nkrumah viewed universities, alongside multinational corporations and institutions like the civil service as 'part of the apparatus of imperialism ... which had to be decolonised' (Ajayi et al. 1996, p. 95). His theory of neo-colonialism emphasised the dangers of 'political independence minus economic independence' and the danger of the creation of new local elites more allied with the West than with Ghana, through institutions such as the university (Nkrumah 1962, quoted in Ahlman 2017, p. 157). Nkrumah demanded universities demonstrated their allegiance to independent Ghana rather than continuing to reproduce a colonial mentality and pushed for the replacement of foreign academics with Ghanaians (Nkrumah 2009 [1964]; Ndlovu-Gatsheni 2017).

Julius Nyerere, the first Prime Minister and President of independent Tanzania, was also a central thinker on the role of African universities (Ndlovu-Gatsheni 2017). Like Ngũgĩ, Nyerere had attended Makerere, though in the 1940s before it became a university college. From 1963, Nyerere was the first Chancellor of the University of East Africa (which included Makerere as a constituent college, alongside the University Colleges of Dar es Salaam and Nairobi). Nyerere's vision was of a university that took 'an active part in the social revolution', prioritising local needs and relevance, whilst maintaining links to an international community of knowledge (Nyerere 1966, p. 219–220). In Tanzania, Ghana, and across Africa after independence, concerns grew that universities were creating a new elite, out of touch with the majority of the people, rather than a socially engaged workforce contributing to national development. Fanon (2019 [1961], p. 99) argued that:

> In an underdeveloped country, the imperative duty of an authentic national bourgeoisie is to betray the vocation to which it is destined, to learn from the people, and make available to them the intellectual and technical capital it culled from its time at the colonial universities.

However, instead of fulfilling this duty, Fanon argued that the educated elites aimed for personal enrichment. Nkrumah, Nyerere, and a host of other African leaders, academics and activists therefore saw the contested future of the university as central to the broader politics of decolonisation and independence.

Universities in Africa were the focus of three impulses in the era of independence. The first was the Africanisation of the staff; replacing Western, mainly British, European, and American lecturers with Africans across universities but also other institutions such as national banks, the military, and the civil service (Stockwell 2018). However, this process did not guarantee more than superficial change: as Fanon (2019 [1961], p. 103) argued early on, in *The Wretched of the Earth*, 'The colonized bourgeoisie ... frantically brandishes the notions of nationalization and Africanization', seeking out the professional positions, privileges and profits previously in the hands of the colonisers for themselves, rather than bringing about revolutionary change. For Ngũgĩ, the importance of class and capital was often overlooked in the era of independence, with imperialism 'far too easily seen in terms of the skin pigmentation of the coloniser.' (Ngũgĩ 1993, p. 62).

A second impulse was towards the Africanisation of the curriculum to produce knowledge and expertise needed in post-colonial Africa. Describing a specific controversy at the University College Nairobi, Kenya, in 1968, around the content of the English programme, Ngũgĩ made it clear that the arguments went much wider than the specific proposals he and his colleagues had brought forward:

> the debate, in other words, was about the inherited colonial education system and the consciousness it necessarily inculcated in the African mind. What directions should an education system take in an Africa wishing to break with neo-colonialism? What should be the philosophy guiding it? How does it want the 'New Africans' to view themselves and their universe? From what base: Afrocentric or Eurocentric? ... who should be interpreting the material to them: an African or non-African? If African, what kind of African? One who has internalized the colonial world outlook or one attempting to break free from the inherited slave consciousness? (Ngũgĩ 1986, p. 101–102)

These mid-twentieth century debates were not in themselves entirely new but echoed long-standing ideas which emerged at the end of the nineteenth century about the need for universities to preserve the 'African personality' by including African languages, cultures, and oral traditions on the syllabus, and teaching in African languages (Blyden 1872; Casely Hayford 1969 [1911]; see Ajayi et al. 1996; Ndlovu-Gatsheni 2018). If 'decolonization was a struggle by the colonized to reconquer the surface, horizons, depth, and heights of their lives', as Mbembe (2021, p. 44) described it, then this included rehabilitation of 'indigenous forms of language and knowledge'.

Ngũgĩ's (1986) critique of the colonial university and disciplines explored questions of culture not in isolation, but within the context of Marxist critiques of the post-colonial political economy of East Africa. These debates about disciplinary knowledges were not reserved only for the humanities. In 1979 the Nigerian political economist Claude Ake, a University of Ibadan graduate and lecturer at the University of Dar es Salaam 1972–1974 weighed into the matter in relation to social science, also drawing on Marxist critiques. In a book uncompromisingly titled *Social Science as Imperialism*, Ake argued that

> Western social science scholarship on developing countries is imperialism in the sense that (a) it foists, or at any rate attempts to foist on the developing countries, capitalist values, capitalist institutions, and capitalist development; (b) it focuses social science analysis on the question of how to make the developing countries more like the West; and (c) it propagates mystifications, and modes of thought and action which serve the interests of capitalism and imperialism. (1982, p. xiii)

For Ake, it was by the late 1970s 'becoming increasingly clear that we cannot overcome our underdevelopment and dependence unless we try to understand the imperialist character of Western social science and to exorcise the attitudes of mind that it inculcates' (1982, p. xiv). Africanisation of knowledge demanded new subjects, questions, theories and methods.

Alongside the Africanisation of staff and curricula, a final – and connected – debate focused on the institutional transformation of universities on the continent, aiming to make them *African* universities rather than universities *in* Africa (Sicherman 2005). This included structural changes – increasing independence in awarding degrees, changes in faculty structure and so on – as well as debates about the relationship between scholars and society (Mazrui 1978). An early aim of post-colonial universities was to satisfy what were known as the 'manpower' [*sic.*] demands of newly independent countries (that is, primarily, in the form of public administration, but also other forms of expertise such as for those taking geography, teaching, planning, surveying, and land-use allocation) (Mkandawire 1995). Another was to produce applied research to inform government decisions on national and local development. As Daniel Clayton (2021, p. 13) has highlighted, 'decolonisation raised a very basic set of geographical questions about land, resources, territory, identity and development' and the discipline of geography was therefore also seen as useful for supporting the project of independence. Training and data production were understood as central to decolonisation as a political and economic project: producing independence and self-reliance through knowledge and knowledgeable people. But as Mahmood Mamdani (2011, n.p.) has argued, based on his experience as a staff member at Makerere University in the 1980s and 1990s, and more recently, such applied research could also be understood as a replication of earlier dynamics of the 'native informer' producing data but not new theorisations or knowledge (Mamdani 2011). As such Paulin Hountondji (1990) has argued this type of activity could be one element of a broader intellectual extraversion leading to dependency. For him, a commitment to applied research betrayed the insidious inculcation of coloniality in the minds of African researchers where their labour and fields of research were only relevant to making African lands productive for extraction (Hountondji 1990).

The Kenyan scholar Ali Mazrui noted that in the period directly after independence, many in Africa grappled with the place of the university on the continent, and 'that uneasy feeling that the beast in our midst is foreign in origin and too rational in its supra-social tendencies' (Mazrui 1978, p. 217). Mazrui was a political scientist at Makerere College (later University) in the 1960s and early 1970s, and this feeling of which he wrote shaped his time at Makerere and his scholarship on education, colonialism, and nationalism. Mazrui's unease reflected the fact that universities were understood as Western institutions – not only in structure, staffing, and curricula, but also in their commitment to detached, supposedly neutral scholarship. As 'detachment' became 'suspected of disengagement', there were increasing demands for universities to support socially committed scholarship relevant to postcolonial development (Mazrui 1978, p. 212; see also Mamdani 2019). However along with demands for relevance came questions over academic freedom and the extent to which African scholars could be socially committed whilst critical of government policies (Mazrui 1978; Ngũgĩ 1986).

Taken together, these discussions over Africanisation of staff, curricula, and institutions concerned the extent to which it was possible, and desirable, to transform African universities, and for what purpose. In these debates we can see different understandings of decolonisation as political, economic, and epistemological project. And whilst the first two elements often dominated, African scholars in the period of constitutional decolonisation grappled with how to both provincialize Europe *and* deprovincialise Africa, 'moving the centre' from which ideas and knowledge should be viewed *and* created to Africa (Ndlovu-Gatsheni 2018; Ngũgĩ 1993; Chakrabarty 2009). Ndlovu-Gatsheni (2018, p. 3) argues that this amounts to a demand for 'epistemological decolonisation' and 'cognitive justice': 'the right to think, theorize, interpret the world, develop own methodologies and write from where one is located'. The next section demonstrates how attention to these histories can make three important analytical contributions to understanding decolonisation as historical process and political project.

## Contributions

### *Historicising Current Debates*

In this book, we recentre the sometimes overlooked but important ideas and practices which emerged from a generation of politicians, scholars, and activists grappling with many of the issues central to coloniality in the mid-twentieth century in Africa. This is our first contribution. Whilst these vibrant debates dominated African universities in the 1960s and 1970s, they seem to have been forgotten by many outside the continent advocating for decolonisation today. For once again over the last decade, decolonisation has once again become a popular term. It has taken on greater and more diverse meanings than just the constitutional moment in high politics, when flags were changed, and a country became politically independent. In the context of higher education, decolonisation has been used to discuss questions about the coloniality of Western ideologies, methodologies, and institutions, and about the whiteness of the university and individual disciplines, including geography (Tolia-Kelly 2017; Bhambra et al. 2018; Ndlovu-Gatsheni 2018; Radcliffe 2022). Publications have proliferated – including from those critical (e.g., Táíwò 2022) – as have sustained, creative, and sometimes violent protests (and their suppression) (Elliott-Cooper 2017; Ahmed 2020).

Specific South African campaigns such as #Rhodesmustfall and the connected #Feesmustfall have been influential world-wide, spawning a range of associated movements, including in the UK (Elliott-Cooper 2017; Gebrial 2018; Jansen 2019). In the UK, Oxford University's own #Rhodesmustfall movement argued for 'critically examining the power struggle that underpins hegemonic knowledge production, and the material structures that make this possible', through work on iconography, curriculum and representation (Gebrial 2018, pp. 20; 23).

Epistemic freedom often forms an increasingly important part of contemporary movements to transform the university in the twenty-first century. Whilst mid-twentieth century debates focused on the primacy of political, and then economic decolonisation, seeing the former as the pre-requisite for any other transformations, Ndlovu-Gatsheni (2018, p. 5) argues that:

> In the co-constitution of political, economic, cultural and epistemological decolonisation, epistemic freedom should form the base because it deals with the fundamental issues of critical consciousness building, which are essential pre-requisites for both political and economic freedom.

Ndlovu-Gatsheni (2018) was writing in the aftermath of the #Rhodesmustfall and #Feesmustfall campaigns in South Africa in 2015–2016, whilst based at Witwatersrand University, formerly a white dominated institution under apartheid. In these protests in South Africa, the meaning of decolonisation, a key buzz-word, was often contested. For Jansen, protests demanded 'the quest for racially and culturally inclusive campuses and more specifically to the transformation of campus symbolism, the university curriculum, institutional cultures and the professoriate' (2019, p. 51). But Ndlovu-Gatsheni (2018, p. 189) and other commentators saw in them a much more fundamental epistemological and political critique of the philosophical, economic, and cultural underpinnings of the university, 'rethinking and redefining the university as a truly African public institution'. In both understandings of the movement, questions of staffing, the student body, the curricula, the campus, relations with society and government were under debate, whilst for some the status and divisions of disciplines and knowledges, and the university itself, were also up for grabs.

These activist campaigns have been matched by an upswell of books and articles engaging with the ideas and practices of decolonisation in the university (Moosavi 2020). Much of the published work – nearly half of publications – has emerged from a South African context (Adefila et al. 2021). More recent research tends to move beyond theoretical interventions, instead focusing on providing summaries of the state of university education today and overviews of progress in decolonisation initiatives (e.g. in a South African context Ammon 2019; Lebelo et al. 2021; on South African geography specifically see Knight, 2018) and on specific experiments in decolonial curriculum reform (often undertaken by the authors) and their relative success (e.g. in Zimbabwe, Gukurume and Maringira 2020). Scholars have also explored current attempts to create new institutions either within or beyond traditional university settings, such as an 'indigenous pluriversity' within a Columbian institution (Padilla 2021) or the Pan Afrikan Marcus Garvey University in Uganda (Mwesigire 2016; see also Schildermans 2021 on university experiments in a Palestinian context).

Much of the recent theorising about decolonisation within and beyond geography has taken its cue from Latin American contexts and discussions of coloniality. Coloniality is defined as the enduring colonial matrix of power in which coloniality and modernity are inextricably linked (Quijiano 2000; Mignolo and Walsh 2018; on the use of these approaches in geography see Radcliffe 2017; 2022; Stanek 2019). These discussions describe how 'modernity/coloniality has worked and continues to work to negate, disavow, distort, and deny knowledges, subjectivities, world senses, and life visions', whilst decoloniality tries to construct 'paths and praxis toward an otherwise of thinking, sensing, believing, doing and living' (Mignolo and Walsh 2018, p. 4). But some scholarship on decoloniality has not engaged in detail with African scholarship, nor with the period of constitutional decolonisation in Africa (Daley and Murrey 2022). This is in part because of the regional context from which decolonial debates emerged, and also because these histories and ideas are not always understood as relevant to contemporary contexts. For example, whilst acknowledging the legacies of non-alignment and the Bandung Conference – in which African and Asian leaders explicitly engaged with many of the issues central to coloniality – Mignolo and Walsh (2018, p. 4) do not see these legacies as 'the central foundation' of the decolonial project. They argue that

> Decolonisation during the Cold War meant the struggle for liberations of the Third World and, when successful, the formation of nation-states claiming sovereignty. By the 1990s, decolonisation's failure in most nations had become clear; with [the] state in the hands of minority elites, the patterns of colonial power continued internally (i.e. internal colonialism) and with relation to global structures. At that moment coloniality was unveiled. (Mignolo and Walsh 2018, p. 5–6)

In focusing on continuities, this and many accounts of coloniality today skip over the optimism of the period and processes of decolonisation, as well as the geographical and historical specificity of colonial experiences (Craggs 2014). As Davies argues, the work of some decolonial theorists narrate histories of the world in which 'The scale of the ambition is matched by thinly referenced, vast geopolitical claims. The history of decolonization becomes merely a state project which ran its course.' (2021, p. 400). Without attention to the complex, particular histories of colonialism and decolonisation in Africa, and to the agency of African scholars, politicians, and others, we risk 'turning colonisation into an eternal category – a form of ontology – instead of a historical one.' (Leonard 2022, n.p.; Táíwò 2022).

As the term decolonisation has become popular, Leon Moosavi (2020) has described a 'decolonial bandwagon', as scholars and managers latch on to the term, but it is applied superficially, in ways that can be ineffective or even harmful.

Moosavi (2020, p. 332) lists 'reducing intellectual decolonisation to a simple task; essentialising and appropriating the Global South; overlooking the multifaceted nature of marginalisation in academia; nativism; and tokenism' as some of the key challenges facing the project of decolonisation today. Whilst Moosavi (2020) is cautiously supportive of decolonial movements in universities today, others are much more critical. Olúfémi Táíwò (2022) is one such critic, arguing that decolonisation debates have failed to take African agency seriously.

This book documents the experiences and contributions of African geographers engaging in important, complex, and ambiguous ways with decolonisation as constitutional, state-building, and epistemological project. In doing so we pay close attention to the concerns outlined above and provide one potential way forward for studying and practicing decolonising scholarship. In examining continuities between the colonial university, the post-independence university and contemporary universities in Africa and beyond, the book is able to explore the ongoing coloniality of geography (Radcliffe 2022) and trace some of the specific legacies from these earlier periods which have shaped today's disciplinary landscape. Our book aims to embed current debates, which sometimes risk being taken over by Western – and often white – academics (Jazeel 2017; Noxolo 2017a; Táíwò 2022), within the experiences, scholarship, and demands of African scholars.

Moreover, motivated by recent calls for 'historical geographies of, and for, the present' which are 'inspired by direct engagement with problems in the present and intend to do something about them' (Van Sant et al. 2020, p. 169), we demonstrate how historical work might inform more progressive and decolonial, current and future academic practices. Unlike Táíwò (2022) we do not see any contradiction between advocating for decolonisation and taking African agency seriously, rather, we see taking African agency seriously as central to making contemporary decolonisation a viable and emancipatory project.

## *Decolonising Geography's Histories*

A second key contribution of the book is to decolonise geography's own histories, by taking seriously the period and process of constitutional decolonisation, examining the role of geographers within that process, and producing more diverse and inclusive disciplinary stories.

Over the last 40 years there has been a growth in histories of geography that highlight the peopled and always political nature of the discipline (e.g. Buttimer 1983; Livingstone 1992; Barnes 2008; Maddrell 2009; Farish 2010; Clayton and Barnes 2015; Albuquerque and Martins 2018; Schelhass et al. 2020). Since the 1990s, scholars influenced by postcolonial theory have carefully deconstructed the relationships between geography and empire highlighting links between the discipline and exploration, resource exploitation, and colonial control and violence (Bell et al. 1995; Driver 2001; Ryan and Naylor 2009). Though Ngũgĩ was most interested in literature and language, he noted 'the [often racist and colonial] images children encountered in literature were reinforced by their study of geography and history, science and technology where Europe was, once again, the centre.' (Ngũgĩ 1986, p. 93). For Ngũgĩ, the discipline of geography was, then, part of a wider 'colonial alienation', the 'disassociation of the sensibility [of a colonised person] from his natural and social environment' (1986, p. 16–17).

Despite this attention to geography and its role in empire building, however, the relationships between geography and decolonisation have for the most part been overlooked. This book responds to this lacuna, shifting attention to the intersections between the period of constitutional decolonisation in the mid-late twentieth century and the discipline of geography, and contributing to a small but emerging body of work on this period (Power and Sidaway 2004; Clayton and Bowd 2017; Clayton and Kumar 2019; de Suremain 2019; Ferretti 2019, 2020; Sarmento 2019; Clayton 2020), and on geography in Africa (e.g. Areola and Okafor 1998; Visser et al. 2016; Daya 2022).

Our account highlights the role of geographers, and the discipline of geography as a whole, in the processes of constitutional decolonisation and postcolonial development and state-building. If we rely on published research, we may conclude that there were few connections between geography and decolonisation. As Clayton (2020, p. 5) notes, 'the term appears only a handful of times in the titles of articles in leading Western geography journals' between 1945–1980. The same is true of the pages of geography journals published in the decolonising world. However, though often obscured in more traditional disciplinary histories, geographers were deeply involved with decolonisation. They worked for the late colonial state, surveying, administering forestry and agriculture, overseeing new development projects and censuses, and they worked for new independent governments in similar roles. They delineated post-colonial boundaries (Fitzpatrick 2019). They worked in the geography departments of colonial university colleges, and in the new universities of recently independent countries. They taught about the (former) colonies in British universities and worked in new institutes specializing in development and area studies (Craggs and Neate 2019). And they wrote about the dilemmas facing newly independent countries through questions of development and modernisation, under-development and neo-colonialism (Clayton 2020; see for example Blaut 1970; Darkoh 1981; Mabogunje 1980; Slater 1974; Soja 1968). In South Africa in the 1980s, the term decolonisation began to be used to name a process required to transform a discipline shaped not only by colonialism but also apartheid (Crush et al. 1982). Paying attention to this work highlights how new sub-disciplines – such as development geography – were shaped (Power and Sidaway 2004), how geographers engaged with ideas and practices of decolonisation (Ramutsindela 2022), and how academic networks and professional norms were produced and contested through this period.

Crucially, the book shifts the focus to the (post-)colonial world, and its relations with and beyond the West, and away from Europe and North America. The book highlights the role and agency of academic geographers from Africa in the discipline in the second half of the twentieth century. We argue that to overlook the contributions of African geography and geographers is to reinforce colonial-era visions of the agency and creativity of academics from the Global South. In placing African geographers centre stage, we intervene in ongoing debates about the exclusionary nature of many histories of geography. Important work by Mona Domosh (1991), Gillian Rose (1995), and Avril Maddrell (2009) has highlighted women's often overlooked roles in geography over two centuries. Work has also begun to unveil the 'hidden histories' of indigenous guides, translators, scholars, explorers, surveyors, and cartographers in the production of geographical knowledge under colonial rule (Driver 2012). Despite these interventions however, our histories of geography remain overwhelmingly western and white (Craggs 2019). Moreover, many of the experiences of those racialised as non-white or indigenous within geography continue to be those of exclusion (Pulido 2002; Peake and Koboyashi 2002; *Area* special issue, ed. Noxolo 2017b).

Whilst histories of geography have unveiled the discipline's colonial past, these critical accounts have rarely addressed the discipline's continued coloniality (Radcliffe 2022). Likewise, those advocating for decolonisation now have rarely engaged in any detail with historical approaches to the discipline, and especially its connections to decolonisation ideas and practices (Radcliffe 2022; though see Jazeel 2019). The book is a timely intervention in recent debates about decolonisation in geography which brings these histories into conversation with present day disciplinary struggles (*Transactions of the Institute of British Geographers* special issue, ed. Radcliffe 2017, 2022; *Area* special issue, ed. Noxolo 2017b; de Leeuw and Hunt 2018; Stanek 2019; Daya 2022).

Academics pursue careers over decades, and their ideas and connections draw upon long years of professional experience stretching from the period of constitutional decolonisation itself, through to the recent past. As we confront questions about ongoing and unfinished decolonisation within geography, the university, and wider society, these longstanding personal experiences of empire and decolonisation provide one important connection between past and present. As Barnes argues, '"now" is produced in large part by actors, scripts, props, ensembles, and scenes from the past' (2002, p. 509). The colonial present is produced (in part) by the colonialism of the past and the processes of constitutional decolonisation; understanding geography's entanglements with these helps us to understand the contemporary legacies of these processes. If 'the writing of certain kinds of pasts is legitimated by, and legitimates, only certain kinds of presents' (Rose 1995, p. 414) then it is essential to produce accounts which reflect the historical and contemporary diversity of geography (as well as highlighting its lack of diversity in many places).

Historical accounts such as this one, which focus explicitly on academic practice, might also offer lessons for how we approach the labour of academic geography today, cognizant of the impacts and implications of our approach to colleagues, students, research and publication. Such thinking is important in light of the arguments raised by the contributors to these debates who ask all of us to take decolonisation seriously as a radical practice today (Esson et al. 2017; Tolia-Kelly, 2017; Noxolo 2017a; Ahmet 2020; Esson 2020; Esson and Last 2020; Oswin 2020; Daya 2022; Radcliffe 2022). Clayton (2020, p.16) argues that 'a decolonial agenda in geography will be diminished if it becomes disconnected from the history of decolonisation and its everyday geographies (including geographers' lives) and convened in a more rarefied or theoretically cloistered politics of knowledge.' In drawing out the connections between engagements with decolonisation in the past and present, between academic theorising and more mundane forms of scholarly labour, and about the everyday experiences of geographers' working lives, the book contributes to calls from Noxolo (2017a), Esson et al. (2017), Daya (2022), and Radcliffe (2022) to think carefully and ethically about the consequences of decolonisation work today. The next section sets out the third major contribution of our book: how our biographical methods can contribute to the historiography of decolonisation.

## *Professional Lives and Histories of Decolonisation*

A recent flurry of scholarship in the history of decolonisation signals a 'vibrant and productive period' in this interdisciplinary field (Milford 2017, n.p.). We argue that focusing on the entanglements of professional lives and the end of empire provides important new directions for the interdisciplinary historiography of decolonisation. First and foremost, this focus highlights the specificity of this period, the opportunities it offered, and the changes that were felt. As Milford et al. (2021, p. 394) put it:

> The era of decolonisation opened up new forms of political community and new terms of engagement with the wider world; it was not merely a new stage in a history of external dominance indistinct from the colonialism that preceded it or the marginalisation that followed in the age of global markets and capital.

Olukoshi et al. (2020) argue persuasively that instead of writing of this period as one of continuity and failure, there is a need to reclaim Africa's early post-independence history as a moment of clarity of purpose and creativity. In the book, our biographical approach offers the opportunity to explore the legacies of empire as lived, experienced, and practiced through professional lives, and to recapture the agency of those working for decolonisation. Conceptualising decolonisation through professional lives provides new sources and spaces – cultural, educational, disciplinary, professional – through which to explore these processes.

Second, many careers were shaped in important ways by decolonisation, which created new opportunities, including for unprecedented mobility, as well as closing down others. As Mkandawire (1995, p. 75) suggests, in the context of African academic trajectories,

> since independence there have been at least three generations of indigenous researchers in Africa. Each has witnessed changes in their countries' economic fortunes and political trajectories, as well as cultural and societal transformation. All of these factors have impinged on the nature and meaning of their academic careers.

These experiences of decolonisation shaped many professional realms – from the discipline of geography which is our focus here – to the related realms of urban planning (Craggs and Neate 2017) and international development (Kothari 2006a, 2006b). A focus on professional lives demonstrates how disciplinary and other sorts of professional knowledge were shaped through decolonisation.

Third, the everyday labour of these careers was not only influenced by decolonisation but was itself part of the process, performing shifting power relations, materialising new norms and priorities, and contributing to the crucial but often overlooked affective, social, and cultural realms of this geopolitical transition (Bailkin 2012). Here we follow recent accounts in understanding decolonisation as a process involving not only politicians and diplomats, but everyone – in the post-colonies and the former metropole – through the realms of work, culture, and education (Schwarz, 2011; Bailkin 2012; Craggs and Wintle 2016; Livsey 2017).

*Decolonising geography?* contributes to recent scholarship which explores questions of expertise and professional knowledge, as well as of the role of cultural, professional, and educational institutions, in decolonisation (Craggs and Wintle 2016; D'Auria 2016; Eagleton 2016; Mew 2016; Waters 2016; Stockwell 2018). The university has been a productive institutional site through which to understand empire and decolonisation beyond the realm of high politics (Pietsch 2013; Jons, 2016; Elliott-Cooper 2017; Livsey 2017; Surman 2018; Sharp 2019). Shifting the frame from an imperial and national one as utilised by Tim Livsey (2017) in his account of Nigerian universities, decolonisation and development, or an institutional one as deployed by Tamsin Pietsch (2013) in her account of selected British and 'British world' universities in the nineteenth and earlier twentieth centuries, here we show how a biographical approach with a disciplinary framing can unveil further, and different, intersections between the academy and (the end of) empire. The book contributes to a vibrant set of literatures surrounding western disciplinary knowledges – from anthropology and sociology to urban planning and history – and their relationships with empire and decolonisation (Bhambra 2007; Chakrabarty 2009; McIntyre 2009; Bailkin 2012; Home 2013; Steinmetz 2016). Providing an account that goes beyond the contributions and experiences of Western professionals to explore also the professional lives of colonial subjects and post-colonial citizens, the book demonstrates how disciplinary knowledges and academic labour were deeply intertwined with decolonisation.

Finally, a focus on professional lives helps to address longstanding concerns in postcolonial history to view the 'core' and 'periphery' within one analytic lens (Stoler and Cooper 1997). Porter (2004) has long argued that empire (and decolonisation) had little impact on society and culture in Britain. However, this minimal impact thesis has been hotly disputed (e.g. by Ward 2001), and more recently in powerful accounts by Schwarz (2011), Bailkin (2012), and Buettner (2016) who have traced the long lasting and often surprising impacts of imperialism in Europe, in everything from ideas of whiteness, to the shape of the welfare state. Contributing and extending this body of work, this book underlines the social and cultural impact of decolonisation in Britain as the end of empire intersected with, and impacted upon, the development of universities, disciplines and research in the UK. In connecting developments in places often conceptualised separately – like Europe and the post-colonial world – biographical approaches help to expand the range of spaces in which knowledge and expertise is often understood to be created.

In developing our approach, we have been influenced by the work of a range of scholars who have explored colonialism and its aftermath through individual lives. Lambert and Lester's (2006) account of imperial 'careering' showed the possibilities of following professional lives in order to develop a networked and peopled account of colonial governance and has been used by historians in a wide range of colonial contexts within and beyond Africa (Dhupelia-Mesthrie 2016; Hofmeyr et al. 2016; Matsuda 2016; Vahed 2016; Alavi 2019; Rogers 2019; Minawi 2020). In the context of decolonisation, Hodge (2010, p. 24), adapting Lambert and Lester's phrase, has provided similar insights through his account of the 'post-colonial careering' of colonial development experts whose careers form an 'important thread of continuity across the seemingly fundamental rupture of decolonisation and independence' (see also Home 2013; Craggs and Neate 2017; Muschik 2018). Kothari's interviews with colonial administrators who became expatriate development professionals highlight the ways in which ideas and practices were translated into different political worlds: not only did individuals remain involved in development, but 'colonial forms of rule and governance' and the 'performance of expertise and authority' endured, as did ways of imagining colonial space (Kothari 2006a, p. 118; 2006b). Recent work has also highlighted the utility of examining the experiences and contributions of individual diplomats and politicians in and between India, East Africa, and the UK, as they navigated the new spaces and networks opened up in decolonisation (Guyot-Réchard 2022; Milford and McCann 2022).

Biographical approaches have also been applied to academic lives. Jeremiah Arowosegbe (2008) focuses on the Nigerian scholar Claude Ake, who began his studies as an undergraduate student at University College Ibadan in the late 1950s (when Nigeria was still a British colony), before studying and working across the UK, US, and East and West Africa. Arowosegbe's (2008, p. 7) account demonstrates that 'Ake's intellectual shifts are not properly accounted for, unless we critically interrogate, narrativise and analyse his life, scholarship and career in relation to the material conditions and the complex social history, which facilitated such a transformation.' Al Bulushi (2022) provides an even more fine-grained approach in his account of scholars working at Dar es Salaam at the same time as Ake. Al Bulushi's approach – what he calls 'an intellectual geography' – blends 'personal stories of everyday life … political praxis, and classroom pedagogy, all of which are grounded in particular places'. In doing so, he is able to draw out how 'the realities of survival sometimes brought to the foreground contradictions and chance encounters that were just as important as movement manifestos and explicit organisational plans to the trajectory of national independence.' (Al Bulushi 2022, p. 2–3). Our account brings together material conditions, ideas, scholarship and geopolitical transformation in a similar way, to explore the careers of geographers - many of whom were Ake's colleagues - in the UK, Nigeria, and Tanzania.

'Writing biography involves mediating between two competing questions. On the one hand is "what can we really know of a life?". And on the other, "what can a life reveal to us about something else?"' (Hodder 2017, p. 455). We are concerned primarily with the latter (though they intersect), with biography as 'method', to navigate and understand the complicated and entangled histories of a discipline and its practitioners in and beyond the continent of Africa. Academic careers are the key focus and a strategy for engaging with these vast histories to tell stories about decolonisation 'that are bigger, bolder, and broader, and to do so in more interesting ways' (Hodder 2017, p. 457).

Unlike most of the work described above, our approach has been to shift the emphasis from individuals in isolation, instead focusing attention on whole cohorts of professionals. Careers are not pursued alone, but rather alongside colleagues and collaborators. Such an understanding helps to highlight the everyday embodied and social nature of working lives as they are worked through 'chats over coffee or lunch, drinks in the bar … through friendship and conflict …' (McFarlane 2011, p. 145–147). An approach that focuses on a range of people, and their relationships, rather than following one individual provides further opportunities to be inclusive in writing about the history of decolonisation. Our approach aims to reconsider 'whose life matters and therefore whose life should be written about and remembered', and in doing so to write different people into the history of geography (McGeachan 2018, p. 143). Through the use of biographical approaches, the book challenges how certain histories have been told, and through this, illuminates other lives, histories, and knowledge-making practices that otherwise remain in the margins (see also Bressey 2013). In the next section, we set out how we researched the professional lives that are at the heart of our account.

## Biographical Methods

The book follows a number of different academic careers, both extraordinary and more ordinary, in order to explore the tensions, conflicts, and excitement within academic lives lived against the backdrop of decolonisation. We focus both on the British geographers who founded many of the first geography departments in late-colonial Ghana, Nigeria, South Africa, Tanzania, and Uganda, and more substantively, on their African colleagues who have played a key role in transforming them in the decades since. The individuals who are part of our account were chosen because they represented a range of experiences. We examine the working lives of some of the most prominent and pioneering African geographers – such as the Nigerian Akin Mabogunje, the Kenyan Simeon Ominde, and the Tanzanian Adolfo Mascarenhas – alongside others who followed in their footsteps or worked alongside them in less prominent roles. In South Africa we focused primarily on those working for the end of apartheid within the universities assigned under the repressive regime to 'Coloured', 'Indian' and 'Black' students. Our approach has not been to focus on high profile individuals only; indeed, we have been careful to also highlight the contributions of those whose labour is often overlooked: university teachers, administrators, and consultants as well as those with prolific publication histories (for similar arguments in other contexts, see Roy 2012 on technocrats; Larner and Laurie 2010 on engineers, and Lagae and De Raedt 2013 on committee members and brokers). Decisions also reflected source material available – on which more detail is given below.

In exploring the experiences and working lives of these geographers we keep in mind Frederick Cooper's (2019, p. 51) warning that those we are writing about 'cannot be reduced to stick figures in a drama with two actors, colonizer and colonized, or a story with one plot line: the struggle for the nation.' Rather, their geographical careers intersected with and contributed to decolonisation and post-colonial development whilst at the same time fulfilling (or not) personal ambitions, providing a salary (usually), providing career satisfaction, frustration, and fun, and leading to long lasting friendships and animosities.

## *Sources*

Decolonisation is not immediately visible in the publications of geographers in the twentieth century, but if you look carefully at other sorts of materials, 'it can be found in the margins and folds of other debates and concerns, by other names' (Clayton 2020, p. 5). The often subtle manifestations of decolonisation as theorised and practiced by academic geographers necessitated drawing not only on published research but on a range of archival and published material, as well as oral histories. These materials provide substantial evidence of what Lorimer and Spedding (2002, p. 295) have called 'geography's more quotidian history'. Personal papers, including diaries and correspondence provide evidence of academic networks and experiences. Institutional records (including those of universities and geographical societies in Africa and the UK) describe hiring and promotion decisions, curriculum and teaching changes, and relationships with government, business and civil society. Autobiographies and obituaries provide details of academic careers, travel, and experiences. Academic publications illustrate developing research interests and modes of publication as well as shedding light on the academic infrastructure of conferences, research groups and exchanges linking institutions across and beyond Africa. Oral histories provide insight into the lived experiences of establishing and maintaining an academic career.

Across these sources, we looked for evidence of the everyday work of academic geography, which included working with others in developing, maintaining, critiquing, and transforming geography departments. Despite consulting a wide range of materials, one area that remained less visible was trade union activity, which was sometimes difficult and controversial in colonial and post-colonial African university contexts (McQuinn 2022). This is a notable absence. Nevertheless, we tried to read against the grain for small and hidden acts of resistance to the colonial university and discipline within the sources we had access to, in written complaints and humorous stories shared, in silences and understatements (see Guha, 1999 [1983]). Complaints in particular, as Sara Ahmed (2021, p. 28) has convincingly demonstrated, can teach us about power: 'We learn *about* institutions from the wear and tear of coming up against them.'

Whilst as Sou (2021, p. 2–3) notes, 'theories of resistance often presume acts are explicit, collective, and spectacular', they can also be 'individual, relatively safe and seemingly invisible acts of opposition to power' (see also Noe 2022). Sou writes in the context of North-South development partnerships in the present but similar dynamics, of professionals working together against the backdrop of uneven power relations in Western institutions were also present in African universities in the mid-twentieth century.

We consulted university and geography department archives in Ghana, Nigeria, South Africa, and the UK.[1] Whilst university records were often formally archived (though in varying states of organisation and repair), collections in geography departments tended to be informal and accessed as a result of the goodwill of academic and professional services staff. As has recently been highlighted, documents, student work, and other materials produced at departmental level are often archived only patchily, with records existing only as the result of diligent work by historically minded staff, or a lack of pressure on departmental space (Bruinsma 2021). We also accessed the records of the Inter-University Council for Higher Education Overseas;[2] the Royal Geographical Society, UK; the South African Geographical Society and South African Society for Geography;[3] and the International Geographical Union (IGU).[4] Aspirations for further visits to Tanzania and Uganda were curtailed by the global Covid-19 pandemic and maternity leave, which had significant impacts on the authors' mobility and capacity for research, underlining the importance of understanding academic labour as contingent and gendered. However, the quantity and heterogeneity of the materials and collections that we were able to consult still provided valuable insights into how everyday labour intersected with intellectual and geopolitical developments.

The book also draws substantially on autobiographical accounts and oral history interviews, as well as materials shared by interviewees (CVs, photos, personal papers). The Oxford Development Records Project contains sometimes extensive autobiographical reflections on the experiences of British academics working in African universities in the 1940s, 1950s, and 1960s, and was a key source for the experiences of this earlier generation of geographers working in Africa, most of whom are now dead.[5] Two of the first generation of African academic geographers, the Nigerian Akin Mabogunje and the Ghanaian George Benneh wrote autobiographies containing important reflections (Mabogunje 2011; Benneh 2016).

Oral histories and other personal testimony can provide useful insight into academic lives (Barnes 2002; Tolia-Kelly 2017; Craggs and Neate 2020). A total of 43 conversations with geographers working in Africa between the period 1948–1998 took place between 2016 and 2021 mostly in the form of face-to-face interviews in Ghana, Nigeria, South Africa, and the UK. Others were carried out virtually, or through email correspondence. Some correspondents did not have regular or reliable internet access which ruled out virtual interviews, others simply preferred to write. This breaks down as six Nigerians, two Ghanaians, two Tanzanians, 19 South Africans and 14 British geographers.[6] We thank all interviewees who were willing to be named in our acknowledgements. The Geography Departments at the University of Ibadan, University of Ghana, University of the Western Cape, and the University of KwaZulu Natal facilitated access to archival collections and provided contacts for interviews. We were welcomed in each of these departments, and colleagues' and students' critical engagements with early findings presented in seminars in Nigeria and South Africa were invaluable in developing the research.[7]

Almost all the geographers interviewed were based in Africa for several years, if not full careers. However, as seen in the breakdown of interviewees, coverage was not even between countries, and only five women were interviewed. Contacting interviewees was sometimes difficult and usually required access through their former departments (where they were now retired). Those we were able to contact were likely to be those who spent long or at least successful parts of their career at their institutions and who left on good terms. This inevitably shapes the narrative presented, with an emphasis on insights from those who continue to value their association with the departments they worked in (and were valued in return).

Talking about professional lives, particularly with those reflecting in retirement, can lead to heroic narratives and the downplaying of difficult periods and failures as career trajectories are smoothed out in the subsequent narration, and disappointment, anger, or sadness are sometimes edited out. Nevertheless, asking about everyday experiences and triangulating interview discussion with archival and secondary material provides both nuance to heroic stories and insight into negative aspects of academic life – particularly around colonialism, racism, and apartheid. We are white British researchers so some of these issues could be difficult to raise, though we followed our interviewees' lead in relation to topics for discussion, and found many people were motivated to explore these concerns alongside more positive experiences. The dynamic between interviewers and interviewees was generally that of professionals interacting, and sometimes of junior and more senior colleagues as many of our interviewees were senior, sometimes renowned, mainly male geographers (and we are young(ish) women). Throughout the book interviews are quoted extensively to allow the experiences of geographers to be described in their own words. All research was carried out following the frameworks for conducting ethical research at our respective institutions. All oral histories were conducted with informed consent from participants, and interviewees gave permission to be quoted and named in this book.

We were privileged to have the funding and opportunities to travel extensively due to our positions as British scholars in still (relatively) well financed institutions which, in the twenty-first century academic world still marked by legacies (and continuing realities) of colonialism, exist at the heart of what was the British Empire and still benefit materially from British imperialism.[8] We specify our authorial position here, briefly, in order to acknowledge our positions within this nexus, whilst trying to avoid 'unfettered self-reflexivity' which can work to recentre the emotions and experiences of mostly white researchers (de Leeuw and Hunt 2018, p. 3). We are cognisant of concerns about what might be called '"capture" … the predominant involvement of non-Africans in studying the continent' (Bob-Milliar 2020, p. 2; Obeng-Odoom 2019). British (and more broadly Western) scholars are over-represented in research about Africa, reflecting the still hugely uneven landscape of funding and opportunities as well as the epistemological structures of the academic world (many of which are the subject of this book). We discussed these concerns explicitly with colleagues in Nigeria and South Africa as we developed the research.

This is a story not just about geography in Africa, but also about the discipline in the UK and the connections between these academic worlds. The idea for this book came from a growing awareness of the colonial connections of the university departments where we received our geographical training, and now work. For example, John Pugh, who founded the Geography Department at University College Ibadan, later worked at King's College London where one of us now works. Indeed, we had colleagues in common. This intertwined history of British and African universities is little discussed (at least in British/Western academia) underlining the work that is still required when it comes to 'The task of provincializing, historicizing, and relativising metropolitan ways of thinking about the world and narrating historiographies of non-Western … geographies' (Boyle 2017, p. 49). In exploring the interconnections between African and British geography, and geographers, we hope that the following book can contribute in a small way to the important task of decolonising the discipline, something which should not be the job solely of racialised minorities working in the academy (Radcliffe 2022).

What is presented in the following chapters is by no means complete or exhaustive; some places and people are discussed extensively, whilst others are touched upon more briefly or not at all. Whilst in some ways, the account that follows contributes to the agenda of diversifying geography's histories, in other ways it falls short. Women are often absent, reflecting primarily the historical imbalances in the discipline, but also the fact that some female geographers we hoped to speak to were unavailable for interview having died, or were untraceable after leaving academia. Where possible we have written female geographers, as well as geographers' wives and partners, into the narrative, trying to make visible what Tamson Pietsch (2013, p. 81) has called 'the shadow networks' of the academic word. Nevertheless, the gendered ways in which academic careers and their aftermaths in retirement were structured reinforces the dominance of male voices in our account.

The book also focuses only on the experience of academic faculty members, not students (apart from those pursuing academic careers) or other actors. The experiences of those with wider roles in an academic department, namely support staff such as administrators, cartographers, or drivers, or those working in or doing geography beyond academia, such as teachers, would also make an important contribution to our understanding. Many administrators, as well as schoolteachers, were women, and as has been shown in other contexts, geography at school level is also an important part of the story of geography (Norcup 2015). Our focus on academic geographers who mainly worked in the most prestigious universities in Anglophone Africa, institutions that mainly developed under colonial rule, reinforces an elite version of African geography, rather than seeking a more radical inclusivity. Nevertheless, we argue that there is real value in exploring the sites where many Africans locate the development of the discipline of geography in the continent and of exploring the opportunities, restrictions, and power relations attendant within these spaces.

## *Case Studies*

The book focuses on geographers at universities in Ghana, Nigeria, Tanzania, South Africa, Uganda, and the UK. All of these countries were part of a tight academic network connecting Britain and its (former) colonies in Africa. The institutions in Ghana, Nigeria, and Uganda (The University Colleges of Ghana, Ibadan, and Makerere) were the first in their countries and all set up after World War II as part of colonial development projects as colleges of the University of London. The first university college in Tanzania was also part of this relationship with London, though it was founded later in 1961, and is included as an important centre for innovation in research and teaching. These common foundations provide the rationale for the choice of institutions across East and West Africa.

In addition, the book explores a number of institutions in South Africa which falls slightly outside this pattern. South Africa was dominant in the British African academic world of this era; it was both the site of a more established and powerful university sector, and apartheid politics which reverberated across the continent in this period. These connections to the British colonial academic landscape and to the politics of decolonisation and racial justice in Africa together mean that it is included in the book alongside Ghana, Nigeria, Uganda, and Tanzania. Moreover, despite these differences in origin from the other cases, our book reveals remarkable similarities across the South African and other case studies. Experiences of racism, decolonisation, and state-building were shared across the continent, though with individual variations. Rather than treat South Africa as exceptional, as many accounts do, here it is placed alongside other African countries revealing fascinating and important connections as well as points of rupture. The institutions discussed in any detail are all located on Figure 1.1, whilst the departments that figure centrally are also described below, and detailed in Table 1.1.

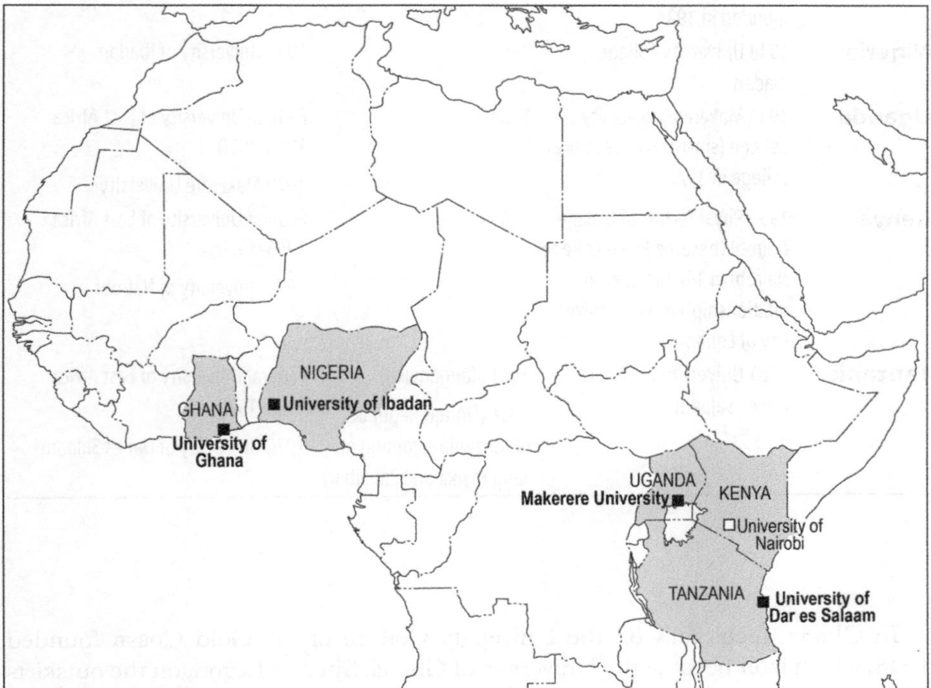


**Figure 1.1** Map of Africa showing case-study institutions (in bold) and other universities discussed in the book. Produced by Philip Stickler.

**Table 1.1** Key higher education institutions and their transition to university status.

| | *University College in 'special relationship' with the University of London* | *Country's Independence* | *Creation of universities with their own degree-awarding powers* |
|---|---|---|---|
| **Ghana** | 1948 University College of the Gold Coast (University College Ghana from 1957)<br>(preceded by Achimota College founded in 1924) | 1957 | 1961 University of Ghana |
| **Nigeria** | 1948 University College Ibadan | 1960 | 1963 University of Ibadan |
| **Uganda** | 1949 Makerere University College (started as a technical college in 1922) | 1962 | Federal University of East Africa 1963-1970;<br>1970 Makerere University |
| **Kenya** | 1956 Royal Technical College Nairobi (became Royal College Nairobi in 1961 in 'special relationship' with the University of London) | 1963 | Federal University of East Africa 1963-1970;<br>1970 University of Nairobi |
| **Tanzania** | 1961 University College of Dar es Salaam | 1961 (Tanganyika)<br>1964 (United Republic of Tanzania – joining of Tanganyika and Zanzibar) | Federal University of East Africa 1963-1970;<br>1970 University of Dar es Salaam |

In Ghana, focus falls on the University College of the Gold Coast, founded 1948, which later became the University of Ghana. Sited in Legon, on the outskirts of the capital Accra, UC Gold Coast was initially a college of the University of London, becoming Ghana's first degree-awarding university in 1961 (Figure 1.2). Ghana itself became independent four years earlier.

Whilst the final transition in 1957 was relatively peaceful, longstanding campaigning for independence had been met by violence and jail terms for many involved – including Kwame Nkrumah – by the British colonial government. The first, very modest, moves to self-government only followed strikes and widespread disturbances in Accra and other major cities in 1948, the same year that the University College was founded. Designed by a British architect on the model of British universities, after independence, Legon soon became the focus of attention from the new Ghanaian government under Kwame Nkrumah (Agbodeka 1998).

**Figure 1.2** 'The Chancellors Procession after his installation'. Inauguration of the University of Ghana 25th November 1961, photographer unknown. Royal Commonwealth Society Collection, Cambridge University Library, Y30448F: 16. Reproduced by kind permission of the Syndics of Cambridge University Library.

Nkrumah's socialist and pan-Africanist ideology meant that he saw the university as a potential site for the production of new interdisciplinary knowledge about Africa in support of his wider visions for pan-Africanism, but was also deeply suspicious of the university's British roots, staffing, and outlook, accusing it of fostering a 'colonial mentality' and a liberal, rather than socialist outlook (Agbodeka 1998, p. 125). Geography was one of the founding disciplines at Legon, and the department still exists today, though it became known as the Department of Geography and Resource Development in 1988–1989, to reflect its expertise in applied research for development (Agbodeka 1998).

In Nigeria, we focus on University College Ibadan (founded 1948), the first higher education institution in Nigeria and also initially a college of the University of London (Figure 1.3). Holding this supreme position, Ibadan remained one of the most prestigious universities in the country through the later twentieth century, though it was subsequently joined by numerous other universities. British colonial policy had divided the country into three large regions – Northern, Eastern, and Western – and political parties from each of these dominated campaigns for independence in the 1950s. Ibadan was in the Western region, and in the 1960s and 1970s, the divisive geographical divisions imposed by the British formed the

**Figure 1.3** Tower Court and Administration Buildings, University of Ibadan, early 1960s, photographer unknown. Rockefeller Foundation Records, NIGERIA, Series 497 University of Ibadan, Box 134, Folder 2563, number 19. Courtesy of the Rockefeller Archive Center.

basis of damaging regional competition and violence in which universities became caught up (Livsey 2017).

The Ibadan campus buildings were also designed by British architects, this time in bold tropical modernist style (Figure 1.3). Again, the Geography Department was founded at the outset and the department continues, under the same title, today. When Nigeria became independent in 1960, it followed a much more Western, capitalist path than Ghana, and initially, under Abubakar Tafawa Balewa, remained close to the UK (Nwosu 1993). A bloody civil war and successive military coups between 1966 and the 1990s meant that universities were often subject to military interference. This also had implications for the sorts of development projects that geographers would become involved in, in the 1960s and 1970s.

In Uganda, we examine experiences at Makerere University College, founded in 1949 (Figure 1.4), though there was already a precursor college at Makerere before this date. Makerere later became part of the federal University of East Africa (in 1963), and Makerere University in 1970. In the seven years as part of the University of East Africa, Makerere was federated with University College Nairobi, in Kenya, which is discussed in the book but is not a central case-study, and University College Dar es Salaam, discussed below. Uganda became independent in

**Figure 1.4** Makerere College, Uganda, 1940s. Pump Park Vintage photography/ Alamy Stock photo.

1962, under Milton Obote, who discussed socialism but had made no real political changes in that direction before he was deposed by Idi Amin in 1971 (Ngũgĩ 1993). Amin's regime ravaged the university as it did the country as a whole, with wholesale expulsions of Ugandan Asians as well as violence and mass killings. By the time Amin was himself toppled in 1979, the university, according to Sicherman (2005), lay almost in ruins, though the institution continued and continues until today. Geography was a founding department at Makerere, and still exists today, though is now called the Department of Geography, Geoinformatics, and Climatic Sciences.

University College Dar es Salaam, Tanzania, was founded in 1961, briefly as a college of the University of London before becoming part of the University of East Africa in 1963. It became an independent university from 1970. The Geography Department was founded in 1965 and continues under that name today. Julius Nyerere, who lead Tanzania from its independence in 1961 (as Tanganyika, becoming the United Republic of Tanzania in 1964) propagated a version of non-aligned African socialism that saw a particular place for the university in post-colonial Tanzania. As such the relationship between the university and the government of the one-party state became both close and sometimes fraught. Nyerere's reputation attracted a global set of scholars (including geographers) excited by his radical vision, but as some intellectuals criticised the government they were increasingly unwelcome, and into the 1970s the university increasingly focused on producing 'manpower' [*sic.*] to drive this Tanzanian vision for development, rather than critical commentary.

The first South African universities were founded in the early twentieth century, and were almost entirely closed to black South Africans, whose only option was the South African Native College opened at Fort Hare in 1916 for black African students (Ajayi et al. 1996). From 1959, the university system in South Africa was formally incorporated into the apartheid regime, with restrictions on who could work and study at different universities. Entrance to the established universities was limited to white students, whilst Fort Hare was co-opted into the national government system and dedicated to Xhosa students. New universities were founded, originally as University Colleges of the University of South Africa (Unisa), for 'Indian' and 'Coloured' students, as defined in the 1950 Population Registration Act, in Durban (University of Durban Westville, UDW, initially University College, Durban) and Cape Town (University of the Western Cape, UWC) respectively (Visser et al. 2016). Universities serving the newly designated black Bantustans ('Homelands' in apartheid government parlance) were also founded after the Extension of Universities Act in 1959. In the book we focus primarily on the contributions of those at the University of Durban Westville and University of the Western Cape. However, we also discuss more briefly Fort Hare, the University of the North, and the University of Bophutswana (set aside for various 'black African' groups as defined by the apartheid regime), the supposedly 'open' but in practice almost entirely white Universities of Witwatersrand, Natal, and Cape Town, and the most prestigious Afrikaans speaking institution (also exclusively white), the University of Stellenbosch.

## Structure

The book is structured thematically, but is also broadly chronological. It begins with the foundation of geography within colonial university colleges in the 1940s in the next chapter (Chapter 2). Here we meet the British geographers who staffed these departments and explore their motivations for taking up these roles and the impact they had on shaping geography as a discipline in Africa. Uniquely in the book, this chapter primarily foregrounds the voices and practices of British geographers, though there is also reflection on the experiences of African students and politicians engaging with the colonial era institutions. This focus reflects British geographers' influential role designing everything from departmental buildings to new syllabuses, and setting the tone for teaching methods, staff-student relations, and important areas for research. The chapter argues that this early generation embedded a colonial vision for the discipline in African universities, even as their own motivations and career histories varied substantially, spanning wartime service and colonial surveying, conscientious objection, paternalist multi-racial liberalism, and even radical anti-racism.

Chapter 3 explores constitutional decolonisation and its impacts on geography departments in the 1950s and 1960s. This period was dominated by debates over Africanisation – the replacement of Western and mainly British institutional structures, staff, and curricula with African ones. We explore the work of the first generation of African geographers as they advocated for, benefitted from, and struggled with how to produce an African geography within an African university against the colonial structures in which they studied and worked (both in the UK and Africa). This chapter demonstrates both the close connections between Africanisation as a project in the 1960s and twenty-first century debates about decolonising disciplines/universities, but also some differences in demands and outcomes. It also highlights the complexity and ambivalence of Africanisation as a project involving academic research and new conceptualisations, but also new curricula, staff, institutional structures, and campus relationships.

Chapter 4 explores the shifting international networks in which African geographers were embedded, as British relationships waned somewhat but others, notably with the United States and, to a lesser extent, the Eastern Bloc, came to the fore in the 1950s and 1960s. The chapter demonstrates the impact of the Cold War on African geography providing huge amounts of funding and travel opportunities but also restricting the room for manoeuvre of African governments negotiating independence. It argues that this geopolitical context enabled both African geography's embrace of new quantitative techniques led by Nigerian scholars, and, in juxtaposition and often in explicit critique of quantitative modernisation work, the emergence of new interdisciplinary critical development scholarship in the creative space that was the University of Dar es Salaam in Nyerere's non-aligned Tanzania.

Chapter 5 explores how geography became bound up in the project of postcolonial national development in the 1960s and 1970s as African universities, and individual disciplines, were required to demonstrate their relevance after independence. The chapter focuses on the work of geographers both in training students to be development experts to fulfil what were known as the 'manpower' [*sic.*] needs of newly independent countries, and in undertaking development work themselves, in the fields of urban, rural, and regional planning, resource and land use evaluation, and census taking. We argue that geographers contributed substantially to the multifaceted project of knowing, planning, and shaping the postcolonial state through their contributions to 'national development', but that these contributions could sometimes be ambivalent, reinforcing inequalities and creating new, often technocratic, market oriented, top-down modernisation projects. Moreover, in working explicitly with governments, the influence of (some) African geographers could be amplified at the same time as their space to criticise these same governments was curtailed.

Chapter 6 focuses on the varied work of geographers in apartheid South Africa. It demonstrates how geographers were implicated in materialising the apartheid project of separate development on the ground, as well as in reproducing apartheid and colonial assumptions within their work and teaching. The main focus of the chapter is the specific anti-apartheid and decolonising work of South African geographers, primarily in the 1980s and 1990s, when an increasing number of black geographers began to get jobs in South African universities, and when geographers began to engage more critically with apartheid within and outside the academy. Examining campus spaces as sites for the reinforcement and transgression of apartheid racial categories, anti-apartheid activism, and an increasing critical scholarly engagement with apartheid geographies, as well as structural changes to South African geography's institutions, the chapter highlights the wide variety of academic labour under the banner of anti-apartheid work. In doing so, it makes clear the need to approach decolonising work in a similarly holistic matter.

Chapter 7 explores the intertwined legacies of decolonisation in African and British geography. It begins with the experiences of the British geographers who returned to Britain in the 1950s and 1960s, as well as those – both African and British – forced out of their roles in the 1970s. The second part of the chapter shifts the focus to later decades of the twentieth century, highlighting the ongoing connections between British and African geography departments into the 1980s and 1990s. It shows how these could be used creatively to mitigate the challenges of extreme austerity and precarity in African departments, but also had a varied impact in the UK. It argues that, through scholarly activity such as textbook publication, PhD supervision, and the growth of the new geography of developing areas/development, and Area Studies, experience in decolonising Africa was an important influence on British geography. But it argues that we have to look far beyond these aspects to new student communities, attitudes, stories, racism, friendships, and solidarity too, to fully grasp the legacies of empire and its endings in African and British geography.

The final chapter (Chapter 8) begins by summarising our key contributions to historical debates about decolonisation, before exploring a key question that runs through the entire book – to what extent was geography decolonised in the fifty years between 1948–1998? The book concludes by asking what lessons it might be possible to learn from these earlier experiences for broader projects to decolonise the university in the twenty-first century. Looking historically provides resources for change in the future, as well as clear evidence of the need to understand decolonisation past and present as a complex, multifaceted process requiring diverse forms of labour, not all of which are often recognised.

## Notes

1 In Ghana, the University of Ghana Archives, Geography Departmental records (informally archived in the department). In Nigeria, the University of Ibadan Archives, and Geography Departmental records (informally archived in the department). In South

Africa: the University of Cape Town Special Collections; University of the Western Cape Records and Archives Department, the University of the Western Cape Special Collections, and Geography Department records (informally archived in the department); University of KwaZulu-Natal Documentation Centre (Westville Campus, Durban) the University Archives, Pietermaritzburg Campus; Geography Department archives (informally archived in the department, Pietermaritzburg campus); University of Stellenbosch University Archives. In the UK King's College London Archives; the University of London Archives; the University of Liverpool Archives and Special Collections.

2 Housed at the UK National Archives, catalogue reference series BW90.

3 Housed at the University of Stellenbosch University Archives.

4 The IGU records are held in Germany, at the Archiv für Geographie, Leibniz-Institut für Länderkunde.

5 The Oxford Development Records Project, Bodleian Library, holds substantial material from founding members of the Geography Departments at Ibadan and Makerere, John Pugh (MSS. Afr. s. 1825 (92) BOX LVII) and Kenneth Baker (MSS Afr. S. 1825 (3) BOXES I-II), as well as material from David McMaster (Makerere), Michael Kenneth Barbour (MSS.Afrs.1825 BOX III) and BW 'Dick' Hodder (MSS Afr. S. 1825 (51)) (both Ibadan).

6 With permission, we also draw on several interviews carried out by Joanne Sharp with geographers who worked at the University of Dar es Salaam in Tanzania in the 1960s and 1970s.

7 Decolonisation, post-colonial careering, and the practice of post-war geography, Department of Geography Seminar Series, University of Ibadan, 4 July 2017; Decolonising disciplines past and present: Disciplinary histories and the experience of the end of empire in Africa, 1948–1998, School of Built Environment and Development Studies Seminar Series, University of KwaZulu Natal 29 October 2019; Decolonising disciplines past and present: Disciplinary histories and the experience of the end of empire in Africa, 1948–1998, Department of Geography, Environmental Studies and Tourism Seminar series, University of the Western Cape, 17 October 2019.

8 Whilst our institutions Manchester Metropolitan University and King's College London have yet to do so, several British universities are now researching these histories in relation to slave ownership. See for example: Bristol University to confront its links with the slave trade, *The Guardian*, 5 May 2019. https://www.theguardian.com/education/2019/may/05/bristol-university-slave-trade-history [accessed 26 November 2021].

## References


Adefila, A., Teixeira, R.V., Morini, L. et al. (2021). Higher education decolonisation: #whose voices and their geographical locations? *Globalisation, Societies and Education* 1–15. https://doi.org/10.1080/14767724.2021.1887724.

Agbodeka, F. (1998). *A History of University of Ghana: Half A Century of Higher Education (1948–1998)*. Accra: Woeli Publishing Services.

Ahlman, J.S. (2017). *Living with Nkrumahism: Nation, State, and Pan-Africanism in Ghana*. Athens, Ohio: Ohio University Press.

Ahmed, A. (2020). #RhodesMustFall: how a decolonial student movement in the global South inspired epistemic disobedience at the University of Oxford. *African Studies Review* 63 (2): 281–303. https://doi.org/10.1017/asr.2019.49.

Ahmed, S. (2021). *Complaint!* Durham: Duke University Press.

Ahmet, A. (2020). Who is worthy of a place on these walls? postgraduate students, UK universities, and institutional racism. *Area* 52 (4): 678–686. https://doi.org/10.1111/area.12627.

Ajayi, J.F.A., Lameck, G., and Johnson, G.A. (1996). *The African Experience with Higher Education*. Athens, Ohio: Ohio University Press.

Ake, C. (1982). *Social Science as Imperialism: The Theory of Political Development*. 2e. Ibadan: Ibadan University Press.

Al Bulushi, Y. (2022). Dar es Salaam on the frontline: red and black internationalisms. *Third World Thematics: A TWQ Journal*. https://doi.org/10.1080/23802014.2022.2141851.

Alavi, S. (2019). Slaves, Arms, and political careering in nineteenth-century Oman. In: *Asia Inside Out: Itinerant People* (ed. E. Tagliacozzo), 179–200. London: Harvard University Press.

Albuquerque, S. and Martins, L. (2018). Place, gender and the making of natural history: Hannah im Thurn in British Guiana, 1895–1897. *Journal of Historical Geography* 62: 1–14. https://doi.org/10.1016/j.jhg.2018.03.006.

Ammon, L. (2019). *Decolonising the university curriculum in South Africa: a case study of the university of the free state. Unpublished Masters Dissertation*. Linæus University.

Areola, O. and Okafor, S.I. (ed.) (1998). *50 Years of Geography in Nigeria: The Ibadan Story*. Ibadan: Ibadan University Press.

Arowosegbe, J.O. (2008). Decolonising the social sciences in the global South: Claude Ake and the praxis of knowledge production in Africa. *Asc Working Paper Series*, (79), Leiden University. Retrieved from: https://hdl.handle.net/1887/13015 (accessed 11 January 2022).

Bailkin, J. (2012). *The Afterlife of Empire*. Berkeley: University of California Press.

Barnes, T.J. (2002). Performing economic geography: two men, two books and a cast of thousands. *Environment and Planning A* 34: 487–512. https://doi.org/10.1068/a3440.

Barnes, T.J. (2008). History and philosophy of geography: life and death 2005–2007. *Progress in Human Geography* 32: 650–658. https://doi.org/10.1177/0309132507086883.

Bell, M., Butlin, R., and Heffernan, M. (ed.) (1995). *Geography and Imperialism, 1820–1940*. Manchester: Manchester University Press.

Benneh, G. (2016). *My Time, My Nation*. Accra: Winmat Publishers Limited.

Bhambra, G. (2007). *Rethinking Modernity: Postcolonialism and the Sociological Imagination*. Basingstoke: Palgrave Macmillan.

Bhambra, G., Gebrial, D., and Nişancıoğlu, K. (2018). *Decolonising the University*. London: Pluto Press.

Biko, S. (1978 [1969–1972]). *I Write What I Like: A Selection of His Writings Edited by Aelred Stubbs C.R.* London: Heinemann.

Blaut, J. (1970). Geographic models of imperialism. *Antipode* 2: 65–85.

Blyden, E. (1872). *The West African University*. Freetown: "Negro" Printing Office.

Bob-Milliar, G.M. (2020). Introduction: methodologies for researching Africa. *African Affairs* 1–11. https://doi.org/10.1093/afraf/adaa011.

Boyle, M., England, K., Farish, M., et al. (2017). Introduction: Geography and geographers: Anglo-American human geography since 1945. *The AAG Review of Books* 5 (1): 48–61. https://doi.org/10.1080/2325548X.2017.1257291.

Bressey, C. (2013). Geographies of belonging: white women and black history. *Women's History Review* 22: 541–558. https://doi.org/10.1080/09612025.2012.751767.

Bruinsma, M. (2021). The geographers in the cupboard: narrating the history of geography using undergraduate dissertations. *Area* 53: 67–75. https://doi.org/10.1111/area.12653.

Buettner, E. (2016). *Europe after Empire: Decolonization, Society and Culture*. Cambridge: Cambridge University Press.
Buttimer, A. (1983). *The Practice of Geography*. London: Longman.
Casely Hayford, J.E. (1969 [1911]). *Ethiopia Unbound*, 3e. London: Frank Cass.
Chakrabarty, D. (2009). *Provincializing Europe*. Princeton: Princeton University Press.
Clayton, D. (2013). Militant tropicality: war, revolution and the reconfiguration of 'the tropics' c.1940–c.1975. *Transactions of the Institute of British Geographers* 38: 180–192.
Clayton, D. (2020). The passing of "geography's empire" and question of geography in decolonization, 1945–1980. *Annals of the American Association of Geographers* 110 (5): 1540–1558. https://doi.org/10.1080/24694452.2020.1715194.
Clayton, D. (2021). Historical geography I: doom, danger, disregard – towards political historical geographies. *Progress in Human Geography* 45 (6): 1692–1708. https://doi.org/10.1177/03091325211011664.
Clayton, D. and Barnes, T.J. (2015). Continental European geographers and World War II. *Journal of Historical Geography* 47: 11–15. https://doi.org/10.1016/j.jhg.2014.12.003.
Clayton, D. and Bowd, G. (2017). *Impure and Worldly Geography: Pierre Gourou and Tropicality*. London: Routledge.
Clayton, D. and Kumar, M.S. (2019). Geography and decolonization. *Journal of Historical Geography* 66: 1–8. https://doi.org/10.1016/j.jhg.2019.10.006.
Cooper, F. (2019). *Africa since 1940: The past of the Present*. Cambridge: Cambridge University Press.
Craggs, R. (2014). Postcolonial geographies, decolonization, and the performance of geopolitics at Commonwealth conferences. *Singapore Journal of Tropical Geography* 35 (1): 39–55. https://doi.org/10.1111/sjtg.12050.
Craggs, R. (2019). Decolonising the geographical tradition. *Transactions of the Institute of British Geographers* 44 (3): 444–446. https://doi.org/10.1111/tran.12295.
Craggs, R. and Neate, H. (2017). Post-colonial careering and urban policy mobility: between Britain and Nigeria, 1945–1990. *Transactions of the Institute of British Geographers* 42: 44–57. https://doi.org/10.1111/tran.12147.
Craggs, R. and Neate, H. (2019). Post-colonial careering and the discipline of geography: British geographers in Nigeria and the UK, 1945–1990. *Journal of Historical Geography* 66: 31–42. https://doi.org/10.1016/j.jhg.2018.05.014.
Craggs, R. and Neate, H. (2020). What happens if we start from Nigeria? diversifying histories of geography. *Annals of the American Association of Geographers* 110 (3): 899–916. http://doi.org/10.1080/24694452.2019.1631748.
Craggs, R. and Wintle, C. (ed.) (2016). *Cultures of Decolonisation: Transnational Productions and Practices, 1945–70*. Manchester: Manchester University Press.
Crush, J., Reitsma, H., and Rogerson, C. (1982). Decolonizing the human geography of Southern Africa. *Tijdschrift voor economische en sociale geografie* 73: 197–198. https://doi.org/10.1111/j.1467-9663.1982.tb00955.x.
d'Auria, V. (2016). More than tropical? modern housing, expatriate practitioners and the Volta river project in decolonising Ghana. In: *Cultures of Decolonisation: Transnational Productions and Practices 1945–1970* (ed. R. Craggs and C. Wintle), 196–221. Manchester: University of Manchester Press.
Daley, P.O. and Murrey, A. (2022). Defiant scholarship: dismantling coloniality in contemporary African geographies. *Singapore Journal of Tropical Geography* 43: 159–176. https://doi.org/10.1111/sjtg.12422.

Darkoh, M.B.K. (1981). Geography and Society. *Journal of Eastern African Research and Development* 11: 32–61.

Davies, A. (2021). Walter D. Mignolo and Catherine E. Walsh, On Decoloniality: concepts, analytics, praxis (Durham, NC, and London: Duke University Press, 2018), pp. xii + 291, $99.95, $26.95 pb and E-book; £77.00, £20.99 pb and E-book. *Journal of Latin American Studies* 53: 399–402. https://doi.org/10.1017/s0022216×21000274.

Daya, S. (2022). Moving from crisis to critical praxis: geography in South Africa. *Transactions of the Institute of British Geographers* 47: 91–15. https://doi.org/10.1111/tran.12459.

De Leeuw, S. and Hunt, S. (2018). Unsettling decolonizing geographies. *Geography Compass* 12 (7): 1–14. https://doi.org/10.1111/gec3.12376.

de Suremain, M.A. (2019). Between political reserve and scientific change: French geographers and decolonisation in tropical Africa, 1945–1967. *Journal of Historical Geography* 66: 9–19. https://doi.org/10.1016/j.jhg.2018.11.002.

Dhupelia-Mesthrie, U. (2016). Betwixt the oceans: the chief immigration officer in Cape town, Clarence Wilfred Cousins (1905–1915). *Journal of Southern African Studies* 42 (3): 463–481. https://doi.org/10.1080/03057070.2016.1168626.

Domosh, M. (1991). Towards a feminist historiography of geography. *Transactions of the Institute of British Geographers* 16: 95–10.

Driver, F. (2001). *Geography Militant*. Oxford: Blackwell.

Driver, F. (2012). Hidden histories made visible? reflections on a geographical exhibition. *Transactions of the Institute of British Geographers* 38: 420–435. https://doi.org/10.1111/j.1475-5661.2012.00529.x.

Eagleton, C. (2016). Designing change: coins and the creation of new national identities. In: *Cultures of Decolonisation: Transnational Productions and Practices 1945–1970* (ed. R. Craggs and C. Wintle), 222–244. Manchester: University of Manchester Press.

Elliott-Cooper, A. (2017). 'Free, decolonised education': a lesson from the South African student struggle. *Area* 49: 332–334. https://doi.org/10.1111/area.12375.

Esson, J. (2020). 'The why and the white': racism and curriculum reform in British geography. *Area* 52: 708–715. https://doi.org/10.1111/area.12475.

Esson, J. and Last, A. (2020). Anti-racist learning and teaching in British geography. *Area* 52: 668–677. https://doi.org/10.1111/area.12658.

Esson, J., Noxolo, P., Baxter, R., Daley, P. and Byron, M. (2017). The 2017 RGS-IBG chair's theme: decolonising geographical knowledges, or reproducing coloniality? *Area* 49: 384–388. https://doi.org/10.1111/area.12371.

Fanon, F. (2019 [1961]). *The Wretched of the Earth*. Milton Keynes: Diana Publishing.

Farish, M. (2010). *The Contours of America's Cold War*. Minnesota: University of Minnesota Press.

Ferretti, F. (2019). Rediscovering other geographical traditions. *Geography Compass* 13 (3): 1–15. https://doi.org/10.1111/gec3.12421.

Ferretti, F. (2020). History and philosophy of geography I: decolonising the discipline, diversifying archives and historicising radicalism. *Progress in Human Geography* 44 (6): 1161–1171. https://doi.org/10.1177/0309132519893442.

Fitzpatrick, H. (2019). Imagining and mapping the end of an empire: Oskar spate and the partition of India and Pakistan. *Journal of Historical Geography* 66: 55–68. https://doi.org/10.1016/j.jhg.2019.07.005.

Gebrial, D. (2018). Rhodes Must Fall: Oxford and movements for change. In: *Decolonising the University* (ed. G. Bhambra, D. Gebrial, and K. Nişancıoğlu), 19–36. London: Pluto Press.

Guha, R. (1999 [1983]). *Elementary Aspects of Peasant Insurgency*, 2e. Durham: Duke University Press.

Gukurume, S. and Maringira, G. (2020). Decolonising sociology: perspectives from two Zimbabwean universities. *Third World Thematics: A TWQ Journal* 5 (1–2): 60–78. https://doi.org/10.1080/23802014.2020.1790993.

Guyot-Réchard, B. (2022). Stirring Africa towards India: apa pant and the making of post-colonial diplomacy, 1948–54. *The International History Review* 44 (4): 892–913. https://doi.org/10.1080/07075332.2022.2093941.

Hodder, J. (2017). On absence and abundance: biography as method in archival research. *Area* 49 (4): 452–459. https://doi.org/10.1111/area.12329.

Hodge, J.M. (2010). British colonial expertise, post-colonial careering and the early history of international development. *Journal of Modern European History* 8: 24–46. https://doi.org/10.17104/1611-8944_2010_1_24.

Hofmeyr, I., Dhupelia-Mesthrie, U., and Kaarsholm, P. (2016). Durban and Cape Town as port cities: reconsidering Southern African studies from the Indian Ocean. *Journal of South African Studies* 42 (3): 375–387. https://doi.org/10.1080/03057070.2016.1174514.

Home, R. (2013). Knowledge networks and post-colonial careering: David Oakley (1927–2003). *ABE Journal* 4. https://doi.org/10.4000/abe.3388.

Hountondji, P. (1990). Dependence in Africa today. *Research in African Literatures* 21 (3): 5–15.

Jansen, J. (ed.) (2019). *Decolonisation in Universities: The Politics of Knowledge*. Johannesburg: Wits University Press.

Jazeel, T. (2017). Mainstreaming geography's decolonial imperative. *Transactions of the Institute of British Geographers* 42 (3): 334–337. https://doi.org/10.1111/tran.12200.

Jazeel, T. (2019). *Postcolonialism*. London: Routledge.

Jöns, H. (2016). The University of Cambridge, academic expertise and the British empire, 1885–1962. *Environment and Planning A: International Journal of Urban and Regional Research* 48: 94–114. https://doi.org/10.1177/0308518×15594802.

Knight, J. (2018). Decolonizing and transforming the Geography undergraduate curriculum in South Africa. *South African Geographical Journal* 100 (3): 271–290. https://doi.org/10.1080/03736245.2018.1449009.

Kothari, U. (2006a). From colonialism to development: reflections of former colonial officers. *Commonwealth and Comparative Politics* 44: 118–136. https://doi.org/10.1080/14662040600624502.

Kothari, U. (2006b). Spatial practices and imaginaries: experiences of colonial officers and development professionals. *Singapore Journal of Tropical Geography* 27: 235–253. https://doi.org/10.1111/j.1467-9493.2006.00260.x.

Lagae, J. and De Raedt, K. (2013). Editorial: global experts off radar. *Abe Journal* 4: 2–13. https://doi.org/10.4000/abe.3384.

Lambert, D. and Lester, A. (2006). *Colonial Lives across the British Empire: Imperial Careering in the Long Nineteenth Century*. Cambridge: Cambridge University Press.

Larner, W. and Laurie, N. (2010). Travelling technocrats, embodied knowledges: globalising privatisation in telecoms and water. *Geoforum* 41: 218–26. https://doi.org/10.1016/j.geoforum.2009.11.005.

Lebelo, R.S., Moloi, K., and Chitumwa, C.C. (2021). Decolonising the postgraduate diploma in higher education curriculum at one university of technology in South Africa. *South African Journal of Higher Education* 35 (1): 82–95. https://dx.doi.org/10.20853/35-1-4406.

Leonard, R. (2022 17 June), Olúfẹ́mi Táíwò's "against decolonisation" *Areo Magazine* Retrieved from: https://areomagazine.com/2022/06/17/olufemi-taiwos-against-decolonisation (accessed 4 July 2023).

Livingstone, D. (1992). *The Geographical Tradition.* Oxford: John Wiley and Sons.

Livsey, T. (2017). *Nigeria's University Age: Reframing Decolonisation and Development.* Basingstoke: Palgrave Macmillan.

Lorimer, H. and Spedding, N. (2002). Excavating geography's hidden spaces. *Area* 34 (3): 294–302.

Mabogunje, A.L. (1980). *The Development Process: A Spatial Perspective.* London: Hutchinson.

Mabogunje, A.L. (2011). *A Measure of Grace: The Autobiography of Akinlawon Ladipo Mabogunje.* Ibadan: Book Builders.

Maddrell, A. (2009). *Complex Locations: Women's Geographical Work in the UK 1850–1970.* Oxford: Royal Geographical Society-Institute of British Geographers/ Wiley-Blackwell.

Mamdani, M. (2011). The importance of research at a university. *Pambazuka News*, Retrieved from: https://www.pambazuka.org/resources/importance-research-university (accessed 4 July 2023).

Mamdani, M. (2019). Decolonising universities. In: *Sharing Knowledge Transforming Societies* (ed. T. Halvorsen, K. Skare Orgeret, and R. Krøvel), 48–67. Cape Town: African Minds.

Matsuda, H. (2016). Japanese "new imperial history" in comparative perspective: the cases of Okinawan and Taiwanese migrations. *Journal of History for the Public* 13: 194–199.

Mazrui, A.A. (1978). *Political Values and the Educated Class in Africa.* Berkeley: University of California Press.

Mbembe, A. (2021). *Out of the Dark Night: Essays on Decolonization.* New York: Columbia University Press.

McFarlane, C. (2011). *Learning the City: Knowledge and Translocal Assemblage.* Oxford: Wiley Blackwell.

McGeachan, C. (2018). Historical geography II: traces remain. *Progress in Human Geography* 42: 134–147. https://doi.org/10.1177/0309132516651762.

McIntyre, W.D. (2009). *The Britannic Vision – Historians and the Making of the British Commonwealth of Nations, 1907–48.* Basingstoke: Palgrave Macmillan.

Mew, S. (2016). Managing the cultural past in the newly independent states of Mali and Ghana. In : *Cultures of Decolonisation: Transnational Productions and Practices 1945–1970* (ed. R. Craggs and C. Wintle), 177–195. Manchester: University of Manchester Press.

Mignolo, W. and Walsh, C. (2018). *On Decoloniality: Concepts, Analytics, Praxis.* Durham: Duke University Press.

Milford, I. (2017). Zoom in, zoom out, change lens: new directions in the historiography of decolonisation. *Journal of Colonialism and Colonial History* 18 (2). https://doi.org/10.135/cc.2017.0026.

Milford, I. and McCann, G. (2022). African internationalisms and the erstwhile trajectories of Kenyan community development: Joseph Murumbi's 1950s. *Journal of Contemporary History* 57 (1): 111–135. https://doi.org/10.1177/00220094211011536.

McQuinn, M. (2022). African trade unions: an introduction. *Global Labour Journal* 13 (2): 157–170. https://doi.org/10.15173/glj.v13i2.5194.

Milford, I., McCann, G., Hunter, E., and Branch, D. (2021). Another World? East Africa, decolonisation, and the global history of the mid-twentieth century. *The Journal of African History* 62 (3): 394–410. https://doi.org/10.1017/S0021853721000566.

Minawi, M. (2020). *The Ottoman Scramble for Africa*. Stanford: Stanford University Press.

Mkandawire, T. (1995). Three generations of African academics: a note. *Transformation* 28: 75–83.

Moosavi, L. (2020). The decolonial bandwagon and the dangers of intellectual decolonisation. *International Review of Sociology* 30 (2): 332–354. https://doi.org/10.1080/03906701.2020.1776919.

Muschik, E.-V. (2018). Managing the world: the United Nations, decolonization, and the strange triumph of state sovereignty in the 1950s and 1960s. *Journal of Global History* 13: 121–144. https://doi.org/10.1017/S1740022817000316.

Mwesigire, B. (2016). Decolonising Makerere: on Mamdani's failed experiment. *African Arguments*, Retrieved from: https://africanarguments.org/2016/08/decolonising-makerere-on-mamdanis-failed-experiment.

Ndlovu-Gatsheni, S.J. (2017). The emergence and trajectories of struggles for an 'African university': the case of unfinished business of African epistemic decolonisation. *Kronos* 43 (1): 51–77. http://dx.doi.org/10.17159/2309-9585/2017/v43a4.

Ndlovu-Gatsheni, S.J. (2018). *Epistemic Freedom in Africa: Deprovincialization and Decolonization*. London: Routledge.

Ngũgĩ wa Thiong'o. (1986). *Decolonising the Mind: The Politics of Language in African Literature*. Portsmouth: Heinemann.

Ngũgĩ wa Thiong'o. (1993). *Moving the Centre: The Struggle for Cultural Freedoms*. Portsmouth: Heinemann.

Nkrumah, K. (1962). *Guide to Party Action: Address by Osagyefo Dr. Kwame Nkrumah at the First Seminar at the Winneba Ideological school on 3rd February 1962*. Accra: Central Committee of the Convention People's Party.

Nkrumah, K. (2009 [1964]). *Consciencism: Philosophy and Ideology for De-Colonization and Development with Particular Reference to the African Revolution*. New York: Monthly Review Press.

Noe, C. (2022). Intellectuals at the Hill: scattered pieces of defiant African scholarship. A commentary on Patricia Daley and Amber Murrey's 'Defiant scholarship: dismantling coloniality in contemporary African geographies'. *Singapore Journal of Tropical Geography* 43: 177–179. https://doi:10.1111/sjtg.12425.

Norcup, J. (2015). *Awkward geographies? An historical and cultural geography of the journal contemporary issues in geography and education (CIGE). (1983–1991)*. PhD thesis. University of Glasgow.

Noxolo, P. (2017a). Introduction: decolonising geographical knowledge in a colonised and re-colonising postcolonial world. *Area* 49: 317–319. https://doi.org/10.1111/area.12370.

Noxolo, P. (ed.) (2017b). Special section: decolonising geographical knowledge in a colonised and re-colonising postcolonial world. *Area* 49: 317–334.

Nwosu, N.I. (1993). The dynamics of Nigeria's decolonization policy in Africa. *Transafrican Journal of History* 22: 74–86.

Nyerere, J.K. (1966). *Freedom and Unity: Uhuru Na Umoja: A Selection from Writings and Speeches 1952–65*. Dar es Salaam: Oxford University Press.

Obeng-Odoom, F. (2019). The intellectual marginalisation of Africa. *African Identities* 17 (3–4): 211–224. https://doi.org/10.1080/14725843.2019.1667223.

Olukoshi, O., Hormeku-Ajei, T., Balaji, A., and Nayar, A. (2020). Reclaiming Africa's early post-independence history. *Africa Is a Country*, 202. Retrieved from: https://africasacountry.com/2020/07/reclaiming-africas-early-post-independence-history (accessed 20 December 2022).

Oswin, N. (2020). An other geography. *Dialogues in Human Geography* 10 (1): 9–18. https://doi.org/10.1177/2043820619890433.
Padilla, N.L. (2021). Decolonizing indigenous education: an indigenous pluriversity within a university in Cauca, Colombia. *Social & Cultural Geography* 22 (4): 523–544. https://doi.org/10.1080/14649365.2019.1601244.
Peake, L. and Kobayashi, A. (2002). Policies and practices for an antiracist geography at the millennium. *The Professional Geographer* 54: 50–61. https://doi.org/10.1111/0033-0124.00314.
Pietsch, T. (2013). *Empire of Scholars Universities, Networks and the British Academic World, 1850–1939*. Manchester: Manchester University Press.
Porter. (2004). *The Absent-minded Imperialists*. Oxford: Oxford University Press.
Power, M. and Sidaway, J.D. (2004). The degeneration of tropical geography. *Annals of the Association of American Geographers* 94: 585–601. https://doi.org/10.1111/j.1467-8306.2004.00415.x.
Pulido, L. (2002). Reflections on a white discipline. *The Professional Geographer* 54 (1): 42–49. https://doi.org/10.1111/0033-0124.00313.
Quijano, A. (2000). Coloniality of power and eurocentrism in Latin America. *International Sociology* 15 (2): 215–232. https://doi.org/10.1177/0268580900015002005.
Radcliffe, S. (2017). Decolonising geographical knowledges. *Transactions of the Institute of British Geographers* 42: 329–333. https://doi.org/10.1111/tran.12195.
Radcliffe, S. (2022). *Decolonizing Geography: An Introduction*. Oxford: Wiley.
Ramutsindela, M. (2022). Epistemic injustice in geography. A commentary on Patricia Daley and Amber Murrey's 'Defiant scholarship: dismantling coloniality in contemporary African geographies'. *Singapore Journal of Tropical Geography*. https://doi.org/10.1111/sjtg.12426.
Rodney, W. (1990 [1975]). *Walter Rodney Speaks: The Making of an African Intellectual*. Trenton: Africa World Press.
Rogers, R. (2019). Conversations about the transnational: reading and writing the empire in the history of education. In: *The Transnational in the History of Education* (ed. E. Fuchs and E. Roldán Vera), 101–124. Basingstoke: Palgrave Macmillan.
Rose, G. (1995). Tradition and paternity: same difference? *Transactions of the Institute of British Geographers* 4: 414–416. https://doi.org/10.2307/622970.
Roy, A. (2012). Ethnographic circulations: space–time relations in the worlds of poverty management. *Environment and Planning A: Economy and Space* 44 (1): 31–41. https://doi.org/10.1068/a44180.
Ryan, J.R. and Naylor, S. (2009). *New Spaces of Exploration*. London: I.B. Tauris.
Sarmento, J. (2019). Portuguese tropical geography and decolonization in Africa: the case of Mozambique. *Journal of Historical Geography* 66: 20–30. https://doi.org/10.1016/j.jhg.2018.11.002.
Schelhass, B., Ferretti, F., Novaes, A.R., and Schmidt Di Friedberg, M. (eds.) (2020). *Decolonizing and Internationalising Geography: Essays in the History of Contested Science*. Cham: Springer Nature Switzerland AG.
Schildermans, H. (2021). *Experiments in Decolonizing the University: Towards an Ecology of Study*. London: Bloomsbury Publishing.
Schwartz, B. (2011). *The White Man's World*. Oxford: Oxford University Press.
Sharp, J. (2019). Practising subalternity: postcolonial Tanzania, the Dar School and pan-African geopolitical imaginations. In: *Subaltern Geographies* (ed. T. Jazeel and S. Legg), 74–93. Athens: University of Georgia Press.

Sicherman, C. (2005). *Becoming an African University: Makerere, 1922–2000*. Trenton: Africa World Press.

Slater, D. (1974). Contribution to a critique of development geography. *Canadian Journal of African Studies / Revue Canadienne Des Études Africaines* 8 (2): 325–354. https://doi.org/10.1080/00083968.1974.10804439.

Soja, E.W. (1968). *The Geography of Modernization in Kenya: A Spatial Analysis of Social, Economic and Political Change*. Syracuse: Syracuse University Press.

Sou, G. (2021). Aid micropolitics: everyday southern resistance to racialized and geographical assumptions of expertise. *Environment and Planning C: Politics and Space*. https://doi.org/10.1177/23996544211048196.

Stanek, M.B. (2019). Decolonial education and geography: beyond the 2017 royal geographical society with the institute of British geographers annual conference. *Geography Compass* 13: e12472. https://doi.org/10.1111/gec3.12472.

Steinmetz, G. (2016). Social fields, subfields and social spaces at the scale of empires: explaining the colonial state and colonial sociology. *The Sociological Review Monographs* 64 (2): 98–123. https://doi.org/10.1111/2059-7932.12004.

Stockwell, S. (2018). *The British End of the British Empire*. Cambridge: Cambridge University Press.

Stoler, A.L. and Cooper, F. (eds.) (1997). *Tensions of Empire*. Berkeley: University of California Press.

Surman, J. (2018). *Universities in Imperial Austria 1848–1918: A Social History of A Multilingual Space*. West Lafayette: Purdue University Press.

Táíwò, O. (2022). *Against Decolonisation: Taking African Agency seriously*. London: Hurst Publishers.

Tolia-Kelly, D.P. (2017). A day in the life of a geographer: 'lone', black, female. *Area* 49 (3): 324–328. https://doi.org/10.1111/area.12373.

Vahed, G. (2016). Family, gender, and mobility among passenger migrants into colonial natal: the story of Moosa Hajee Cassim (c.1840s–1921). *Journal of Southern African Studies* 42 (3): 505–522. https://doi.org/10.1080/03057070.2016.1171557.

Van Sant, L., Hennessy, E., Domosh, M. et al. (2020). Historical geographies of, and for, the present. *Progress in Human Geography* 44 (1): 168–188. https://doi.org/10.1177/0309132518799595.

Visser, G., Donaldson, R., and Seethal, C. (ed.) (2016). *The Origin and Growth of Geography as a Discipline at South African Universities*. Stellenbosch: Sun Press.

Ward, S. (ed.) (2001). *British Culture and the End of Empire*. Manchester: Manchester University Press.

Waters, R. (2016). Henry Swanzy, Sartre's zombie? black power and the transformation of the Caribbean Artists Movement. In: *Cultures of Decolonisation: Transnational Productions and Practices 1945–1970* (ed. R. Craggs and C. Wintle), 68–95. Manchester: University of Manchester Press.

# Chapter Two
# 'New, Interesting, and Even Exciting Opportunities': Geography and the Founding of Colonial Universities in Africa[1]

## Introduction

Soon after the publication of the Commission on Higher Education in the Colonies, the so-called Asquith Report, in 1945, The British geographer S.J.K (Kenneth) Baker, then working at the University of Liverpool, sat down eagerly to read it.[2] This was a report that would have significant impacts on the provision and shape of higher education in the British colonies through the establishment of new university colleges in the Gold Coast (Ghana), Nigeria, and Uganda. It also had a dramatic impact on Baker's career trajectory. As an individual with 'a long-standing concern for international affairs and race relations', as well as a research specialism in population mapping in East Africa, Baker was inspired by the Report to seek a position in one of the new African university colleges (Baker 2015). He applied for the first vacancy he saw advertised, at Makerere College in Uganda.[3] Baker's application was successful, and in January 1947 he arrived in Uganda where he began the task of establishing the institution's geography department. Baker is an example of a British geographer who helped to realise late-colonial visions for the expansion of higher education in Africa. In some ways, Baker is exceptional – his career at Makerere spanned twenty years – a length of time few other British geographers spent at a single African university. However, as we will see in the rest of this chapter, he was part of an identifiable

*Decolonising Geography? Disciplinary Histories and the End of the British Empire in Africa, 1948–1998*, First Edition.
Ruth Craggs and Hannah Neate.

cohort of British geographers (almost all of whom were men) who were tasked by the British government in the late 1940s with establishing geography as a founding subject at the select band of prestigious new African colonial university colleges.

This chapter focuses on developments in geography in British colonies in Africa in the 1940s and 1950s through the foundation and staffing of geography departments in the new university colleges in the Gold Coast, Nigeria, and Uganda. It also examines the role of the University of London – to which these institutions were affiliated as constituent colleges – and the South African university landscape, which was an important part of this British (African) academic world. Unlike the rest of the book, British voices dominate this chapter. This reflects one of its key arguments: that geography in the new colonial university colleges in Africa was dominated by British colonial perspectives. At independence, this bequeathed African countries geography departments dominated by British staff and British visions for the discipline of geography which reinforced colonial ideas about African land, people, and climate, and about development as (Western) modernisation.

The chapter develops two other important arguments that run through the book. The first is that a biographical focus can highlight the contingencies shaping geography's history. Opening up the histories of geography beyond those of published research, this chapter explores why British geographers moved to colonial university roles, and demonstrates the importance of exploring the interconnections between (geo)political and economic transformations and the consequences for academic job markets. An explicit attentiveness to biographical detail, the mundane, and the everyday adds to the rich work within historical geography on life-writing and the interlinked nature of the 'intimate' with 'large-scale histories and geographies' (Ogborn 2005, p. 382; Hodder 2017; Jones 2018).

A further contribution of this chapter is to the broader interdisciplinary historiography of twentieth century empire and decolonisation. University colleges in Africa were explicitly a part of the colonial development project which aimed to recast empire after World War II, to ready colonies for (eventual) independence, and to provide an educated and anglophile elite with whom the British could continue to work after decolonisation (Livsey 2017). Often also demanded locally, universities were visible and important sites through which colonial development was materialised, practiced, and experienced. Focusing on the roles and experience of geographers within this broader process brings to the fore the lived experience of the fraught and often ambivalent project of late colonial development. It also allows spaces for African agency to be recognised, through the voices and experiences of colonial students and early staff members in these departments, as well as the nationalist leaders campaigning for further access to higher education.

The chapter serves an introduction to the geography departments and university colleges (and later universities) that are a key focus of the book. It begins with the establishment of new university colleges in Nigeria (Ibadan), the Gold

Coast (Legon) and Uganda (Makerere) in the 1940s as an act of late-colonial development caught up in the wider politics of empire and decolonisation. The chapter moves on to explore the colonial networks that underpinned the foundation of these university colleges. This feeds into discussion of the early faculty including an exploration of the motivations of geographers working in these colonial locations and the challenges associated with the establishment of new departments. Geography was a central academic discipline adopted from the outset in these new colonial university colleges and the teaching and research of the discipline is explored next, with discussion turning to how the research conducted and the curriculum delivered at the university colleges closely mirrored those of British geography, whilst often ignoring the politics of (de)colonisation. The chapter ends by drawing attention to working and associational life on campus, examining the informal labour of producing the elite space of the university, including the gendered and racialised aspects of this.

## Asquith Colleges and the 'Imperial Family of Universities'

Ajayi et al. (1996, p. 54) describe the 1940s as an optimistic period when expectation had been raised in British colonies in Africa about 'major structural changes – constitutional, economic and social reforms, especially in the area of education'. Reflecting African demands and as part of this vision for post-war change, Sir Cyril Asquith led a Commission on Higher Education in the Colonies; this resulted in the 1945 Asquith Report, a comprehensive review of higher education in the British colonies that provided a blueprint for expanding universities, based on a British model, to colonial subjects. It outlined proposals for these new colonial universities to be 'part of the imperial family of universities' (Maxwell 1980, p. 20) whilst also underpinning wider post-war plans of facilitating the colonies towards self-government. It joined several other Commissions focused on higher education in East Africa, West Africa, and the West Indies.

A key recommendation of the Asquith Commission was the establishment of new colonial university colleges in 'special relationship' with the University of London. This special relationship involved the establishment of new university colleges tied to the University of London, who awarded degrees, agreed the curricula (based on those taught in London), appointed faculty and moderated final exams. The university colleges were designed from the outset as residential institutions in the mould of Oxford and Cambridge, with an emphasis on Liberal Arts and Science, and as such, were positioned as elite centres of learning whose purpose was to educate future leaders in key roles in the civil service and other leadership positions (Mngomezulu 2004). The anticipated number of students was modest to begin with, reflecting a wider colonial belief that a small number of select universities serving vast territories would suffice for the anticipated development in Africa. This educational project was funded by Colonial

Development and Welfare Funds (Stockwell 2008). Five new university colleges were established in the late 1940s, either from scratch or building on existing educational institutions: the University College Ibadan, the University College of Khartoum, the University College of the Gold Coast, Makerere University College, and the University College of the West Indies. The University of Malaya was also founded as part of this process. This resulted in the creation of prestigious new higher education institutions that sat alongside universities founded earlier in India and Ceylon (Maxwell 1974).

The newly formed University Colleges followed a British model with emphasis on 'transplantation, not adaptation' (Ajayi et al. 1996, p. 56). Walter Rodney (1990 [1975], p. 36) described the process derisively as 'the University of London establishing virtually a colonial out-house' in these territories. However, their existence was also the fulfilment of decades of agitation for higher education provision from Africa's educated elites, especially in West Africa (Livsey 2016). African elites, often themselves educated in Britain, articulated demands for not only the presence of universities in Africa, but also for the qualifications offered at these institutions to hold equivalence with their British counterparts (Mngomezulu 2004; Livsey 2016).

The 'Asquith Colleges' were an articulation of significant plans to expand British colonial education on the African continent and wider colonies in the twentieth century. They were not, however, the first colonial universities in Anglophone Africa. Fourah Bay College was established by the Church Missionary Society in 1827 and affiliated with Durham University from 1876 until the mid-1960s, eventually becoming the University of Sierra Leone. Gordon Memorial College, founded by Lord Kitchener in 1898, was the forerunner of the University (College) of Khartoum in Sudan (Mngomezulu 2004). There was also an important pre-colonial history of higher education in Africa, with universities and other institutions of higher education in Cairo, Fez, and Timbuktu, for example, though these looked very different to the colonial universities imported on a Western model (Ajayi et al. 1996; Ndlovu-Gatsheni 2017).

The experience at the Asquith colleges was broadly comparable to other colonial colleges in Anglophone Africa, and beyond, where their early years was dominated by British appointments and 'circulatory migration' before later 'indigenisation' (Pietsch 2013, pp. 70–71, 78; Agbodeka 1998; Stockwell 2008). With different racial dynamics, there were also parallels with the earlier experiences of the new university colleges in the settler colonies of Australia, Canada, and New Zealand which remained intimately tied to the UK university system until the 1950s (Pietsch 2013). In South Africa too, there were similar patterns of recruitment and staffing, albeit starting earlier with the founding of the first South African universities around 1916. However, these early universities were more or less closed to the non-white South African population, who had to attend the South African Native College opened at Fort Hare in 1916 for black African students (Ajayi et al. 1996).

## Geography at the Asquith Colleges: Colonial Networks

Geography was one of the original subjects offered to students at the Asquith colleges. At Ibadan in Nigeria, geography was offered on the recommendation of the Elliot Commission on Higher Education in West Africa, which saw the subject as one of a suite of arts disciplines that could be offered 'to provide a training in clear and objective thought'.[4] This emphasis on providing high-level academic training in order to equip graduates with necessary skills to take on future high-level professional and leadership roles was common across the new university colleges. For example at Makerere University College (est. 1949) the first subjects taught were English, History, Social Studies, Geography, Mathematics, Art, Physics, Chemistry, and Biology.[5] Debates concerning 'manpower' [*sic.*] needs and the importance of technical training grew in importance as countries transitioned towards independence, however, the university colleges that were established in the late-1940s were based on an elitist British model of higher education that bestowed great prestige upon these new institutions.

In 1948, University College Ibadan (UCI) took in its first cohort of students studying geography as part of a general degree (alongside two other subjects); cohorts of geography students followed soon after at the University of the Gold Coast, and Makerere University College (Mabogunje 1998). Although the plans for the Asquith Colleges were shaped significantly by African demands, in the early years 'mechanisms for staff appointments favoured British candidates' who were 'often selected through informal British academic networks that encompassed Canada, Australia and New Zealand as well as Britain itself' (Livsey 2017, p. 43). Indeed, appointment of geographers to the newly established geography departments at the Asquith colleges was reliant on British colonial institutions and the informal networks that underpinned them. As Pietsch has noted for an earlier period, appointments can be seen as 'a power-laden, historically contingent … world of access and exclusion' (Pietsch 2013, p. 61). The Inter-University Council (IUC), a body established in 1946 by the UK government to work with British universities 'to assist the advancement of higher education in developing countries', handled requests for staffing.[6] Whilst the final decision to appoint candidates was with the appointing university, it is clear from the archives that a few British geographers – notably R.O. Buchanan of the London School of Economics and Robert Steel of Oxford and Liverpool – wielded significant influence in this process across the new geography departments at Ibadan, Makerere, and the Gold Coast.[7]

Steel – lecturer in colonial geography at Oxford, and later professor at Liverpool – began his career-spanning interests in Africa with fieldwork in Sierra Leone in the 1930s. This led to his membership of the Ashanti Survey team, a social and political survey supported by the Colonial Research Council, colonial government and Ashanti Confederacy Council in the Gold Coast (Prothero 1998). His influence, in both Britain and Africa, was shaped by his roles on committees of learned societies, including the Institute of British Geographers, the Geographical Association, the Inter-University Council, and the Royal African Society. R.O. Buchanan was born in New Zealand and came to Britain in 1925 to study at the London School of Economics (LSE), where he was later chair of geography. Buchanan was one of the founders of the Institute of British Geographers and, like Steel, served on several committees throughout his career including the Association of Commonwealth Universities, the Geographical Association, and the British Association for the Advancement of Science (Wise 1983). These connections, combined with the hierarchical and informal processes of academic appointments across Britain and the wider colonial world, put both in pivotal positions for hiring decisions in the newly established universities in Africa, and for geographers seeking to (re)join UK academia after periods in Africa (which we discuss in Chapter 7).

Steel regularly received requests for suggestions for appointments from contacts working in African universities.[8] Many geographers who worked in the new university colleges were appointed on his recommendation: some had direct academic links (e.g. Ernest Amano Boateng, University College Gold Coast, studied under Steel at Oxford), in other instances he was referee (e.g. for Terry Coppock, visiting scholar at Ibadan 1964–1965), or provided advice to more junior colleagues (e.g. Dick Hodder, lecturer at Ibadan 1956–1962, who Steel advised on how to make the necessary connections to apply).[9] He was on the selection panel for the chair of geography and head of department role at Ibadan, appointing Michael Barbour to the position in 1962.[10] Steel's archives abound with informal letters reflecting the continuing value of 'writing to friends' in the appointments process across the former colonies in Africa, South East Asia, and the Caribbean until the 1960s (Pietsch 2013, p. 82).

## Early Faculty

In an address to the Geographical Association in 1947 Alexander Carr-Saunders, chairman of the Inter-University-Council 'pointed to the new, interesting and even exciting opportunities' that now existed for geographers in colonial universities. These new institutions offered 'the attraction and responsibility attached to pioneering' he noted, continuing 'We have not failed [in the past] as a nation to find men to shoulder our national responsibilities overseas' (Carr-Saunders 1948, pp. 1,4,7). The implication was that working in these academic posts should be understood in similar terms: as service in less developed places with the added excitement of being a colonial pioneer. These were gendered opportunities: it was men who (in this account) had taken on these responsibilities in the past and would do so again. The 1945 commission which had recommended the founding of University College Ibadan and the University of the Gold Coast was also explicit about this: 'Any institution or institutions in West Africa will have to ... rely on European staff for the most part upon young men who will go out to West Africa to seek new opportunities'.[11] This sense of pioneering, in institutional and research terms, is present in the accounts of the early faculty at the new university colleges.

In the late 1940s a career in geography overseas could be framed within a broader tradition of colonial service, embracing both the excitement and responsibility associated with empire roles. Indeed, previous roles in colonial or military service such as surveying, cartography, forestry, agriculture, or resource management were often seen as useful precursors to working in newly established geography departments (see also Hodge 2007). For others, political or religious convictions, including experience in missionary service underpinned motivations for working in Africa in the 1940s and 1950s (see Williams 2001).

In the context of the University College of the Gold Coast, Makerere University College, and University College Ibadan the responsibility of setting up new geography departments rested on the shoulders of a small number of key staff who were tasked with establishing the departments both physically, in terms of new estate and equipment, and intellectually. Students studying geography at University College Ibadan were taught initially by the British duo of Keith Buchanan and John Pugh, teaching human and physical geography respectively. Buchanan (Figure 2.1) came from a lectureship at Pietermaritzburg in South Africa, travelling north with his family because 'they could not accept the attitude of the Nationalist Government [newly elected in South Africa] towards the non-whites'.[12] Buchanan was a colourful character who refused to conform to the expectations of senior university administrators. According to Akin Mabogunje (2011, p. 79), who as an undergraduate was taught by Buchanan, 'He was very unorthodox in his attitude to many aspects of campus life and he used to jump into the class through the window nearest his office. He was also heavily-bearded and loved wearing floral shirts over his short trousers.'

Buchanan's clothing choices obviously made a strong impression on everyone as they were also mentioned by his colleague Pugh who recalled he would wear shirts 'made from cloth from the local market with a pattern of rosebuds or Crown birds'. Pugh also noted that Buchanan 'had a wide range of acquaintances in Ibadan Town ... and never hesitated if he saw a party of drummers heading somewhere – he would follow.'[13] This engagement by a British academic with local communities outside of the university was very unusual in this period. His lectures touched on issues of colonialism and development, and showed signs of the more radical critiques of his later scholarship (Power and Sidaway 2004). Eventually the University College Ibadan principal deemed that Buchanan's clothing provided a visual representation of his wider inappropriateness for the job, leading to his departure in 1951 for London School of Economics, and then what was to become the University of Wellington in New Zealand.[14]

**Figure 2.1** Keith Buchanan. Image Services, Victoria University of Wellington. Portraits – Buchanan [Geography Professor], 1958–1959. Tapuaka – Heritage Collections & Archives, Te Herenga Waka, ref. VUWI00108.

Buchanan's story – of radical political commitment and local engagement – is not however representative of the early faculty at Ibadan. If we turn to John Pugh, Buchanan's first colleague, a different story emerges. Following an undergraduate degree at Cambridge, Pugh arrived in Nigeria as a result of wartime service in the Royal Engineers and later the Colonial Office, where his role was to carry out mining surveys. Pugh remained in Nigeria following the war to continue surveying for the colonial government, and in this role he was involved in both the primary triangulation of the country, and in teaching Nigerian surveyors at the Survey School at Oyo.[15] It was from here that he was asked to join the new geography department. Pugh therefore arrived in Nigeria courtesy of the war, and remained as part of the colonial service. He did not actively make the decision to travel, nor was his move prompted by political commitments. Rather, like many in this period, his career trajectory was shaped by war and then by the on-going opportunities that colonial service offered. In the early years, the department had close links to the Nigerian Colonial Service, and to geographical knowledge – through survey and agriculture – for colonial development. Pugh's trajectory highlights how geography in these colonial universities was made in part not through some 'pure' academic expertise but also through the wider visions for the discipline of explorers, travellers, and surveyors.

In Uganda the task of establishing geography in the colonial university college was taken on by S.J. (Kenneth) Baker. Makerere College was established in 1922 as a technical college located on the outskirts of Kampala. As the only significant educational institution offering advanced education in East Africa, Makerere became an Asquith College, and from this point was positioned as a central hub for the development of higher education across the region. The establishment of geography as an academic discipline in Uganda gained momentum in January 1947, when aged 49 Baker arrived at Makerere from his previous position at the University of Liverpool. Baker served Makerere in both departmental and high-level administrative roles until his retirement and return to the UK in 1968. He was awarded an OBE in 1966 for services to Education in Uganda and long-standing efforts to promote Anglo-Ugandan cooperation.

For Baker the University of Liverpool provided 'a training ground for international citizenship'; during his time there, he was member of the Student Christian Movement, East and West Friendship Committee, the League of Nations Union, and the International Society (later the Cosmopolitan Club) (Baker 2015, p. 2). It also embedded him to the networks that facilitated the career moves of numerous British geographers to African colonial universities.

Unlike many others whose first experience of Africa or international travel came through wartime service, Baker was a conscientious objector and therefore was not posted overseas.[16] His specialist interest in East Africa developed during his time as a student and then member of staff at Liverpool, where he was 'responsible for the courses on Africa' (Baker 2015 p.2). In addition to teaching and research, Baker also became a 'guide, philosopher and friend to a constant stream of African students' studying at Liverpool (Baker 2015, p.2). He later reflected:

> For a number of years I had acted as tutor to overseas students in the University of Liverpool and at the time of my appointment to Makerere I was chairman of the University Cosmopolitan Club. During the war of 1938–1945 I had close connections with the Students Department of the British Council and with the Colonial Office Welfare Department, and many contacts with the West African and West Indian population of Liverpool.[17]

During his early years at Makerere, Baker found working there 'delightful' as he enjoyed the climate and situation of the campus, as well as the low staff to student ratio: at that time there were 'rather less than 250 students in the College'. At that point he also commented, 'I imagine, perhaps wrongly, that it has not yet caused any reduction in my intellectual output!'[18] In the longer term, and in common with other geographers who focused more on the administrative aspects of setting up and running departments, Baker produced little in the way of published research. Despite this, Baker stayed at Makerere for close to two decades,

**Figure 2.2** S.J.K. Baker teaching in the geography department, Makerere College. Oxford Development Records Project, The Bodleian Libraries, University of Oxford, ODRP MSS Afr. S. 1825 (3) BOX I, photograph 54.

driven by an 'active desire to contribute to the development of higher education in the colonial territories'.[19]

With Pugh and Buchanan at Ibadan, and Baker at Makerere, we see examples of British geographers who worked in new African universities because of existing colonial links, political convictions, as well as broader social commitments. At the University College of the Gold Coast a series of British geographers were also integral to establishing geography as an academic presence. Their careers, in contrast, demonstrate how, for some, African universities were stop-off points in the wider context of longer (and sometimes controversial) academic careers.

Geography at the University College Gold Coast was established in 1948, and to begin with this was driven by W.J. (Bill) Varley, who was recruited to set up the department and lead on the development of the University College more broadly. In this strategic role Varley 'worked tirelessly alone for nearly two years, except for the active support of his wife who set up the first photography unit of the college' (Agbodeka 1998, p. 59–60). In the context of a very male-oriented discipline in the 1940s and 1950s here we see an important glimpse of the supporting role that Varley's wife played in establishing the department. This was a common though not always acknowledged pattern. Varley was joined by T.E. Hilton and J.H. Hubbard, two further male British geographers in 1950.

Varley studied under Fleure as an undergraduate at Aberystwyth, University of Wales, where he also completed his MA, followed by a period at the University of Liverpool. By the time he got to University College Gold Coast he was 'one of the most powerful professors of the early College' (Agbodeka 1998, p. 59–60). However, Varley went on sick leave in 1956, and then made an abrupt departure 'sometime in May 1957' linked to rumours that he had falsified his doctoral degree (Agbodeka 1998, p. 122).[20] When asked at the airport to explain his departure Varley is reported to have exclaimed 'Ten years has been enough'![21] Varley is the subject of a series of intriguing files in the National Archives relating to communism in Africa and his 'connection with subversive factions in the Gold Coast'.[22] However, rather than linking Varley with any political activities the files mainly indicate that he had a tendency to drink a lot and stretch the truth.

Following Varley's departure, a professor of geography role was advertised by the Inter-University Council, with applications sent to London, not the Gold Coast. As was often the case, Robert Steel was nominated to the board of electors and this resulted in the appointment of Professor E.H.G. Dobby in 1958. Dobby moved from the University of Malaya where 'he started and was largely responsible for building up the Department of Geography' (Hodder 1982, p. 510). His experience in founding a geography department in another colonial university, along with proven administrative and academic credentials (e.g. establishing what is now known as *The Singapore Journal of Tropical Geography*) provided a relevant basis for the appointment to the role at what by then was University College Ghana, following the country's independence in 1957. However, Dobby's role in Singapore wasn't always without controversy; he departed following his presentation of a speech critical of the older university college alumni.[23] Varley and Dobby are examples of how controversial moments could punctuate the academic careers of geographers, as well as, for Varley at least, how colonial positions could give substantial power to those whose qualifications may not have been accepted in the UK.

The archival record also reveals pragmatic decisions regarding the opportunities and challenges of positions in the new university colleges, which tend to dominate over any sense of mission or political engagement. As Dick Hodder (Ibadan 1956–1962) recorded, for example, 'I had no religious or political motivation in going abroad, nor did I have any strong interest in development. Indeed, having once decided to enter academic life, my motivation in going was that at 28 and with a wife and child I could not face going to Hull on £450 a year'. Initially working in Singapore at the University of Malaya, Hodder 'became keen to move to an African university in 1954-5' to develop his CV for a move back to the UK: 'I felt that to widen my field experience in this way would give me a better chance of competing in the British academic market'.[24] Michael Barbour (Ibadan 1962–1972) also took a role in Africa as a way to progress his career. As he recorded, 'Job in Khartoum [1948–1956] seen as a means of entry without D.Phil. etc. to academic life – intention to try in due course for jobs in UK or Commonwealth'.[25]

Hodder, Barbour, and Dobby were not alone in working in multiple colonial contexts. For instance, nearly forty percent (seven of eighteen) of the British geographers who worked at Ibadan had previously worked in other universities across Africa, South-East Asia, and New Zealand. Movement within African institutions and on to Australia, New Zealand, and Canada was common, painting a picture of academic mobility facilitated through the British World that drew on long-standing colonial connections (Pietsch 2013).

Dobby left Ghana for the University of Cape Town, South Africa, where he joined the Scot William Talbot in a wider institutional landscape still dominated by British connections. The founding geographers in South African departments were all from Europe: Petrus Serton appointed in 1920 to Stellenbosch (from the Netherlands, see Figure 2.3); and John Wellington, Frances Plummer, and Talbot, all British, at Witwatersrand, Pretoria, and the University of Cape Town respectively (Barnard 2016). As in the Asquith Colleges, they 'brought their own myths, histories, and professional training' from Europe, and founded a discipline that closely followed the shape of British, and to a lesser extent, French, Dutch and German geography (Barnard 2016, p. 1). Most oversaw departments that were almost exclusively white (Figure 2.3).

**Figure 2.3** Geography students with Petrus Serton centre, front row, at the University of Stellenbosch in 1953. Stellenbosch University Archives, Foto Versameling Watson-Lockley, reference 16933/24414.

For these mainly British men working in African geography departments, these choices to move continents, sometimes more than once, were taken primarily to build careers. Often accompanied by their families, a position in Africa could be a temporary or permanent move, up the career ladder and occasionally away from controversy. These personal and professional choices shaped the making of geography in Africa.

## Building a Department

While the prospect of helping to set up a new department was in some respects exciting it also presented challenges, for example, overseeing the specifications and requirements for a fully functional modern geography building. For Baker at Makerere University College to begin with he was allocated space within pre-existing campus buildings – 'two pleasant and well furnished rooms in the top floor of the Main Building, one used as a Studio and the other as a Lecture Room' (seen in Figure 2.2), though a departmental building would follow as part of later expansions of the institution.[26]

By contrast, University College Ibadan was built from scratch which involved appointing internationally renowned British architects Maxwell Fry and Jane Drew to design the campus, including the halls of residence and faculty buildings (Figure 2.4). Before the campus was built UCI was based at the 'Temporary Site' with the geography department housed in the geography 'lecture hut'. Given Pugh's stout booted background this was something that he took in his stride. However, he noted how these conditions were not to everyone's taste:

> Staff accommodation was in the wooden huts of the old military hospitals forming the Temporary Site. To those of us accustomed to life in tents and mud huts it was luxurious, with electric light (spasmodically), running water (usually) and furniture other than folding camp equipment. Some of the newcomers from Europe, however, found it rather primitive, with wooden shutters instead of glass windows and with dry "thunderbox" toilets.[27]

Fry and Drew had a formidable international reputation as specialists in tropical modernism, an architectural style that bought together aspects of international modernism with a design ethos that included adaptations to suit tropical climates. However, as Livsey (2014, p. 673) has noted, Fry and Drew 'were part of the colonial establishment, sharing colonial-era attitudes', including about the relationship between climate, race and character, and about the role of modern architecture in shaping modern citizens.

**Figure 2.4** Tedder Hall, University of Ibadan, early 1960s, photographer unknown. Rockefeller Foundation Records, NIGERIA, Series 497 University of Ibadan, Box 134, Folder 2563, number 3. Courtesy of the Rockefeller Archive Center.

For John Pugh, it was not an easy working relationship:

> [T]here was considerable dissatisfaction with the architects, Maxwell Fry and Jane Drew. When in Ibadan, either together or singly, they stayed in the Principal's house, and their dealings with academic staff were conducted in a manner which could be described as both condescending and conceited, as though only architects could know what was good for academic layout.[28]

Indeed, during an exhibition of proposed designs that took part in the Geography 'lecture hut', Jane Drew alienated her audience by insisting on a creating a large 'floral arrangement' in a tank that had been 'constructed for experiments and demonstrations of wave and stream action.' The fact that this damaged 'a major piece of academic equipment' did not go down well.[29] Fry and Drew were not popular with academic staff who issued a vote of no confidence as a means of expressing their frustrations with the arrogant approach the architects were taking (Livsey 2014). The final result, however, was an impressive modern university campus:

> Fry and Drew's university buildings rose in dazzling concrete: halls of residence, faculties, places of worship, a theatre and library, and the infrastructure to serve them. They were spectacular. Even Nigerian nationalists, not easily impressed by the achievements of colonial development, were momentarily stunned. (Livsey 2014, p. 667)

As at Ibadan, establishing an exclusive, modern campus was a key priority in the Gold Coast, and Varley oversaw the move of University College Gold Coast from Achimota College to a new campus site at Legon in 1952, including the design and construction of a new geography building. It was specified that the building should have two lecture rooms, a library, a 'study for honours types', two practical rooms, a drawing office, a dark room, an instrument room, a room for photogrammetry, a model room, and a meteorological records room. There were also discussions about the need for air conditioning.[30] Despite the aspirations for a well-equipped cutting-edge departmental building, translating this into a feasible design was a challenge. As Varley remarked in a letter to the Vice-Principle of the College in December 1955: 'I'm afraid I have to say that the plans submitted by the architect for the school of geography are so silly that I suggest that it would be kind to pretend they don't exist.'[31] Revised plans were eventually approved in November 1956. The departments of geography and geology were (and remain) sited on 'the main avenue' on the campus East Hill, a site which was of 'great importance in the approach from the main entrance to the Dodoma Road' (Figure 2.5).[32] Early British faculty were key in materializing the very physical shape

**Figure 2.5** The geography department building, University of Ghana (now renamed geography and resource management) in 2017.

of geography departments, in doing so, concretising visions for the discipline in the mould of mid-century British geography.

On arrival at University College Ghana, Dobby was circumspect about some of the decisions that had been made by his predecessor Varley relating to kitting out the department:

> two station wagons are held by the department … I am inclined to think that greater control could be exercised with greater economy, … previous estimates have made generous provision of capital and recurrent expenses for transport yet I am not satisfied that serious academic research has been consistently served. Travel is not of itself research.[33]

His concerns reveal that investment in estates and equipment did not guarantee academic outputs. This also related to the challenges faced by the small number of staff and the impacts of this. University College Ghana compared poorly with other similar departments: 'Apart from the Professor, four teachers are now on the establishment, a total smaller than at either Ibadan or Makerere, it is fewer than most of the London colleges catering for honours degrees'.[34] Dobby was keen to highlight that it took longer to prepare for teaching in a new place without resources and that teaching hours were longer. This raised concerns about the ability to function as an academic department: 'the scantiness of staff in part explains the paucity of research which has been put out by members of the department.' Calls were made at this time for new lecturers in climatology, historical geography and applied geography 'branches necessary to the balance of our teaching and at the same time offering wide scope for research of practical value to Ghana'.[35]

Dobby had left University College of Ghana by the end of 1959, serving his post as Professor of Geography for only one year before taking up a post at the University of Cape Town, South Africa.[36] Whilst we have no evidence for why Dobby moved on so quickly, it is possible that these material conditions, along with the rapid political changes in Ghana and at the university from the late 1950s combined to influence his decision. The pull of a well-resourced South African department, in an institution where policies of Africanisation did not apply (quite the opposite) may have been too great. What we can say for sure is that Dobby's career trajectory highlights the generally peripatetic nature of British geographers who worked in the new university colleges in Africa. After his departure T.E. Hilton was temporarily left in charge of a department in an interesting state. As he recorded, revealing animosity between the two British lecturers, 'what equipment we have is mouldering quietly in a garage for which we have no vehicle, Dobby having got rid of that as well'.[37] Hilton was, however, the last British academic to serve as head of geography. As will be discussed in Chapter 3, Africanisation took place more quickly at the University College Ghana than either Ibadan or Makerere with the first Ghanaian Chair of Geography, Ernest Boateng, appointed in 1960.

## Teaching and Researching Geography

Staff and buildings were only part of setting up new departments; there were also decisions to be made about the shape and content of the curriculum and degree qualifications to be offered. Geography was taught as part of general degree courses in the early university colleges, rather than to honours level. It wasn't until 1957 that Makerere university college was able to admit their 'first group of students to the newly-established three-year Honours course in geography'. Providing an insight into the exclusivity of Honours courses Baker recorded that: 'There are only half-a-dozen of them; but, more important than numbers, I think that they will all measure up to the requisite standard'.[38] A BSc in Geography followed in 1961.

Baker provided a rare personal insight into what teaching geography in Uganda meant to him in the early years of the university college:

> We try to help our students to understand their own East African region and the problems and opportunities presented by it; to appreciate forms of regional experience other than their own; and to view with disciplined imagination the terrestrial whole to which the varied regions of the world all belong. In brief, it is our view that in Africa as elsewhere in the world good geographers make good citizens.[39]

From the outset Baker was committed to providing a 'balanced and reasonably comprehensive general treatment of the subject'; his was a moral, liberal, vision for geography.[40] Other British academics didn't necessarily share this outlook, but all brought with them not only the university culture but also the disciplinary traditions of geography in the UK. Their research interests shaped what areas of geography were taught at university level in Ghana, Nigeria, and Uganda and the structures of teaching reflected those of the University of London. But the experience of working in Africa also shaped the research interests of those same geographers, and the travel of ideas and knowledge was multi-directional. Many of those working at Ibadan became 'experts' in the geography of Nigeria for example, converted by their experience into West African, tropical, or, later, 'development' specialists (Farmer 1983).

At University College Ibadan, several faculty members arrived with little knowledge of Nigeria or even West Africa, and were sent on a tour of the country in the department's vehicle so they would be in a better position to teach about its human and physical geography.[41] Mansell Prothero, arriving in Nigeria as an expert in UK land use (and promptly being sent on tour), was transformed into a migration and health expert, a shift that lasted his whole career. His obituary recorded:

> The excitement of being in that young Geography Department at Ibadan in the early 1950s … led in particular to his studies of seasonal labour migration from Sokoto, NW Nigeria, and of Land use in Soba (in Economic Geography, 1957) that were early major conceptual markers in African geography.[42]

Prothero travelled with a young Nigerian student, Akin Mabogunje, who acted as his research assistant and contributed to the development of this expertise (Mabogunje 2011).

Some, like Pugh, gained their PhDs whilst at Ibadan, meaning that their doctorates focused on Nigeria (Pugh's on the country's geomorphology). Later publications based on doctoral research often followed. Pugh published on everything from 'fringing pediments and marginal depressions' to 'river captures' and 'sand movement' in Nigeria throughout the 1950s and 1960s.[43] Others like Bill Morgan and Dick Hodder brought an interest in Africa to Nigeria, but developed this substantially at Ibadan. Morgan's PhD had explored Malawi (then Nyasaland) and his years in Nigeria cemented his position as an Africanist. Hodder's time in Nigeria refocused his interests in economic development in the tropical world from Asia (where he had been lecturing before Ibadan) to Africa, where they remained throughout his career.[44]

Studies describing the impacts of colonisation on agriculture and settlements but ignoring the politics of (de)colonisation were common throughout the 1950s. Those focusing on contemporary issues framed them through notions of 'development' but were silent on the politics of this and took for granted who would lead this process. For example, at Ibadan Roland Moss described land-use mapping as 'a basic necessity for all development' (1959, p. 8). In this, he followed in the tradition of 'land utilisation' pioneered by Dudley Stamp in Britain in the 1930s, where these surveys were seen similarly as valuable for 'providing a cartographic base upon which efficient and orderly land-use planning could proceed' (Rycroft and Cosgrove 1995, pp. 91, 93; Stamp 1938). Mapping continued to be connected to the exploitation of resources (Haines 2016).

Research also often described and evaluated colonial development schemes, though not often from a critical perspective. For example, T.E. Hilton, working at the University College Ghana, published on 'Land planning and resettlement in Northern Ghana' in 1959, and Bill Morgan, of the University College Ibadan, published from his PhD on 'The Lower Shire Valley of Nyasaland [later Malawi]: A Changing System of African Agriculture' in 1953. Both papers outline concerns about overpopulation and soil degradation, dominant anxieties in the late colonial state (Beinart 1984), and about African agricultural practices which were seen as either inefficient or damaging. The papers demonstrate common tropes about African farming, for example this, from Hilton: 'Progress is made in the face of tradition ... they greatly appreciate the improvement in water supplies and the opportunities of earning wages, but they do not always take advantage of all that is done for them, e.g., irrigation schemes' (1959, p. 234). Hilton in particular is unquestioning of the value of colonial interventions (which were often continued after independence). He described the interventions following a 1953 law change which provided powers to enforce actions 'to improve the land and prevent *misuse*' without unpacking these underlying assumptions (Hilton 1959, p. 233; emphasis added). In line with other research papers of the time neither include any sense of the political economy of late colonial agricultural development: the desire for cash crops for export, white settler schemes and the associated land dispossession, and infrastructure projects leading to forced resettlement (Beinart 1984).

The methods, approaches and curriculum used by faculty at the university colleges closely mirrored those of British geography, a pattern observed across colonial university colleges and disciplines (see for example Mazrui 1978; Kaweesi 2021 on Makerere). Given the colonial context and staffing, it was, as an early Ibadan student Akin Mabogunje noted, 'not surprising that the British tradition in geography formed the foundation of much of the training of geographers in the country' (2004, p. 63). Initially research reflected the regional tradition of description and synthesis still dominant in Britain in the mid twentieth century (Johnson 1991). This included a focus on land utilisation as well as on describing landscape change over time, themes which dominated the work of two titans of British geography: H.C. Darby and Dudley Stamp. Local variants of these traditions emerged that produced descriptive regional work with an emphasis on historical studies and, increasingly, an interest in population, economic, urban and agricultural change, land use mapping and the development of infrastructure (Morgan 1957; Prothero 1957; Hodder 1962–1963; Gleave 1965). As we highlight in Chapter 5, research focusing on land use, mapping, population and resources reinforced utilitarian visions of African land and resources and contributed to the imagining of geography as an applied subject to be harnessed for colonial and later national development.

The research and other experience of British staff at Ibadan found their way into the curriculum. Pugh's earlier career as a surveyor translated directly into the geography teaching at Ibadan, where, as a third year, Mabogunje and his classmates assisted in a guinea-worm eradication project by using their survey skills to produce village maps. In this way, Nigerian geographers in training were putting their skills into practice in support of late-colonial development projects, albeit in this case led by the Nigerian Professor of Preventative and Social Medicine at Ibadan, Oladele Ajose (Mabogunje 2011, p. 79). Whilst efforts were made by early Faculty to localise content of the syllabus by including Nigerian and African examples (see Chapter 3), many of the courses still focused on the regional geography of Britain, Europe, and North America (Areola and Faniran 1998). And though the maverick Buchanan's lectures sometimes critiqued colonialism, the majority of content reflected colonial, developmentalist visions for land and people in Nigeria, in which western expertise, including largescale environmental interventions, were key for development, and land was for utilisation, of value economically above all else.

## Campus Relations

Life as an academic at the University Colleges in the Gold Coast, Nigeria, and Uganda meant being part of a colonial and post-colonial elite, an atmosphere and position that many of the university's architects, administrators, academics, and students were keen to protect (Livsey 2017). The development of these university campuses as elite spaces involved not only the construction of iconic university buildings but also the everyday conduct of the lecturers and their families, which worked to reinforce the divisions between the university colleges, including their new students, and the cities they were part of (Livsey 2014). This was perhaps why Keith Buchanan's behaviour at Ibadan was seen as so threatening: by meeting people in Ibadan town he unsettled the important division between the elite status of the university and the city after which it was named.

Of the jobs available overseas, some were regarded as a superior opportunity. Ibadan was one such example: on arrival there in 1963 to take up post as head of department, Barbour wrote approvingly to Steel that it compared positively to his previous appointment in Khartoum, Sudan, in terms of both 'morale' and 'material' conditions.[45] Indeed, reflecting towards the end of their careers, both Barbour and Hodder argued that in the early 1960s Ibadan was a substantial centre for research, with Barbour stating it was 'as serious re: academic life as UCL [University College London]'.[46] This atmosphere attracted and contributed to the careers of those such as Mansell Prothero, Dick Hodder, John Pugh and Bill Morgan who returned to positions at Liverpool, Birmingham and the University of London. In addition, many young geographers who would go on to have a high profile in the UK and US were attracted to Ibadan for visiting positions (Areola and Okafor 1998).

Ibadan also had good benefits, high pay and good support for research.[47] This was something also evident in other university colleges, which, via the Inter University Council, employed staff on British terms which enabled 'expatriate' staff to have British pensions, and earnings sweetened by what one geographer working in Africa referred to as 'a modest topping up of the salary'.[48] Leave every year for 'expatriate' staff, provided to prevent academic isolation, was seen by British faculty as central to the 'intellectual vigour and resilience' of the institutions, though it was not initially extended to local faculty, something which caused some considerable discontent (Livsey 2017).[49] To these institutional benefits could also be added the lifestyle possible for department members in the new university colleges.

The vast majority of staff in geography departments in this period were men, and in many cases male academics travelled with their families to the new university colleges. Jane Barbour, who had worked as a meteorologist during the war, later published work on Nigerian textiles based on research she conducted whilst accompanying her husband to Ibadan.[50] She also contributed to the publication of the *Nigerian Geographical Journal*, preparing the indexes.[51] In this she was not alone, as W J Talbot, the Head of Department at the University of Cape Town 1936–1973 recalled, in his valedictory comments on retirement:

> Fortunately, the Head of the Department had married another Berkeley geographer who, in an unofficial and honorary capacity, generously lent invaluable help in many ways. Besides a 1935 V.8 for field work, she had brought a

> second-hand office typewriter as part of her dowry and, as honorary secretary in an era when no department in the Arts and Science faculties were deemed to need any clerical assistance whatever .... Typed reams of duplicated material for class use, including tables of climatic data for every continent that are still in use ... [In 1946] Mrs Talbot ... began her intermittent career as a "temporary lecturer" in the Department. Ineligible being the wife of the professor, for a permanent appointment in her husband's Department, she has nevertheless stood by through the years as a reserve member of the team, ever ready and competent to step in a carry the ball whenever needed in every emergency and equally prepared to step aside in deference to University policy when a competent alternate eligible for permanent appointment appeared.[52]

Whilst Jane Barbour's work was recorded in an acknowledgement, and Talbot was explicit about his wife Anna-Maria's role, much other informal academic labour by wives and other family members at African colonial university colleges is not recorded. This pattern of men feeling free to travel to take up academic roles, and bringing partners with them, continued into the 1970s and beyond. Men's career development often meant women travelling too.

Much of the labour of producing the elite space of the university was also done by women. In addition to undertaking paid work and some of the unglamorous administration associated with the university, they created family homes, looked after children, and held key roles in university societies and associational life (Pietsch 2013). For example, alongside her job as a school teacher on campus at University College Ibadan, Cambridge geography graduate Margaret Pugh found time to grow flowers and tropical fruits in the garden of the family's staff accommodation, and to star in annual opera performances staged by staff. Her gardening was part of a wider transformation of the Ibadan campus into a typical British-style university landscape (Figure 2.6), and her musical roles contributed to an anglophile social landscape to match.

The campus at Makerere was similarly transformed. Randall Baker (no relative of Kenneth Baker), a Welsh geography graduate who went to Makerere for post-graduate study in the mid-1960s recalls in his autobiography how the university campus projected its elite status:

> everything was a case study in order and control. Everything that could be trimmed was. It was all as clean as Singapore ... We drove up and past the stately administration building and the modern library ... My new home was a corner room in Livingstone Hall ... I observed on a short walk around my immediate surroundings that all the residential buildings had European, and often Scottish, names which seemed, to me, rather curious for an African university, but not for a colonial institution. (Baker and Polonskyi 2015, pp. 34–35)

The social calendar at Ibadan included drinks receptions and parties, including one in which the young unmarried geographer Mansell Prothero was 'made up ... as a Red Indian brave' for a fancy dress party (this included providing 'thin pyjama trousers and

**Figure 2.6** Tower court and part of Mellanby Hall, University of Ibadan, photographer unknown. Rockefeller Foundation Records, NIGERIA, Series 497 University of Ibadan, Box 134, Folder 2563, number 1. Courtesy of the Rockefeller Archive Center.

a feathered head-dress' and applying 'war-paint (lipstick) on his chest').[53] The same party also led to controversy, as one British academic chose as his costume local Yoruba robes, offending some of the Nigerians present.

Though in the early years before the permanent university buildings were completed conditions could be basic, by the late 1950s they had improved significantly.[54] Soon after arriving in 1962, Barbour wrote to Steel that 'Life is full and quite jolly here. Jane and I are just getting into regular tennis fours, which is nice, and there is good swimming nearby; very soon there will be a College pool that we can use, staff only'.[55] The social scene at the other Asquith colleges was similarly British in flavour, as a brief description of the interests of geographer Bryan Langlands, at Makerere from 1953–1976 illustrates: 'he was a Patron of the Rugby Club at Makerere University, a member and later patron of the Scottish Dancing Club, and he was also a good actor. He was nicknamed Ratty for his memorable performance in Toad of Toad Hall' (BakamaNume and Mutabuza 1997, p. 7). Student life on campus at Makerere in the early years was also heavily influenced by British expectations and norms. As the historian Bethwell Ogot recalled: 'The social life of students was varied and full. The English department staged Shakespeare plays annually for first year students ... There were also ballroom dancing lessons, which had been started at the College in 1947' (2003, p. 51).

These practices highlight both the active labour involved in the making of an elite university space, and the tensions within that project, not only between the university community and the local population, but also within the university's staff body, where British lecturers dominated throughout the 1950s, and casual racism was often observed (Livsey 2017). These tensions are not often recorded explicitly in the archival records, nor in the celebratory accounts of departmental histories such as that produced by to celebrate 50 years of the Geography Department at Ibadan (Areola and Okafor 1998). They can sometimes be glimpsed however, such as in retrospective reflections like those of John Pugh, founding member of the department at Ibadan, who recalled that 'with some [Nigerian] staff colleagues, one was not able (allowed?) to forget that one was white'.[56]

Relations between British staff and Nigerian students could sometimes be close, only to come up against the racism of the colonial authorities. An example of this is when as an undergraduate in the early 1950s, Akin Mabogunje accompanied the British lecturer Mansell Prothero to Zaria in the north as a research assistant on a land-use survey project. As Mabogunje (2011, p. 81) recalled in his autobiography:

> My most memorable recollection of this trip was the encounter with British colonial civil servants in northern Nigeria. The relaxed relationship between my lecturer and myself was almost an anathema to them. Indeed, while they were prepared to entertain him in their sitting room, I was not welcome and walked out. Mr Prothero soon joined me.

Here a British colleague stood in solidarity with a Nigerian facing racism from the colonial state (though crucially the younger and more junior Mabogunje was the one who lead the protest). Prothero and Mabogunje remained friends for decades after this, with the former writing a tribute to Mabogunje published more than 50 years later.[57] Such stories of colonial racism are much more prominent in Mabogunje's autobiography than they were in our interviews with him, providing a glimpse into the way that lives are narrated for different audiences.

According to the British geographer Michael Chisholm, who was a visiting lecturer in the department at Ibadan 1964–1965, though 'at an interpersonal level in the department ... we all got along perfectly well' some British lecturers failed to make close relationships with Nigerian colleagues, struggling with different cultures of hospitality, and socialising in different spaces. A weekend away (visiting an inselberg at Idanre), alongside a Nigerian colleague Reuben (known as Ken), Udo was for Chisholm 'actually rather special' because it 'was the only time that I recollect that there was that kind of association'.[58] Social life on campus revolved around a staff club with swimming pool and restaurant, which were still in the early 1960s distinctly white. Udo remembers that 'the club was actually a white club when I came here [in 1963], there were very few Nigerians'.[59] Chisholm provides further detail:

> [O]n a Sunday, at about midday you could get that day's airmail edition of the *Sunday Times*. ... The only day of the week that you could read an English newspaper on the same day it was published. And the Staff Club was the social focus for academics, but it was more the expatriates who were there rather than the indigenous.

Though there was no colour bar at the club or the university throughout its history, the academic staff (for whom the club was reserved) remained dominated by non-Nigerians well into the 1960s. Moreover, the social atmosphere offered by the club was distinctly Western, with British newspapers and gin and tonic flowing at the bar.[60] The persistence of such forms of segregated associational life even after independence highlights the making of the university campus as an elite and still racialised space, though by the 1970s the club was increasingly utilised by Nigerian academics.

Other traditions from elite British universities had also been adopted at the Asquith Colleges. At Makerere 'brilliant red academic gowns, required for dinner until the late 1960s, were an outward and visible sign of the gold standard' (Sicherman 2005, p. 40). Early students at the university colleges were part of a small elite and often experienced undergraduate life as one of fun, freedom and opportunity. The Ghanaian geographer George Benneh, recalled how his time at University College Ghana in the mid-1950s was 'an experience worth going through': living conditions were good and 'each student had a room and a balcony' and there were 'excellent meals'.[61] Although it was a privileged experience for students studying at the university colleges, there were also tensions, discrimination, and racism in what remained throughout the 1950s colonial institutions, with overwhelmingly white and male, and mainly British, academic staff (Livsey 2017).

If for academic staff at Ibadan, the Staff Club was a decidedly white space, at Makerere for Randall Baker it was the dining hall where he first became aware of the 'ethnic composition of Uganda', 'for the meals fell into two categories: African and South Asian' (Baker and Polonskyi 2015, p. 53). As a white student, and later staff member, Baker was able to move and mix between the two, though Mills (2006) notes that 'liberal' lecturers were in a minority in halls where a high table hierarchy was in operation, and some would not mix across racial lines. Mills argues that at Makerere:

> It was not just the African students who were engaged in a form of colonial mimicry. Academics and students alike were between two if not three worlds – the racialized legacy of 'native' education in colonial Africa, the imagined potential of the Oxbridge ideal for new campus universities, and the rapidly changing politics of African societies and states heading towards independence... As the social climate changed in the post-independence era, these universities inevitably began to be criticized for their elitism and irrelevance to African 'needs'. (Mills 2006, p. 251)

## Conclusion

At the geography departments that were founded in the late-1940s, staffing, curriculum, research and conduct on campus bore the indelible mark of Western (specifically British) ideas about academic life, the purpose of university departments, and the instruments that were needed in order to ensure these colonial institutions had a footing on a wider international stage. Through the appointment of British lecturers in geography, the construction of new departmental buildings, the 'special relationship' with the University of London, and adoption of curricula that reflected ideas about what was appropriate for British rather than African geography students, the Asquith colleges all shared very similar characteristics, even when allowing for differences and occasional idiosyncrasies when it came to staffing. Many South African departments, founded earlier in the twentieth century, were also part of this wider British academic world that also embraced universities in Australia, Canada, and New Zealand.

Ajayi et al. (1996, p. 67) summarise the introduction of these higher education institutions as follows:

> The Colonial University Colleges were established partly in response to African demand, especially in West and Southern Africa, partly in response to the demand from European officials and settlers, especially in Central and Eastern Africa, but generally in pursuit of laid down colonial policies and objectives... The Colleges introduced a new age of higher education in Africa. As such they were welcomed and widely appreciated.

However, they go on to say that, despite this initial welcome, 'Eventually they were recognised as part of the colonial legacy put in place in the period of decolonization.' (Ajayi et al. 1996, p. 67). Criticisms fell into four main areas, all of which related to the inadequacy or irrelevance of the British model of higher education introduced. Critics argued, first, that they 'deliberately set out to train an elite that would merely step into the shoes of outgoing colonial rulers and become exploiters of their own people' (Ajayi et al. 1996, p.74). Second, they argued that the curricula were too narrow, and not necessarily aligned with needs for development. Third, they felt that the British degree structures, focusing on specialist and abstract knowledge, might not be best fitted for providing the manpower needs for national development. Finally, they felt that '[t]he Ivory Tower nature of the Colleges underlined their separation from the wider community' and that there was a need to address this through both institutional and a physical relocation, through extension classes and the shifting of campuses to the centre of cities for example (Ajayi et al. 1996, p. 75). Whilst this chapter has foregrounded the careers and impact of British geographers in shaping the institutionalisation of geography in Africa universities, a key theme of the rest of the book is how African geographers attempted to reckon with this colonial legacy as it pertained to the discipline of geography. As such, the following chapters include a wider diversity of voices and contributions, with a focus on African perspectives.

As we have seen in this chapter, buildings, staffing, curricula, disciplinary norms, university relations with wider society, international networks and the very idea of the university and the academic were shaped in a British mould (Mazrui 1978; Ake 1982; Livsey 2017). As Walter Rodney put it, reflecting on his time at the University College of Dar es Salaam 'both the academic program and the physical environment within the university – the administration, everything – were built and structured towards forcing one to behave along certain orthodox, conventional, bourgeois academic lines.' (1990 [1975], p. 37). Some of these things were easier to change than others. As Chapter 3 shows, Africanisation of staff could happen relatively quickly, though the speed of change was slower in the colonial university colleges than those founded anew in post-colonial Africa, because contracts, staff in senior positions and embedded power relations could hamper change. But Africanisation of the geography curricula was a more slippery subject. How was geography to be Africanised? What sort of geography was best suited to independent Africa? These debates, and the changes made to what was taught, and how, are the focus of Chapters 3, 4, and 5, which examine early attempts at localisation of content, the introduction of new methods influenced by international disciplinary developments, and the perceived need for 'relevance' and 'applied' knowledge and skills in geography.

Buildings were quite permanent, and many of the original designs introduced on the behest of the first British staff in colonial university colleges remain in use today, for example at the University of Ghana. However, as noted in Chapters 4 and 5, sometimes the impetus of key individuals and new funding streams – such as those of the American Foundations like Rockefeller – could change the spaces available to geography staff and students for their work. The networks of the British colonial academic world were (and remain) remarkably tenacious. Nevertheless, as Chapter 4 shows, these were challenged and joined by other networks, including those of US academia and that of the Eastern Bloc, influenced by the Cold War and soft power diplomacy in African higher education. The Western idea of the university, and its relation to the society in which it was based, remained stubbornly persistent throughout the period, though it was hotly contested (Nyerere 1966; Mazrui 1978; Nkrumah 2009 [1963]). These debates are ongoing today, often taking on the terminology of 'decolonising' universities and individual disciplines (e.g. Bhambra et al. 2018; Ndlovu-Gatsheni 2018). The rest of this book provides a historical account of geography in African universities in the period of decolonisation, and the conclusion asks what such an account might have to contribute to current debates.

## Notes

1 Carr-Saunders, A. 'The teaching of geography in colonial colleges: Address to the Geographical Association', *Geography*, 33 (1948) 1–7, p. 1.
2 S.J.K. Baker 'Memorandum from S.J.K. Baker' Oxford Development Records Project (hereafter ODRP) MSS Afr. S. 1825 (3) BOX II, p. 1.
3 S.J.K. Baker, 'Memorandum' p. 1.
4 *Report of the Commission on Higher Education in West Africa*, London, 1945, 55.
5 Undated speech discussing Baker's early experiences at Makerere, D402/1/2/1 from papers of S.J.K. Baker, University of Liverpool Special Collections and Archives (hereafter SJK) D402/1/2/1.
6 Inter University Council (IUC), Papers of Robert Steel, University of Liverpool Special Collections and Archives (hereafter RS), D91/4/8/17.
7 Professional Papers RS, D91/1.
8 David Davies, The University of Zambia, to Robert Steel, 4 December 1969, RS D91/1/4/2iii (part 4 of 14).
9 Terry Coppock to Steel, 1963, RS D91/1/4/2iii (part 4 of 14); Steel to Michael Barbour, RS D91/1/4/2iii (part 1 of 14); B.W. Hodder 'Memorandum', ODRP, MSS Afr. S. 1825 (51), unpaginated.
10 Steel to Barbour, 2 August 1961, RS D91/1/4/2iii (part 1 of 14).
11 *Report of the Commission on Higher Education in West Africa*, 62.
12 John Pugh 'Memorandum', ODRP, MSS Afr. S. 1825 92/Box LVII, p. 5.
13 John Pugh 'Memorandum', p. 32.
14 John Pugh 'Memorandum', p. 32.
15 John Pugh 'Memorandum'.
16 Undated newspaper clipping. SJK D402/1/1/2.
17 S.J.K. Baker, 'Memorandum'.
18 Undated speech SJK D402/1/2/1.
19 S.J.K. Baker, 'Memorandum' p. 20.
20 University of Ghana Archives, UG1/3/2/3/56(18).
21 Daily Graphic No. 2060 (Weds 8th May 1957), p. 1.
22 UK National Archives, hereafter TNA, FO 371/90059.
23 Dobby was actually reading the speech on behalf of Vice Chancellor Sidney Caine, who was away. However, he refused to apologise to the Varsity when pressed and subsequently resigned in 1957 with the *Straits Times* describing him as 'a very tired and harassed man'. Though he was persuaded to withdraw this, he resigned finally a year later, in 1958, having been in post since 1939. See 'Dobby may withdraw his notice of resignation, *Straits Times*, 16 April 1957 p. 4; 'Dobby gives notice of resignation.' *Straits Times*, 12 February 1958 p.1.
24 B.W. Hodder 'Memorandum', ODRP, MSS Afr. S. 1825 (51), unpaginated.
25 Kenneth (Michael) Barbour memo, ODRP, MSS Afr. S. 1825 Box III (4).
26 Baker 'Memorandum'.
27 John Pugh 'Memorandum' p. 20.
28 John Pugh 'Memorandum' p. 30.
29 John Pugh 'Memorandum' p. 30.

30 'Letter, Varley to Vice-Principal, 6 December 1955', 'Design and construction of the department of geography', University of Ghana Archives, UG/1/3/3/2/100 [4203/38].
31 Letter, Varley to Vice-Principal.
32 Letter to Vice-Principal from architects, Harrison Barnes and Hubbard, Design and construction of the department of geography', University of Ghana Archives, UG/1/3/3/2/100.
33 E.H.G. Dobby, 1958 'Report of the Department of Geography' University of Ghana Archives, UGI/3/2/3/56 [32] p. 2.
34 E.H.G. Dobby, 1958 'Report of the Department of Geography' University of Ghana Archives, UGI/3/2/3/56 [34] p. 3.
35 E.H.G. Dobby, 1958 'Report of the Department of Geography' University of Ghana Archives, UGI/3/2/3/56 [34] p. 4.
36 E.H.G. Dobby to Registrar, Letter of resignation, University of Ghana Archives, UGI/3/2/3/56 [49].
37 'T.E. Hilton to Provost, Letter, 19 October 1959' University of Ghana Archives, UG/1/3/3/2/100.
38 Letter from Baker to Boesch, 30 July 1956, International Geographical Union Archives, II-2–29 UGANDA.
39 Undated handwritten manuscript 'Makerere College and Geography', SJK D402/2.
40 Undated handwritten manuscript 'Makerere College and Geography', SJK D402/2.
41 John Pugh 'Memorandum'.
42 Obituary, R. Mansell Prothero, 1924–2013, *Royal Geographical Society with Institute of British Geographers*, 1. Retrieved from https://www.rgs.org/geography/news/r-mansell-prothero-(1924-2013)/ last accessed 8 April 2017.
43 John Pugh, 'Curriculum Vitae' Personal papers of John Pugh, private collection, courtesy of Mike Pugh, hereafter JP papers.
44 Dick Hodder: Obituary, *The Times*, 24 November 2006, 79.
45 Barbour to Steel, 1963 RS D91/1/4/2iii (part 1 of 14).
46 Barbour 'Memorandum', p. 7; Hodder 'Memorandum'.
47 Pugh to Sidney Wooldridge, 28 November 1955, JP papers.
48 Interview with Leonard (Len) Berry, 12 March 2021.
49 Hodder 'Memorandum', p. 2; Barbour 'Memorandum'. The term 'expatriate' here refers to non-local staff, in practice mainly British but also from the British World of Australia, New Zealand and Canada, and the US and other parts of Europe. As such, expatriate also often meant in practice 'white' (Livsey 2017).
50 Obituary, Jane Barbour, *The Guardian*, 14 June 2014.
51 Acknowledgements, *The Nigerian Geographical Journal* 5 (1962).
52 Talbot, W.J. (1973) 'Lest we forget …. Annals of the University of Cape Town Department of Geography, 1936-1973' p. vii, ix. Professor William J. Talbot Papers, University of Cape Town Special Collections (Hereafter UCT) BC1079 D.5 Departmental Farewell Speech.
53 Obituary: Margaret Pugh; Pugh 'Memorandum', p.8.
54 Pugh 'Memorandum'.
55 Barbour to Steel, 11 March 1962, RS D91/1/4/2iii (part 1 of 14).
56 Pugh 'Memorandum'.
57 Prothero, M. 'A distinguished life'.*Celebrating Mabogunje at 75*. p. 101.
58 Interview with Michael Chisholm, 14 March 2017.

59 Interview with Reuban K. Udo, 9 July 2017.
60 Interview with Michael Chisholm, 14 March 2017.
61 'Prof Emeritus George Benneh turns 80' *Daily Graphic*, Ghana, 5 March 2014. https://www.graphic.com.gh/features/features/prof-emeritus-george-benneh-turns-80.html [Last accessed 1 December 2021].

## References

Agbodeka, F. (1998). *A History of University of Ghana: Half a Century of Higher Education (1948–1998)*. Accra: Woeli Publishing Services.

Ajayi, J.F.A., Lameck, G., and Johnson, G.A. (1996). *The African Experience with Higher Education*. Athens: Ohio University Press.

Ake, C. (1982). *Social Science as Imperialism: The Theory of Political Development (Second Edition)*. Ibadan: Ibadan University Press.

Areola, O. and Faniran, A. (1998). Undergraduate training at Ibadan. In: *50 Years of Geography in Nigeria: The Ibadan Story* (ed. O.I. Areola and S.I. Okafor), 19–46. Ibadan: Ibadan University Press.

Areola, O. and Okafor, S.I. (ed) (1998). *50 Years of Geography in Nigeria: The Ibadan Story*. Ibadan: Ibadan University Press.

BakamaNume, B. and Mutabuza, M. (1997). Bryan Wooleston Langlands: Ugandan geography and the gospels of St. Langlands, 1953–1976. *East African Geographical Review* 19 (2): 1–12. https://doi.org/10.1080/00707961.1997.9756243.

Baker, R.A. (2015). Samuel John Kenneth Baker 1907–1992. In: *Geographers Biobibliographical Studies Volume 22* (ed. P.H. Armstrong and G.J. Martin), 1–11. London: Continuum.

Baker, R. and Polonskyi, M. (2015). *Cold War in a Hot Place*. Sofia: East-West Publishing.

Barnard, W.S. (2016). *Encountering Adamastor: South Africa's Founding Geographers in Time and Place*, vol. 1. Stellenbosch: African Sun Media.

Beinart, W. (1984). Soil erosion, conservationism and ideas about development: a southern African exploration, 1900–1960. *Journal of Southern African Studies* 11 (1): 52–83.

Bhambra, G., Gebrial, D., and Nişancıoğlu, K. (2018). *Decolonising the University*. London: Pluto Press.

Carr-Saunders, A. (1948). The teaching of geography in colonial colleges: address to the geographical association. *Geography* 33: 1–7.

Farmer, B.H. (1983). British geographers overseas, 1933–1983. *Transactions of the Institute of British Geographers* 8: 70–79.

Gleave, M. (1965). The changing frontiers of settlement in the uplands of Northern Nigeria. *Nigerian Geographical Journal* 8: 127–141.

Haines, E. (2016). Pledging the future': investment, risks and rewards in the topographic mapping of Northern Rhodesia, 1928–1955. *Environment and Planning A: Economy and Space* 48 (4): 648–664. https://doi.org/10.1177/0308518×15594808.

Hilton, T.E. (1959). Planning and resettlement in Northern Ghana. *Geography* 44 (4): 227–240.

Hodder, B.W. (1962–1963). Badagri. I: slave port and mission centre. *Nigerian Geographical Journal* 2: 75–85.

Hodder, B.W. (1982). E.H.G. Dobby. *Transactions of the Institute of British Geographers* 7 (4): 510.

Hodder, J. (2017). On absence and abundance: biography as method in archival research. *Area* 49 (4): 452–459. https://doi.org/10.1111/area.12329.

Hodge, J.M. (2007). *Triumph of the Expert: Agrarian Doctrines of Development and the Legacies of British Colonialism*. Ohio: Ohio University Press.

Johnson, R. (1991). *Geography and Geographers*, 2e. London: Edward Arnold.

Jones, M. (2018). Anne Buttimer's *the practice of geography*: approaching the history of geography through autobiography. *Geografiska Annaler: Series B, Human Geography* 100 (4): 396–405. https://doi.org/10.1080/04353684.2018.1544468.

Kaweesi, E.S. (2021). Hegemonic: the trajectory of political theory at Makerere University College, 1949–1968. *Paedagogica Historica*. https://doi.org/10.1080/00309230.2020.1856155.

Livsey, T. (2014). 'Suitable lodgings for students': modern space, colonial development and decolonization in Nigeria. *Urban History* 41 (4): 664–685. https://doi.org/10.1017/S0963926813001065.

Livsey, T. (2016). Imagining an imperial modernity: universities and the Nigerian roots of colonial development. *Journal of Imperial and Commonwealth History* 44: 952–275. https://doi.org/10.1080/03086534.2016.1210305.

Livsey, T. (2017). *Nigeria's University Age: Reframing Decolonisation and Development*. Basingstoke: Palgrave Macmillan.

Mabogunje, A.L. (1998). The history of the department. In: *50 Years of Geography in Nigeria: The Ibadan Story* (ed. O. Areola and S.I. Okafor), 1–18. Ibadan: University of Ibadan Press.

Mabogunuje, A.L. (2004). Geography in the Nigerian public policy domain: the impact and influence of American geographers. *GeoJournal* 59 (1): 63–67.

Mabogunje, A.L. (2011). *A Measure of Grace: The Autobiography of Akinlawon Ladipo Mabogunje*. Ibadan: Book Builders.

Maxwell, I.C.M. (1974). The changing pattern of universities in developing countries. *World Development* 2: 7–39.

Maxwell, I.C.M. (1980). *Universities in Partnership: The Inter-University Council and the Growth of Higher Education in Developing Countries 1940–70*. Edinburgh: Scottish Academic Press.

Mazrui, A.A. (1978). *Political Values and the Educated Class in Africa*. Berkeley: University of California Press.

Mills, D. (2006). Life on the hill: students and the social history of Makerere. *Africa* 76 (2): 247–266.

Mngomezulu, B.R. (2004). *A Political History of Higher Education in East Africa: The Rise and Fall of the University of East Africa, 1937–1970*. PhD Thesis. Rice University.

Morgan, W.B. (1957). The 'grassland towns' of the eastern region of Nigeria. *Transactions of the Institute of British Geographers* 23: 213–224.

Moss, R.P. (1959). Land use mapping in tropical Africa. *Nigerian Geographical Journal* 3 (1): 8–17.

Ndlovu-Gatsheni, S.J. (2017). The emergence and trajectories of struggles for an 'African university': the case of unfinished business of African epistemic decolonisation. *Kronos* 43 (1): 51–77. http://ref.scielo.org/7tcwpd.

Ndlovu-Gatsheni, S.J. (2018). *Epistemic Freedom in Africa: Deprovincialization and Decolonization*. London: Routledge.

Nkrumah, K. (2009 [1963]). *Consciencism: Philosophy and Ideology for De-Colonization and Development with Particular Reference to the African Revolution*. New York: Monthly Review Press.

Nyerere, J. (1966). *Freedom and Unity: Uhuru Na Umoja: A Selection from Writings and Speeches 1952–65*. London: Oxford University Press.

Ogborn, M. (2005). Atlantic geographies. *Social and Cultural Geography* 6: 379–385.

Ogot, B.A. (2003). *My Footprints in the Sands of Time: An Autobiography*. Victoria: Canada.

Pietsch, T. (2013). *Empire of Scholars Universities, Networks and the British Academic World, 1850–1939*. Manchester: Manchester University Press.

Power, M. and Sidaway, J.D. (2004). The degeneration of tropical geography. *Annals of the Association of American Geographers* 94: 585–601. https://doi.org/10.1111/j.1467-8306.2004.00415.x.

Prothero, M. (1957). Land use at Soba, Zaria Province Northern Nigeria. *Economic Geography* 33: 72–86.

Prothero, M. (1998). Robert W. Steel CBE, 1915–1997. *African Affairs* 97: 397–399.

Rodney, W. (1990 [1975]). *Walter Rodney Speaks: The Making of an African Intellectual*. Trenton: Africa World Press.

Rycroft, S. and Cosgrove, D. (1995). Mapping the modern nation. *History Workshop Journal* 40 (1): 95–105. https://doi.org/10.1093/hwj/40.1.91.

Sicherman, C. (2005). *Becoming an African University: Makerere, 1922–2000*. Trenton: Africa World Press.

Stamp, D. (1938). Land utilization and soil erosion in Nigeria. *Geographical Review* 28: 32–45.

Stockwell, A.J. (2008). 'The crucible of the Malayan Nation': the university and the making of a new Malaya, 1938–62. *Modern Asian Studies* 43: 1149–1187. https://doi.org/10.1017/S0026749X08003752

Williams, D. (2001). *A History of the University College of Fort Hare, South Africa, the 1950s: The Waiting Years*, vol. 4. New York: Edwin Mellen Press.

Wise, M.J. (1983). Obituary: R. Ogilvie Buchanan 1894–1980. *Transactions of the Institute of British Geographers* 8: 112–114.

# Chapter Three
# Shifting the Centre: Africanising Geography in Decolonisation

## Introduction

In 1955 the Kenyan geographer Simeon Ominde returned to Africa following his postgraduate studies in the UK, and secured a post at Makerere University College, in Uganda, where he taught alongside Kenneth Baker who we met in Chapter 2. Whilst at Makerere, Ominde spearheaded a campaign to speed up the 'Africanisation' of the university staff – the replacement of foreign, mainly British, academics and administrators with East Africans. Moving to University College Nairobi, Kenya, as the first Professor of Geography in 1964, Ominde lead an education review for the national government, recommending the overhaul of the school system, including the revision of syllabuses to make them relevant to independent Kenya, and the renaming of schools by replacing British nomenclature with Kenyan. In his inaugural lecture in 1968, Ominde set out his vision for a new African geography based on an 'African philosophy of economic and social development' (1971a, p. 3). Ominde's contributions can all be seen as contributing to the Africanisation, and perhaps the decolonisation, of geography, and education more broadly. Ominde was one of the first generation of African geographers replacing 'expatriate' academics in the geography departments of the continent and his campaigning speeded up this replacement process; he contributed to national policy around curriculum and institutional changes at all levels; and, as a trail-blazing academic, his vision for geography contributed to the reshaping of the discipline in the 1960s.

*Decolonising Geography? Disciplinary Histories and the End of the British Empire in Africa, 1948–1998*, First Edition.
Ruth Craggs and Hannah Neate.

This chapter explores the various strategies for Africanisation highlighted in Ominde's story, moving on from the discussion of the establishment of universities in Anglophone Africa as late-colonial development projects (1940s–1950s) and into the 1960s, when Nigeria, Uganda, Kenya, and Tanzania (along with a host of other African countries) became independent, following Ghana in 1957. This chapter, therefore, charts a shift from the colonial to the postcolonial, into the independence period of African universities. In this era, debates about Africanisation, as well as about inter-territoriality and the possibility of federations were high up the agenda for political elites across many sectors, and played out starkly in higher education, influencing the working lives of geographers in African universities.

In this period, changes were made to geography departments, and the wider universities they sat within, as a result of constitutional decolonisation. This included modifications to the institutional structures of universities, as they became independent universities rather than colleges of the University of London, and as experiments with multi-country federal universities began and ended; new administrative and academic staff; and changes to curricula. There were also a plethora of new universities founded in Africa in the 1960s and 1970s. In some areas change came quickly, whilst in other places things dragged far behind the demands of African politicians, university staff, and students. The term decolonisation was seldom used to discuss the processes transforming the African university sector in this period. However, in many ways the debates and demands of the time echo contemporary debates. First, there were similarities in the focus on the need for change in institutional structure, personnel, and curricula. Second, and as importantly, there were more fundamental demands to recentre the African academic world on the continent rather than the UK (Ngũgĩ 1993). As Ngũgĩ (1993, p. 2) put it, this period in the 1950s and 1960s was one

> when the centre of the universe was moving from Europe or, to put it another way, when many countries particularly in Asia and Africa were demanding and asserting their right to define themselves and their relationship to the universe from their own centres in Africa and Asia.

The 'political struggles to move the centre, the vast decolonisation process changing the political map of the post-war world' extended into debates about universities and individual disciplines too (Ngũgi 1993, p. 3).

Many of these changes reflected debates about the place of the university in African society, and the extent to which these institutions could and should be Africanised. Mazrui argued that in this period African universities were seen as 'both too *foreign* and too *rationalist*.' (1978, p. 202, emphasis in original). They were seen as foreign because they were based on western models, because many of the academic staff working there were from Britain, Europe, and North America well into the 1970s, and because syllabuses were dominated by Western authors, ideas and case studies. There were also questions about whether the rationalist tradition of the detached neutral scholar was appropriate to the needs of post-colonial Africa, where academics and students were required to be socially engaged and support wider state-building processes (Mazrui 1978, p. 202). It is the first aspect, the 'foreignness' of the university and of geography departments – and attempts to challenge this – that are the primary focus of this chapter. But the second aspect, the tradition of scientific detachment, and the way this rubbed up against visions for the university in supporting post-colonial state-building in African countries, is touched upon towards the end. Here discussion turns both to the extent to which new governments attempted to influence university decisions, and the ways in which Africanisation of staff and syllabus contents often stopped short of broader epistemological shifts away from notions of the neutral detached scholar.

The chapter begins with a discussion of African geographers from the colonies travelling to British universities to further their studies with a view to pursuing academic careers as the first generation of African geography lecturers. Their journeys reveal the ambiguities of decolonisation: they were crucial to the process of Africanisation of staff at African universities, but had to travel to the UK to claim their status. In the UK they were welcomed by a network of British geographers many of whom had themselves forged their careers in (or in relation to) colonial Africa. Their scholarships were often part of late colonial development funding, and aimed to produce a new African elite friendly to the UK. The UK provided opportunities but also experiences of racism. For many African students, their time in the UK was framed as necessary but uninspiring and uninteresting: working in Africa was the goal, and when they returned, they drove the change and development of the discipline of geography and the wider university landscape. By revealing the importance of returning to Africa for these students, we begin to unsettle accounts of the relationship between African and UK universities that place British institutions consistently in a hierarchy above the African university colleges. In doing so, we show how the career ambitions of these geographers provided a step towards decolonising by 'Shifting the Centre' to Africa, to use Ngũgĩ's (1993) terms.

The rest of the chapter explores various forms of Africanisation of universities and disciplines on the continent, through institutional, staffing, and curriculum changes. First it explores the changing institutional structures of universities and the broader higher education landscape in the context of constitutional decolonisation, and the impact this had on geography departments. These changes included independence from the University of London, the founding of new post-colonial universities, experiments with federation in East Africa, and, in an already independent but increasingly confident South Africa, the extension of Apartheid to the higher education sector. Following this, the chapter explores the demand for Africanisation of staff at universities – a term used to refer to the indigenisation or localisation of staff – and how this unfolded across different parts of Africa. The expansion of the sector, alongside this policy, meant that this period was one of increasing opportunities for African geographers in many places. Whilst the South African 'Extension of University Education Act' also meant new geography departments in newly established segregated universities, the academic positions were limited primarily to white Afrikaans-speaking staff. By examining the experiences of lecturers involved in the process of staff 'localisation' that began to impact universities particularly after constitutional decolonisation we see how it was felt differently in different locations. In each place, a push for increased numbers of local staff increased opportunities for those scholars whilst displacing others. Academic geographers – such as the Nigerian Akin Mabogunje and the Kenyan Simeon Ominde – took an active role within these wider processes as they pushed for change and negotiated with their often older and more senior British colleagues. Africanisation also had wider meanings in relation to teaching and curricula, and these issues are explored in the final part of the chapter, which examines debates about how geography might be Africanised as a subject in teaching and research in the era of constitutional decolonisation.

## African Geography Students in Britain

Whilst the colonial Asquith Colleges had opened up opportunities for African students to study geography in Africa, initially, only general degrees including geography were available, and students who wanted to continue their university education had to go abroad to pursue honours or higher degrees. After studying for an ordinary geography degree at Ibadan, Akin Mabogunje enrolled at University College, University of London (UCL) for his honours degree (winning the Parry Prize for his performance in the process). He followed this with a Masters and PhD at UCL (the latter at a distance), both supervised by the prominent British historical geographer H.C. Darby (Mabogunje 2011). The Kenyan Simeon H. Ominde initially obtained a Diploma in Education from Makerere College before moving to Britain to study for a Masters in Geography at the University of Aberdeen, followed by another Diploma in Education at the University of Edinburgh (awarded in 1955).[1] From the mid-1950s it was possible to take honours geography degrees at Ibadan, Ghana, and Makerere; however, those studying at higher levels continued to travel beyond Africa until the early 1970s. After obtaining his BA in Geography at University College, Ghana, George Benneh went to the London School of Economics in 1961 to commence postgraduate study.[2] Benneh was awarded his PhD in 1964 based on his study of 'The Agricultural Geography of the Main Forest Region of Ghana'. He was examined by John Pugh and was joined by K. B. Dickson, a fellow Ghanaian geographer, in the pub in London for celebrations afterwards (Benneh 2016). Benneh's PhD viva is indicative of the continuing strength of colonial educational networks into the 1960s: his supervisor Harrison Church had been central in appointing geographers in the colonial university colleges in Ghana, Nigeria, and Uganda, and Pugh was one of those founding staff members at Ibadan before returning to the UK to a position at King's College, University of London. That there was another Ghanaian geography PhD student in London at the time, who would go on to be Benneh's colleague at the University of Ghana, demonstrates the frequency of this route from Africa to London, and back again. Indeed, Mabogunje recalled that whilst he was studying in London there were plenty of fellow Ibadan students who would 'keep running into each other'.[3]

Olusegun Areola, who graduated from Ibadan in 1968 reflected on how a period of study in the Britain in order to obtain higher qualifications was seen as a necessary step for talented (and well supported) students. He remembers that the then head of department at Ibadan, the British Michael Barbour 'had this policy of sending young graduates, especially ones who had a second upper or first class … to a university in Britain to do your PhD.'[4] Mabogunje, Reuban K. Udo, and Johnson Ayoade, early Nigerian staff members at Ibadan, all gained their doctorates from UCL, whilst Areola studied at Cambridge. Long after decolonisation, which for Nigeria came in 1960, British staff at Ibadan influenced where Nigerian students pursued further study, often intervening very directly to secure funded positions abroad for those deemed future stars. This underlines a clear understanding held by British staff that western (and primarily British) universities represented the highest standards. This was also a view shared by Nigerians, and across the other university colleges. Discussions over the founding of universities in West and East Africa from the 1920s onwards highlight demands for British academic qualifications rather than local ones, reflecting understanding of the greater value of these qualifications for demonstrating equivalent status within an unequal colonial system (Mngomezulu 2004; Sicherman 2005; Livsey 2016).

As a first generation of African academics made career decisions, a United Kingdom PhD was in the 1950s a necessity and continued to be a smart choice in the following decade. This reflected a continuing understanding of an international hierarchy placing western institutions above their African counterparts, and structural limitations (few qualified supervisors) that meant there were few opportunities for PhD study in geography at the university colleges until the mid-1960s. By the 1970s increasing numbers of PhD students began to study at what, by then, had become national universities, at new higher education institutions in Africa, as well as in the United States (Areola 1998). However, the British (colonial) connections remained influential.

The influence of British geographers also shaped the curriculum studied and knowledge produced by African geographers not only at the colonial university colleges but through their undergraduate and postgraduate training in the UK. Mabogunje noted, for example, that 'the historical geography angle', imparted by his PhD supervisor H.C. Darby was 'one of the lasting impacts of studying at UCL in my career'.[5] Ghanaian K. B. Dickson also studied under Darby, producing a PhD on 'Cocoa in Ghana'. However, when Areola wanted to study soils in Nigeria for his Cambridge PhD under A. T. Grove,

> unfortunately Professor Barbour [Ibadan Department Head]… said "No money, no money for you to be coming back to Nigeria and roaming about the place and doing fieldwork. Whatever you want to do, do it in Britain."

Areola ended up studying soils in Wales, with repercussions for his future career:

> When you do your PhD abroad, there is a period of adjustment, trying to establish new study areas, and it also affects your writing. Not many people are interested in soils in Montgomeryshire, temperate land, over here. So it was like starting all over, tropical soils.[6]

This was not uncommon. Though her stated goal was to specialise in African geography, the non-availability of a supervisor who was 'conversant with African geography' and a scholarship that did not make any provision for fieldwork in Ghana, lead Elizabeth Ardayfio-Schandorf, from Ghana, to study for a PhD on 'The Development of Canals in Warwickshire, and their Influence on the Social and Economic Geography of the County' at the University of Birmingham (awarded 1974).[7] But moving to England for study also offered opportunities for early African geographers and those they worked alongside. Mabogunje enjoyed long-lasting friendships with fellow geography students at UCL, and with his lecturers, including Eric Brown, who he invited to his wedding in London (Figure 3.1), and Hugh Prince, who taught nineteenth century British historical geography and took Mabogunje and his wife to visit a number of country houses outside London.

**Figure 3.1** The wedding of Akin Mabogunje and Titiola Ogunmekan, London, December 1957. Courtesy of Akin Mabogunje.

Whilst studying, George Benneh moved into 'digs' in Wood Green north London, living with two other Ghanaians, in a first floor flat above a British family. This was an enjoyable period:

> I like playing records, jazz, so I bought a beautiful radiogram, and every month I buy, when I get my allowance, I go and buy long-playing records, and enjoyed it. And of course I also cooked my own meals. And I attended the dances, the Ghanaian dances.[8]

Ardayfio-Schandorf experienced snow for the first time when spending Christmas with another Ghanaian student in Hull.[9]

Despite these pleasures, the UK could also be a hostile place. Whilst memories of racism can be uncomfortable and upsetting, some discussed it explicitly in accounts of their PhD study in London. Ayoade, at UCL from 1968–1971 remembers that:

> There were some of my colleagues, postgraduate students who were very friendly, and I remember one or two of them, I won't mention names, who I think were not too friendly. In fact one of them, even when I greeted him once, he would come to the room where I was with others and he would not greet me. So I stopped talking to him... That was 1969.[10]

Areola was disappointed to find 'that amongst the younger generation especially, people were so ignorant about African countries, they seemed to have some stereotype ideas... I was shocked that that was the type of thing they were teaching the young people'.[11] These accounts chime with the broader experiences of colonial students in the UK, who, in contemporary surveys, reported discrimination in housing, colour bars in pubs and restaurants, prejudice and violence (Stockwell 2008). Alongside this there was a feeling that, as the Guyanese historian Walter Rodney put it, 'one had to vindicate oneself. The colonial had to show that colonized people had the same capacity. And this was a typical response to the colonial student. The moment you go there you are under that load.' (1990 [1975], p. 23).

Just as was the case with the colonial university colleges in Africa (Ajayi et al. 1996), opportunities to study in the UK highlight both the agency of Africans demanding educational opportunities, and the influence of late colonial development policies which aimed to educate a new African elite friendly towards the former colonial power. Walter Rodney understood his time studying for a PhD in history at SOAS, University of London in the early 1960s as part of a process 'to find bright young men ... who would serve the purposes of the Europeans, who would become part of the world of European scholarship, and who would represent, in effect, extensions to the highest levels in universities back in the third world.' (1990 [1975], p. 23–4).

The Colonial Office took a close interest in African students in the UK, seeing them as the leaders of the future. However, the UK government was also concerned about the racism they experienced whilst studying, and that the combination of discrimination and the nationalist politics of their fellow students might turn them against Britain (Adi 1993; Stockwell 2008). Many East African students, including the Kenyan geographer Simeon Ominde, were sent to study outside of London (in this case at Aberdeen) for fear of the radicalising potential of the capital (Ogot 2003; Mngomezulu 2004) and courses designed for colonial students were based at the Universities of Oxford and Cambridge, in part because the Oxbridge college system was seen as better able to keep control of potential nefarious influences on the course participants (Stockwell 2018).

Whilst studying abroad was a necessary step for many early African geographers within the framework of opportunities available in the decolonising state, most did not see Britain as a permanent destination. Studying and working abroad was a means to an end: advancing their careers and securing a permanent position in Africa. Simeon Ominde returned to join the Geography Department at Makerere in 1955 after studying in Edinburgh and George Benneh returned to Ghana in 1964 directly after finishing his PhD to take up a job as lecturer in the Geography Department at the University of Ghana. Similarly, Akin Mabogunje was offered an Assistant Lectureship at Liverpool University following his Masters at University College London in 1958 but turned this down to become the first Nigerian lecturer in the department at Ibadan.

For Mabogunje, his years in London were much less significant than those he had spent as an Ibadan undergraduate, which combined the excitement and freedom of student life and involvement with the national and regional politics of decolonisation. His undergraduate years coincided with the growth of anti-colonial action and the introduction of the Macpherson Constitution (1952), seen as a significant step towards self-government. Anticipating the new constitution, in the Western Region where Ibadan was based, the Action Group, a new left leaning Yoruba dominated political party was formed (Cooper 2019). Mabogunje showed a keen interest in nationalist politics, being first involved with a visit by the anti-colonial nationalist Nnamdie Azikiwe (later the first President of independent Nigeria), to the university campus, and later, playing a key role as Secretary of the university branch of the Action Group which rivalled Azikiwe's party (Mabogunje 2011, p. 349). Mabogunje spent much of his last long vacation 'campaigning in the rural areas of Ibadan', recalling 'it was great fun trying to explain the policies of the Action Group to illiterate farmers and rural dwellers … My last year at the University College, Ibadan was thus very eventful and exciting.' (Mabogunje 2011, pp. 88–89). As he recalled, there was no comparison between this period and his time in London:

> those four years [as an undergraduate] at the University College, Ibadan were some of the most memorable and fruitful years of my life…although I went on to read for

> the B.A. (Honours) in Geography as an undergraduate at University College London, I never really felt the same excitement of life as an undergraduate in those two years… Much of what happened to me as both an undergraduate and a postgraduate student in London, I saw as part of my training for participating in the development of the national and economic entity known as Nigeria. (Mabogunje 2011, p. 91)

Areola, studying in London a decade later in the late 1960s described similar sentiments:

> those of us from Africa had limited time, we wanted to get things done and come back home. Not like nowadays where people study there and they want to stay. In our time we wanted to study over there and come back home… because we were really interested in coming back to teach, in my own department and to contribute.[12]

We see here how the Asquith colleges provided a steady stream of African students who went on to further their education in Britain, thereby becoming part of a late-colonial vision for training new African academic elites. Because the type of education they were part of emphasised British ways of conducting research and reflected what British geographers deemed worthy topics to study this system continued to reflect colonial hierarchies of knowledge making. However, Britain was also decentred by the same group of African students who viewed time in Britain as a means to an end, a stop-off point before returning to their African homes, where the exciting prospect of contributing to newly independent and decolonising countries awaited them.

In the accounts highlighted in this section, the UK was not seen as particularly significant, and certainly not exciting, intellectually, politically or socially. Ibadan, Legon, or Makerere, rather than London, were the places to be and many were anxious to return. Going back to Africa provided opportunities to participate in the politics and development of new nations, and to take advantage of the growing academic opportunities leading up to and following independence. To understand this helps to unsettle accounts of the relationship between the African universities and the UK that suggests that British universities have always been placed above African universities. Western qualifications continued to be understood as necessary, but the most promising academic futures were envisaged in Africa.

These accounts are also a reminder of the optimism felt in the years around constitutional decolonisation in Africa, and the buoyancy of the former Asquith Colleges in these years (Livsey 2017). Indeed, African universities were exciting, well-connected global academic centres in the 1960s. This excitement also captured the imagination of British students. Geographers Randall Baker and Tony O'Connor chose to do postgraduate and PhD study at Makerere. Both described the choice in similar terms: for Baker, Africa was 'the place to be at that time' (Baker and Polonskyi 2015 p. 15), whilst for O'Connor 'in 1960 Africa was THE story'.[13] And Makerere wasn't alone in hosting students from beyond the continent. Students from Britain, Pakistan, and Singapore travelled to Ibadan for postgraduate studies in the 1960s (Mabogunje 2011). Jennifer Bray, a British undergraduate who studied for a Master's degree in Ibadan from 1964 explained:

> The time was Nigeria, four years into independence, pre-oil and pre-war, now awarding its own degrees, a country full of confidence and hope and a university truly international in outlook. I was a 21-year-old on an academic journey, wanting to be part of it all![14]

Lecturers from abroad helped contribute to the dynamism of geography departments throughout the 1960s. At Ibadan they joined a community seen by both Nigerian and foreign lecturers alike as an important centre for research, with 'the quality of discussion and research … quite remarkable'.[15] Foreign staff members and visiting lecturers often returned home with positive 'views of what was happening [in the department that] … made it easy for them to "market" our department in other parts of the world' (Mabogunje 2011, p. 128).

Accounts of decolonisation often rightly highlight colonial continuities that persist today, and the often-violent struggles over power, land, and rights that accompanied late colonial rule and independence in many parts of Africa. However, these were accompanied by positive affective experiences of the end of empire and the novel material and intellectual opportunities it afforded (some) within newly independent nations. Vividly articulated in the accounts of African geographers and those who chose to study in Africa from outside the continent, these experiences are important to acknowledge, and existed alongside frustrating, problematic, and often violent colonial legacies within and without the university.

Nevertheless, despite this optimism for change in the 1960s decade of independence, the colonial universities, and the discipline of geography within them, were not suddenly transformed into decolonised institutions, however we might define this process. It was largely African students who studied in the UK that became the first generation of African academics charged with reshaping African universities in the wake of decolonisation. Knowledge gained abroad would help to incorporate the Asquith colleges within an already existing international system of higher education, rather than to challenge it. Opportunities were primarily available to those who were willing to embrace western visions of academic practice, rather than those who fundamentally disrupted them. And there were institutional structures and vested interests that held back the pace of change. The rest of this chapter explores some of the ways in which geography at African universities changed in the 1960s, and the limits of these changes.

## Shifting Higher Education Structures

The early 1960s saw a substantial shift in the relationships between African higher education and the UK, as the links to the University of London were severed. As the University College of Ghana (in 1961), The University College of Ibadan, and the Makerere University College (both in 1963) became independent institutions, following months or a few years after the countries themselves, they gained greater freedoms to adapt and transform their structures and offerings. At each institution this break was accompanied by the appointment of the first African Principal or Vice-Chancellor: Nana Kobina Nketsia IV in Ghana (albeit temporarily), Kenneth Dike at Ibadan, and Yusuf Kironde Lule at Makerere. The universities (no longer university colleges) could now create new departments and adapt their curricula without reference to London (Livsey 2017). This change was hugely important symbolically, but it also made a substantial material difference to life at these universities, with increasing opportunities for Africans in leadership, teaching, and research roles. It also made a difference to their students. For example, at Ibadan, the early 1960s decision to open up the geography honours degree pathway to all students (where before it had been reserved for a select few) was understood by Akin Mabogunje very much in terms of the decolonisation of student learning.[16] This was because it was tied in to the broadening of access to higher education qualifications, following critique of colonial higher education as focused solely on training a small elite to step into the shoes of the colonial rulers (Ajayi et al. 1996, p. 74).

The shift away from the University of London also offered opportunities for the governments of independent African countries to influence university affairs. In Ghana, Ajayi et al. (1996, p. 86) record that:

> the Pan-African philosophy and commitment to the total liquidation of colonialism in Africa constituted an ideology that placed Kwame Nkrumah and the ruling Convention Peoples Party [CCP] at logger heads with the colonial University College of Ghana.

Therefore, the end of the special relationship offered opportunities to substantially alter the way the university was run. The CCP government argued, in line with Nkrumah's broader arguments about neo-colonialism, that Western education (including that provided at Legon) often produced 'enlightened servants of the colonial administration', or 'abstract "liberal" outlooks' aligned more closely to the imperial powers than to 'the very real problems of colonial domination' (Nkrumah 2009 [1964], p. 3–4). Whilst the theory of neo-colonialism emphasised the dangers of 'political independence minus economic independence' (Nkrumah 1962, quoted in Ahlman 2017, p. 157), it also highlighted the embedding of 'a local elite designed to serve above all the interests of the imperialists' into newly independent countries (Ahlman 2017, p. 158). Liberation therefore required reclaiming 'the psychology of the people, erasing the "colonial mentality" from it', including through shifting the Western (and primarily British) orientation of the University (College) of Ghana (Nkrumah 2009 [1964], p. 106). The Ghanaian government therefore used the change from College to University as an opportunity for change, influencing debates about Africanisation of staff and curricula, and the decisions made about appointments to key roles within the university, something that was not possible previously (Ajayi et al. 1996). They also moved quickly to increase student numbers and the extension work of the university in the community (Agbodeka 1998). As Nkrumah put it:

> We want the University College to cease being an alien institution and to take on the character of a Ghanaian University, loyally serving the interests of the nation and the well-being of our people. If reforms don't come from within, we intend to impose them from the outside. (Nkrumah, cited in Mkandawire 2005, p. 22)

Broadening of access to higher education through the creation of additional student places also led to a clamour for new institutions to join those select few founded by the British in the 1940s. The University of Nigeria Nsukka, 'opened its doors in October 1960 as the climax of independence celebrations' (Ajayi et al. 1996, p. 81). Nsukka was the brainchild of Eastern Region Premier Nnamdi Azikiwe, pan-Africanist and radical anti-colonial nationalist and was portrayed as the epitome of the post-colonial university. Azikiwe had 'warned against the thoughtless imitation of western education', though when Nsukka opened it drew on substantial American assistance and looked much like Ibadan (Azikiwe, cited in Livsey 2016, p. 958). Regional competition also played a substantial role here. In Nigeria, 'each region … aspired to have its own university that would serve the interests of its microcosm' (Mngomezulu 2004, p. 60) and soon after independence, along with Nsukka, three further universities were founded resulting in one institution in each of the three federal regions and the capital, each in a different political context. Each new university was used as an instrument of regional development but also became bound up with damaging regional competition within Nigeria's colonial era federal system bequeathed at independence (Livsey 2017). Nigeria's domestic politics has subsequently spawned multiple new universities; for example, between 1975 and 1982, 22 new universities were founded. Universities were politically popular, and Ajayi et al. (1996, p. 141) note that in the late 1980s, 'The Head of state was known to announce off the cuff the establishment of universities that were not even mentioned in the prepared speeches.' Nevertheless, substantially broadening access to higher education in Nigeria, as well as elsewhere in the continent stands as a 'major achievement of African nationalism' (Ndlovu-Gatsheni 2018, p. 179).

Whilst the pattern in Nigeria was not followed to the same extent in Ghana or Uganda, in these countries too, independence spawned new institutional relationships. In Ghana, the Kwame Nkrumah University of Science and Technology (KNUST) was formally established as Ghana's second national university in 1961, with its name clearly signalling its value as part of a nation-building project. KNUST also signalled its break from the UK in its academic trappings; whilst its earlier incarnation as a technical college had also been affiliated to the University of London, as a full university it did things differently. As Ajayi et al. (1996, p. 88) suggest, it 'did not suffer from any colonial hang-ups: no Latin grace or gowns at dinner; rather national dress on formal occasions; residential hostels rather than collegiate Halls of Residence'.

In South Africa, the 1960s also saw a dramatic growth in the higher education sector. However, whilst the expansion of South African universities paralleled the rapid expansion of higher education in several other newly independent African states, the purpose of these developments, and the opportunities for geography staff and students, were very different. The 1950s saw the intensification of the policy of apartheid as well as a growing white nationalism and confidence which culminated in the country becoming a republic and leaving the Commonwealth in 1960. From 1959, South Africa's universities increasingly became a venue for the reproduction of Afrikaner nationalism and its flagship policy of apartheid, as the university system was formally incorporated into the apartheid regime. This meant restrictions on who could work and study at different universities, and new institutions opened for each of the so-called 'racial groups': Coloured, Indian, and Black African, as defined in the Population Registration Act, 1950 (Dubow 2014). Entrance to the established universities was limited to white students, whilst Fort Hare, the missionary college aimed at black Africans from across Southern Africa was co-opted into the national government system and dedicated to Xhosa students. Each of these new apartheid institutions taught geography, mainly because the subject was on the school curriculum and one of the key roles of the universities was to train teachers for the segregated school system. A number of new Afrikaans-medium white universities were also founded, shifting the balance away from the English-speaking and British world dominated higher education scene which dominated earlier in the twentieth century (Pietsch 2013).

By the 1960s, foreign-born or trained geographers were increasingly uncommon in South Africa as more and more South Africans who had also studied in the country were appointed, especially in the Afrikaner language institutions (Barnard 2016). This huge expansion of the university sector in the 1960s and 1970s provided plentiful new roles for the mainly Afrikaner white lecturers who went on to staff them. However, whilst the expanding higher education sector reflected an increasingly confident South African whose nationalism saw it shifting further away from the UK, this was only an opportunity for some, with most black South Africans excluded.

These opportunities remained restricted into the 1980s: the South African Ronnie Donaldson recalled that in his early career, 'If you had a PhD and you were white, and male, and spoke Afrikaans, you were virtually guaranteed a job.'[17] In echoes of policy across the colonial universities in the British empire in Africa, jobs at the historically black universities (HBUs) were associated with higher salaries (for 'white' staff) than in other institutions to 'compensate' for working in these difficult conditions alongside staff and students of colour. In another echo of the early colonial higher education policy, at these new universities students undertook education at a lower level (with fewer going on to honours and postgraduate qualifications) (Wolpe 1995).

Many in and beyond South Africa protested against the extension of apartheid to the higher education sector, with the English-speaking South African universities – such as the Universities of Cape Town and Witwatersrand – at the forefront of the criticism. Their arguments primarily revolved around the academic freedom to admit who they liked, rather than the fundamentals of discrimination, and their protestations sometimes sounded hollow to Black South Africans (Ajayi et al. 1996, p. 72). The Professor of Geography at the University of Cape Town, William Talbot, was keen to highlight the diversity of the department in the period up until the 'Extension of University Education' Act, with his first group of third year students made up of 'three coloured and seven white students' (all male).[18] However, in 1959 on the eve of the enforcement of 'Bantu' education, there were only 633 non-white students at the University of Cape Town out of a total of more than 5000 (only 39 of whom were black African). The situation at Witwatersrand was worse with even fewer non-white students (Ajayi et al. 1996, p. 71). Thus whilst the 1960s saw a rapid expansion of undergraduate geography programmes, much of this provision was understood by Black communities as both inadequate and ideological (Wesso 1992). Moreover, the lecturing opportunities were restricted overwhelmingly to white Afrikaans speaking geographers. A focus on academic careers underlines the fundamental inequalities in opportunity for lecturers as well as students in South African higher education under apartheid.

In Uganda, Makerere University College, which had been established on the basis that it would serve the whole of East Africa, became subsumed within a new experiment, the University of East Africa, in 1963. The University of East Africa was the educational strand of plans to offer inter-territorial services – 'railways, telecommunications, Airways – co-ordinating tariffs and trans-border trade' – shared between Kenya, Tanzania, and Uganda (Ajayi et al. 1996, p. 92). Proposals for a federation of independent East African states, had gained 'considerable traction among East African leaders and citizens' in the lead up to constitutional independence in these territories (Milford 2020, p. 1326). While ambitions for an East African Federation remained unfulfilled, the University of East Africa did, for a period, function as an inter-territorial 'hard-headed and realistic experiment in regional technical co-operation, as well as a dream in the tradition of Pan-East-Africanism' (Mazrui 1978, p. 240; Ngũgĩ 1993). For East African politicians a regional university was a means of sharing resources, therefore it was also a practical move to ensure educational needs would be met in the period when East African countries were moving towards decolonisation and independence (Mngomezulu 2004). The first Chancellor of the University of East Africa was President Nyerere of Tanzania, a figure who was 'perceived as an epitome of African freedom in East Africa' at the time (Mngomezulu 2004, p. 239).

Makerere University College, as the longest established institution in East Africa was chosen as the original nucleus for the regional university as neither Kenya nor Tanganyika had comparable institutions at the time of the University of East Africa's planning. Makerere was joined by the Royal Technical College (Nairobi) which had been founded in 1954 and became the University College Nairobi in

1964; and the University College Dar es Salaam, founded in independent Tanganyika in 1961 (Morgan 1991) (see Table 1.1 in Chapter 1). The constituent colleges were a means of spreading the sites of higher education across the region, ensuring that funding was not just directed to Uganda. They were also conceived as means of sharing resources, with each college initially designed to offer courses in basic Arts and Sciences with professional courses split between the sites. For example, University College Dar es Salaam offered Law, whilst medicine went to Makerere (Ogot 2003).

Makerere University College, Uganda (Figure 3.2), remained the primary place to study geography (and other disciplines at higher level) across East Africa until the founding of the Geography Departments at Nairobi (1964) and Dar es Salaam (1965), though the subject was taught at Nairobi prior to this, by the British geographer W.T.W. Morgan (Dunbar 1991). In 1956, the Tanganyikan geographer Adolfo Mascarenhas had been planning to go to London to get his degree on the advice of his father. However, he was told by his teacher (an Irish missionary close to future Tanzanian President Julius Nyerere) that because of the strength of his academic performance (the number of places at Makerere were extremely limited) he could apply to Makerere. As he put it: 'there was no

**Figure 3.2** Students in the geography lecture room, Makerere college, 1947. Pictured are Left to right, front row: C.T. Mallinga, from Uganda, F.M. Mbiti, Kenya, R.S. Matano, Kenya; Back row: L–R: L. Kalule, Uganda, J. Kihori, Kenya, and M. Alala, Kenya. Oxford Development Records Project, The Bodleian Libraries, University of Oxford, ODRP MSS Afr. S. 1825 (3) BOX I, photograph 55.

reason to go to London to get a degree ... I could do it from Makerere. I was happy with this'. When Mascarenhas graduated in 1960 he was one of only 54 Tanganyikan graduates in any discipline.[19]

Brief descriptions of the three departments in the federal University of East Africa provide some evidence of the way that national, institutional, and departmental politics could combine to produce different versions of the discipline of geography, whilst also signalling some of the reasons why the federation fell apart in 1970, having lasted only seven years. Despite the plans for Makerere College to serve the whole of East Africa, it had overwhelmingly recruited black Africans, rather than Asian or white students, who preferred to study elsewhere, either in the UK, or at the Royal College (Nairobi), an institution whose foundation was driven by 'the ambition of the European and Asian settlers in Kenya' (Ajayi et al. 1996, p. 92). The British geographer Leonard (Len) Berry described Makerere as 'very set in the traditional academic British setting' and noted that the geography department reflected this broader dynamic.[20] Longer established than Nairobi or Dar es Salaam, for a period spanning twenty years from 1947, Kenneth Baker and his British successor Bryan W. Langlands 'remained as the backbone of the geography department at Makerere' (BakamaNume and Mutabuza 1997, p.3).

In contrast, in Kenya, University College Nairobi's geography department was only founded in 1964, and its founding Chair was the Kenyan Simeon H. Ominde, who became the first East African Professor at University College Nairobi on moving from his lectureship at Makerere. Nairobi 'became a strong department on the basis of local geographers' rather than British staff (BakamaNume and Mutabuza 1997, p.3). In Tanzania, the geography department, founded in 1965 four years after the establishment of University College Dar es Salaam in 1961 was initially headed by another Brit, Len Berry. On arrival Berry was 'given the task of really building the department', but on the understanding that Tanzanians would soon take over.[21]

Although the University of East Africa was greeted with excitement and heralded as an example of pan-East Africanism and regional cooperation, tensions were present even before its inauguration. Developments and expansion at Makerere were put on hold to give Royal College Nairobi and University College Dar es Salaam time to catch up. Despite plans to target professional training at particular sites, there were different 'manpower' [*sic.*] requirements in different territories, and following independence, emphasis shifted to national development which created a demand for localised delivery. Why should Tanzanians go to Uganda to train to be a doctor? Why should Kenyans go to Tanzania to train in law? Many of the intended specialisms across the multi-site university began to break down or were bypassed by local decision-making that was driven by national political demands. Even though the University of East Africa was unified by a common Senate and Council, with a single Vice-Chancellor, this structure, which was oriented towards inter-territorial cooperation, began to look increasingly irrelevant as Uganda, Kenya, and Tanzania began to shift attention to the demands of national development, with different tools and underlying ideologies at play in each country. With Tanzania embracing socialism and the Ugandan Common People's Charter outlining plans for socialist centralized planning, but Kenya opting for a more Western-oriented capitalist approach, it became apparent that the federal system was coming under increasing pressure. Moreover inter-territoriality – initially a colonial era project – placed controls on the numbers of Africans that could attend higher education (Kithinji 2012). National universities would allow the different countries to train their students according to what national governments considered were national priorities (Southall and Kaufert 1974; Mngomezulu 2004). The Federal University of East Africa ended in July 1970 and was succeeded immediately by three national universities: the University of Nairobi in Kenya, Makerere University in Uganda, and the University of Dar es Salaam in Tanzania.

Universities, and their geography departments, were thus shaped in profound ways in the 1960s by the politics of independence. Not only were they considered crucial for national development, they were important symbols: 'a national university was considered as much a hallmark of national independence as having a flag, an anthem, a central bank and a currency' (Mamdani 2018, p. 59). Whilst in the 1940s and 1950s, several different futures were imagined for Africa beyond the nation state, by the end of the 1960s, both imperial and post-colonial pan-African futures for African development had been widely rejected (Cooper 2008; Livsey 2017). These transnational visions were overtaken by primarily national frames for imagining and practising development in Africa, and African universities became increasingly closely connected to *national* development projects as part of this broader process (Livsey 2017).

## Africanisation of Staff

One of the key demands at independence, within universities alongside other public sector institutions such as the civil service was for 'Africanisation' of the workforce; that is, the replacement of foreign (mainly white) lecturers and administrators with Africans. Moves towards 'Africanising' took place at different speeds and with varying degrees of urgency. For example, in independent Ghana, under an increasingly radical Nkrumah, Ghanaian leadership was required for the geography department of the University of Ghana. E. A. Boateng (Robert Steel's former student, alongside Barbour, at Oxford) was duly appointed in 1960, with Steel writing that, 'It was clear to me that [the temporary incumbent, the Brit T.E. Hilton's] appointment to the Chair would have been wrong from the African and student point of view'.[22] Boeteng was one of the first Ghanaians in any discipline to be 'raised to full Professorship' at Legon, and was followed by many more as the now independent university responded to government policy to appoint Ghanaians at the level of Chair (Agbodeka 1998, p. 127). However, even in Nkrumah's Ghana, British expertise was still sought in hiring decisions at the university into the early 1960s. Soon after his appointment Boateng went on a sabbatical to Pittsburgh in the United States and the University's academic board agreed R. O. Buchanan of the London School of Economics (a professor with similar 'king-making' credentials as Steel) to sit on the appointment committee,[23] while Stoughton, the University Principal wrote to Steel again asking for ideas about who might replace Boateng.[24]

Reflecting on Africanisation at the University of Ghana, George Benneh recalled how when he was a student in the mid- late-1950s, there were 'not too many Ghanaians' on the academic staff, but shortly after independence, 'almost all the heads [of department were] by policy Ghanaian'.[25] After the break with the University of London in 1961, the Ghanaian government began wielding increasing power in the university, demanding rapid Africanisation but removing in particular 'expatriates *and Ghanaians* too vocal in their opposition to policies of the government' (Ajayi et al. 1996, p. 87, emphasis added). The main opposition to Nkrumah's CCP government was led by Kofi Busia, a professor at the university, and here we can see the intersections of decolonisation policies such as Africanisation of institutions and positions, with the fight for domestic political power (Agbodeka 1998).

In Nigeria the picture was more mixed. At the University of Nigeria Nsukka, the percentage of Nigerian academics employed increased quickly from 37% in 1962–1963 to 69% in 1966–1967 (Ajayi et al. 1996, p. 81) whereas at Ibadan the pace of change was slower, perhaps reflecting the latter's colonial foundations and legacy of the mainly British faculty. Many Nigerian lecturers at Ibadan were frustrated by the slow pace of change as well as by continuing discrimination, racism and pay disparities between local and international staff. Despite sustained calls for the appointment of more Nigerians to academic roles throughout the late 1940s and 1950s, in 1962 of six lecturers in the geography department, only one – Akin Mabogunje – was Nigerian.[26] When they returned from study overseas, many young Nigerian scholars were better qualified than the existing foreign staff; for example when Reuban K. Udo joined the staff of the geography department in 1963, he was one of only three with PhDs, alongside Mabogunje and the British Bill Morgan. Between 1962 and 1968 staff numbers rose in the geography department, with many of the new appointments Nigerian. By 1968 the staff had grown to eight, and half were Nigerian – R. A. Akinola (urban geography), R. C. Duru (International Atlas of West Africa) and Udo (rural settlement) having joined Mabogunje (urban geography).[27] Despite this, in geography the head of department remained British until 1972. The frustrations of working under a Head of Department who was less well-qualified but continued to have significant power and personal discretion in hiring and promotion decisions was highlighted in contemporary complaints to the Vice-Chancellor from geography staff. Letters also raised serious claims of anti-Nigerian attitudes within the department.[28] Ibadan was not the only place where such grievances were felt.

Despite on-going frustrations however, for Nigerians in this period, there were 'good hopes of rapid promotion'.[29] A background of training in the UK held a particular cachet for Nigerian staff, although the opportunity to be supervised at Ibadan by the rising star Mabogunje also held prestige. A post at Ibadan in the 1960s also came with a high level of kudos as well as financial reward. According to Mabogunje, 'Life at the University College, Ibadan in those days really entailed living in an ivory tower' (2011, p. 135). Udo remembers beginning his first post at Ibadan fresh from studying in London and being bombarded with flyers from motor companies hoping to sell him a car.[30] Further Particulars of a job advertised in 1970 stated that:

> Lecturers' offices are air conditioned... there is a geography laboratory ... The department has a drawing office with a staff of three cartographers, equipped with a Grant Projector and an Apollo dye-line machine... Research in the department is encouraged by a generous annual grant from the University to cover the expenses of fieldwork and excursions. There are also a Peugeot 404 Estate Car and a driver assigned to the Department.[31]

When a new head of the department of geography at Ibadan was appointed in 1962 the search still focused on British academics, rather than Nigerians, and Michael Barbour, a lecturer at University College London, with experience at the University of Khartoum, Sudan, was duly chosen. However, Barbour was also appointed, rather than the temporary incumbent (who was also British), according to Mabogunje, because he 'came with an open mind for changes'. Mabogunje, reflecting on this period, suggests that: 'In effect that transition ... to Barbour would be the beginning of the decolonisation'.[32] Nevertheless, as the 1960s progressed, the position of non-Nigerian heads of department, whatever their outlook, became increasingly incongruous: 'there was some feeling in the university that where you have two professors ... the Nigerian professor should really become head of the department', Mabogunje remembered. Mabogunje was promoted to professor in 1965, but preferred research and international roles to the administrative burden of a departmental headship. Instead the two formed a productive partnership, with Barbour deferring to Mabogunje when necessary; as the latter remembers, 'any major decisions he ha[d] a way of asking my views'.[33] Barbour also described this relationship in positive terms at the time, with the only concern he expressed being that he would have to fill in on Mabogunje's economic geography course whilst the latter was on sabbatical![34]

In his research and writing, which remained in the genre of regional tropical geography, Barbour was hardly central to any debate (let alone critique) of empire and decolonisation. His sometimes antagonistic and sarcastic manner could frustrate colleagues and lead to complaints.[35] However, in his everyday practice, Mabogunje at least understood Barbour as taking an active part in decolonisa-

**Figure 3.3** Photographs of the former heads of the geography department at the University of Ibadan still adorn the wall of the head's office. Source: Ruth Craggs, June 2017.

tion: administering a transforming geography department, overseeing increased access to honours degrees and the Nigerianisation of staff, and working productively with some Nigerian colleagues. Mabogunje finally took over as Head of Department in 1972 (Figure 3.3 shows the portraits of former Department Heads in order, still hanging in the current incumbent's office). After leaving, Barbour stayed in contact with the Department of Geography at Ibadan throughout the rest of his career, returning for their hosting of the Nigerian Geographical Association's Silver Jubilee Conference in 1982 (shown in the image on the front cover of the book), a decade after leaving.[36]

At the University of East Africa, the original impetus was towards 'East Africanisation' as opposed to nationalisation of staff. Kenyan Geographer Simeon H. Ominde was central to this campaign. Liyai (2000, p. 299) argues that Ominde gained a reputation for being a 'freedom fighter in the academic battlefield' when in June 1963 he produced a memorandum entitled *Africanisation of Administrative Posts in Makerere College*, alongside Bethwell Ogot from Kenya (who would later write about Africa's 'cultural dependence' in history, Ogot 1978, p. 33) and the Ugandan Economist Semei Nyanza. In this open attack on university hiring policies, they argued that key administrative posts at Makerere could be carried out by existing African staff with immediate effect. Ogot (2003, p. 121) records that many of the 'expatriate' staff were 'shocked' by these demands and, in a familiar refrain in decolonising Africa, were concerned about the possibility of undermining 'standards' (see also Livsey 2016). While the memorandum focused on administrative roles, it was written with the knowledge that in the geography department there were British staff who held senior academic positions despite having equal or lower qualifications than their African colleagues (Mngomezulu 2004). The campaign went far beyond the university, with a question asked in parliament about it, and was successful; Africanisation of the constituent colleges of University of East Africa beginning in earnest from 1964 (Mngomezulu 2004).

However, the push for *East*-Africanisation did not last long; 'once the heyday of the East African Federation movement was over and tensions developed which threatened to pull the federal university apart, nationalization rather than regionalization of university faculty emerged as a recurrent theme' (Southall and Kaufert 1974, p. 626). Thus by the late 1960s, the demand was for Ugandanisation not East Africanisation at Makerere, and this process was particularly speedy in relation to administrative roles, where Ugandans in post rose from 48% to 81% between 1965 and 1971. For academic staff the change was from 11% to 38% in the same period (Sicherman 2005). The speed of these changes accelerated further under Idi Amin from 1971, though this period also saw the departure of many Ugandan and other East African staff, as a result of the threats and violence against the university from the Amin regime (see Chapter 7).

As aspirations for East African inter-territoriality were superseded by nationalist politics, Ominde's outspoken role in the Africanisation debate at Makerere began to impact his own academic career. In his autobiography, historian Bethwell Ogot, a co-author of the memorandum on Africanisation recalls how, 'Ominde and myself, in particular, were faced with all kinds of threats and insults, demanding our resignation or transfer to Nairobi. Eventually, both of us moved to [University College] Nairobi on promotion in 1964' (2003, p. 121). There were also pull factors: 'great political pressure was put on Kenyan lecturers who were at Makerere to come home to build "their university"' (Ogot 2003, p. 130). Ominde was the first to make such a move and was appointed Professor of Geography in 1964. Campaigning for Africanisation had made building an academic career challenging at Makerere, as it was unpopular with the many British staff who still dominated. The fracturing of pan-East African visions also pushed Kenyans out of Makerere, and towards the new University College Nairobi, as national(ist) visions for higher education began to dominate.

It would be simplistic to suggest that there were always clear-cut lines between British and African academics when it came to debates about Africanisation however. Although Ominde was critical of the glacial speed of Africanisation and the lack of initial policies to support such change at the University of East Africa, in the early 1970s he was also willing to highlight the role that the British Head of Department Kenneth Baker had played in securing roles for East African geographers:

> he unobtrusively demonstrated that a firm foundation for a university department of geography depended on a consistent policy of attracting and recruiting local East African staff. To this end, modifications of the London degree structure in geography had the double aim of safeguarding standards and providing East Africa with university manpower grounded in the discipline. He was among the pioneers in instituting a systematic programme of selecting and training of East African staff. By a regular and sometimes lengthy correspondence he maintained close touch with them at a time when rapid changes were affecting university Education in East Africa. (Ominde 1971b, p. xii)

And whilst Baker's successor Bryan Langlands was criticised for his colonial attitude and tone, the archives show him consistently advocating for senior international roles and representation for East African geographers in the wider geographical community, as well as a strong line against white rule in Rhodesia.[37] In 1976, just as he was leaving Uganda under duress (see Chapter 7), Langlands paid $700 of International Geographical Union fee arrears 'by personal cheque to ensure that Uganda would not be dropped for non-payment of dues.'[38] There was sometimes a tension between the personal, professional, and political: Langlands was committed to developing East African geography, and creating opportunities for East African geographers, but could not – or was not willing to – see that this commitment might lead to the untenability of his own role as Head of Department into the mid 1970s.

By contrast, a commitment to helping to create the conditions for African colleagues to take over is evident from the perspective of Len Berry, who was the first professor of geography at University College Dar es Salaam. Berry joined Dar es Salaam in 1965 from previous positions at Hong Kong and Khartoum, and for him Africanisation was an evolving process:

> during the six years I was there [1965–1971] there was a rapidly escalating desire that the administration should be Tanzanian, which it should be, but in the first couple of years, Wilbert Chagula, the first African Principal had to go away for a few days, and out of three Deans who were Tanzanian and three of us that were expatriate, I was chosen to run the place whilst he was away and no one worried about that. By the end of the period that would have been unthinkable .... So the university was in a really dynamic mode ... and most people were there for a reason not for a job.

This reason most people were there, for Berry, was capacity building for a new nation state, which, for non-Tanzanian staff, was primarily a fixed-term engagement:

> There was a period from 1965 to say 1968, 69, when the external presence was welcomed and regarded as necessary, though had to be watched, but by the time I left, ... I loved it but I felt it was time to leave and let Tanzanians do the job.[39]

Whilst Africanisation provided opportunities for many local geographers, these opportunities overwhelmingly fell to men, rather than women, with the gender imbalance in many geography departments continuing to this day. It was not until the early 1970s that Ghana had a female geography lecturer, Elizabeth Ardayfio-Schandorf.[40] Joining the University of Ghana as one of three female students studying geography in 1966, she returned to take up a position in the department in 1974 following her PhD in the UK. While she did not encounter discrimination, there were logistical challenges, that highlighted the sort of male academic community envisaged by the architects of the early African universities:

> So it was a very cordial atmosphere, you went to lunch together, you chatted together. Yet there were limited facilities set up for females. Almost all the Halls were built for men. With my friends and colleagues all male, being the only female in the Department, I was able to navigate even the male washrooms whilst my male colleagues kept watch at the door [laughter]. For over twenty years I remained the only female geographer in the Department.[41]

Victoria Mwaka, a 'stellar' student in the geography department and the *only* Ugandan to gain her PhD under Langlands' tenure as Head of Department at Makerere, took over as Chair of Geography in 1977. In doing so she became one of the first female Heads of Geography on the continent (BakamaNume and Mutabuza 1997; Sicherman 2005, p. 87).

## Africanisation of Research, Curriculum, and Teaching

Whilst much of the transformation of African universities in the 1960s focused on Africanisation of staff, there was also some interest in the Africanisation of disciplines through teaching, curricula, and research. For some academic disciplines, notably history and English, debates about Africanisation were pronounced. They were linked to calls to reject irrelevant canons that indoctrinated African students into colonial ideologies through curricula that had nothing to do with national and local contexts historically, culturally, or linguistically. From this emerged distinctive schools of African history that embraced methodological innovations, oral traditions and linguistic diversity such as at Ibadan and Dar es Salaam (Falola 2004), as well as fierce debates and rejection of curricula, such as at the English Department at Nairobi, in a campaign lead by Ngũgĩ wa Thiong'o (Sicherman 1998; Ogot 2003; Ngũgĩ 1986). By contrast geography departments were more distanced from these debates and there was less discussion of epistemological transformation or new methods and sources. In general, geographers followed dominant disciplinary approaches and were not party to the same kind of epistemic overhaul that other disciplines demanded.

As the last chapter highlighted, British geographers working in African universities often took advantage of their locations to conduct local research, sometimes leading to PhD theses. Many of these provided descriptive accounts of physical features, or mapped developments in transportation and provided new data related to African landscapes, albeit following the methods and approaches of British geography. Much of the most innovative work was produced by African postgraduate students, especially those who had been able to define their own specialisms despite studying abroad. For example, Akin Mabogunje's PhD, awarded by University College, University of London was adapted to become his first monograph *Urbanization in Nigeria* (Mabogunje 1968), which was one of the first accounts to address *African*, rather than solely *colonial* urbanisation in Nigeria. Beginning in the pre-colonial period, the book is rich in detail about the historical development of Nigerian cities. At a time when, as the book notes, urban theory tended 'to regard urbanization as a peculiarly European and American phenomena', it took seriously urban growth unrelated to the West, and analysed the patterns of life and urban forms of Nigerian cities (Mabogunje 1968, p. 33). Crucially, Mabogunje's study provided a counter-point to the large number of studies, carried out primarily by white European anthropologists and sociologists, focusing on the (mal)adjustment of 'the African' to modern (colonial settler) cities, and to the broader notion that urban spaces were alien to 'traditional' African life (Mabogunje 1968). The book was well reviewed, and remains important today for scholars researching Nigeria's towns and cities.[42] *Urbanization in Nigeria* provided new insights by overturning colonial historical geographies of Africa which started with colonisation, and in doing so, paralleled innovations in the study of history lead by the first generation of historians in Nigeria (Falola 2004).

In his inaugural lecture in 1968, the same year that Mabogunje published his book, Simeon Ominde noted the colonial nature of geography on the continent: 'Like most aspects of the modern intellectual traditions in Africa, geography as a discipline belongs to that phase which might be called a phase of cultural intrusion.' (1971a, p. 3). The lack of trained African geographers meant that there continued to be a reliance on foreign expertise and traditions. But he also articulated his vision for a new African geography:

> as part of the modern intellectual re-awakening of our continent, we are challenged to reflect on geography as an intellectual activity within the context of the African development situation … Here is an opportunity to place in perspective the contribution of Western geographical tradition to the understanding of problems of African development and global relations… if African development is not a blind imitation of the course of events in the countries which have so far nursed the geographical tradition, then there is a need for a redefinition of the goals which would determine the evolving theoretical and practical role of the subject in the solution of our development problems. This is the question of the African philosophy of economic and social development. (Ominde 1971a, p. 3)

Ominde had long been interested in African knowledge and liberation. As an undergraduate at Makerere he had won the Uganda Bookshop Prize for an essay on 'the Meaning of Freedom' (Liyai 2000), as well as the Hancock Memorial Prize (for arts research) for a piece entitled *The Luo Girl from Infancy to Marriage* (Ominde 1952; Atieno Odhiambo 2000). Eisha Stephen Atieno Odhiambo (2000, pp. 8–9) writes that the latter was

> in essence an evocative statement about the validity of African indigenous education … one can read it as a major step in the enterprise of the production of knowledge for the Luo and by the Luo themselves. At one level, *The Luo Girl* was a challenge to the colonial structure of dominance, to the colonial praxis of hegemony over what constituted knowledge and what did not.

In this it was a challenge to colonial anthropology (and geography) 'which had never attributed mind and certainly not autonomy to the African woman in her autochthonous setting' (Atieno Odhiambo 2000, p. 9). Ominde also demonstrated his commitment to pan-African knowledge production through his founding of the population journal *Jimla Mutane* (Odingo 2000, p. 5).

Yet whilst in his inaugural lecture Ominde highlighted the Western roots of the discipline, and the need for an 'African philosophy of economic and social development', his description of the role geography could play articulated later on in the same speech did not depart hugely from that of late colonial development. 'Geography occupies a strategic position in a drive that must be made towards a more rational and fuller use of the resources of the land and manpower to achieve higher standards of living in Africa' Ominde argued (1971a, p. 8). There was one crucial difference however: the explicit rationale for the exploitation of resources was increasing standards of living for citizens in independent Africa, rather than colonial profits. Colleagues reflecting at the end of Ominde's life recorded that he was a 'mainstream scholar', viewing his Western intellectual heritage as 'a major pedagogic resource to be adapted and re-oriented with the purpose of creating an objective understanding of the social reality obtaining in Kenya and in Africa in general', rather than seeking to reject this inheritance altogether (Ojwang 2000, pp. 25; 22).

Ominde's substantial contributions to the study of population between the 1960s and 1980s reflected the broader trends in this field globally in exploring resource crises and overpopulation, and the national-level planning solutions for these issues, rather than more fundamentally questioning the global political economic structures and intellectual framings that produced these 'crises' (Ominde 1981; Greenhalgh 1996). In this Ominde was not unique. Despite the innovation of his first book, Akin Mabogunje sought a geography combining Western and African knowledge, and focused more on practical and applied geographical research for development on a liberal rather than radical model, rather than seeking a 'pure decolonized geography' (Clayton 2020, p. 11; Mabogunje 1980). This highlights the ambiguities of much of the scholarship of the first generation of African geographers. They were central in the institutionalisation of African leadership in the discipline in the continent, open to new approaches which drew on African theories and philosophies, and strongly committed to decolonisation and African nationalism. Because of this nationalism, though, they were also committed to the project of post-colonial national development which often closely paralleled earlier visions for colonial development (Mkandawire 1995). The consequences of this alignment of geographical research with post-colonial national development are explored further in Chapter 5.

Even after independence, teaching was often slow to change. Some British and European academics had little to no interest in updating or changing what they taught. This was especially true of those whose 'tenure of service' went back to 'colonial days' (Mazrui 1978 p. 204). This meant that teaching materials often continued to have affinity to courses verified and taught during the special relationship with the University of London even after that link had expired (Mngomezulu 2004). However, as Ali Mazrui, an academic at Makerere in the 1960s and 1970s noted, Africanisation of staff was

> not by itself a guarantee of accelerated Africanization of the syllabus and curriculum. The local staff, precisely by feeling less vulnerable than some expatriates, with no sense of the guilty outsider, can be more bothered about 'standards' and more resistant to precipitate change. (1978, p. 204–205)

Indeed, this generation of African staff, mainly educated in the UK and with more of a stake in the future of African universities than many of the foreign lecturers, were sometimes reluctant to make the dramatic transformations some demanded to break the Western hold on the university.

As a result, like staffing, Africanisation of the curriculum was not without controversy. Tensions appeared in the context of providing localised content that was relevant to newly independent countries, particularly when what was deemed 'relevant' became linked explicitly to the particular politics of post-colonial governments (Mamdani 1993). Moreover, as Mazrui (1978, p. 207) wryly noted, drawing on his own experience as a Political Science lecturer in Uganda under first Milton Obote and then Idi Amin, Africanisation of syllabi could make them 'all too relevant to the local scene'. This was particularly true in the humanities and social sciences where studies of local politics, democracy, development and resource extraction decisions could be sensitive topics for new governments.

Geography departments saw less of a transformation and more incremental change. When we asked in interviews about the Africanisation or decolonisation of the curricula, geographers answered with details of how the content of courses was changed for relevance to local conditions and requirements.[43] This primarily involved producing local content – including maps, land use surveys, case studies and data sets (for example of local climate through the installation of new weather stations on campus as at Ibadan, Dar es Salaam, and Legon). In 1964, Robert Steel, the British Professor of Geography at the University of Liverpool was invited to the University of Ghana as a visiting professor to help with the reform of the curriculum so it better suited local needs. Steel had played a key role in advising on the staffing the early geography departments in African Universities (see Chapter 2), and had conducted research in the Gold Coast, as part of the Ashanti Survey in the 1940s (Fortes et al. 1947). Nevertheless, the contradiction in drawing on British expertise in order to reframe a curriculum around local content and concerns is stark.

From the early days of geography at Ibadan, being sent out on sorties in the departmental vehicle to better understand Nigeria was an established practice. Generally, there was a distinct need for geographers working in Africa's geography departments to find out, first-hand about their regional and national contexts. This was especially true of the British geographers who were often entirely ignorant of the local African landscapes and societies when they arrived, but also held for African academics. This was both because texts and teaching materials about Africa (in English, and that fitted Western understandings of objective geographical knowledge) were practically non-existent, and because many of the first generation of African scholars hailed from urban areas (where they had been able to access better schooling) and knew little about other rural parts of their countries.[44] Their students too, tended to be from urban areas. Field trips could be crucial in bridging this divide. As Kenneth Baker, of Makerere reflected:

> The field class had a vital role in the curriculum. What better means of ensuring that urban-orientated students came to the realisation of the importance of the rural and agricultural base of their state than to take them on a carefully prepared field course in, say, Bunyoro or Teso? Such a course brought the members face to face with the peasants and showed the cultivators or the fishermen that people at Makerere had a real interest in them and their activities.[45]

Early on Baker instigated a system whereby:

> in the long vacation between the two sessions, also a time when students are not allowed to borrow books from the College Library, we ask them to work out an exercise in local geography; and we ought to add that by this means the staff is learning much about the local regional geography of the territories served by the College.[46]

This involved students carrying out research in diverse languages, though this is not something that is commented on, and the research and teaching materials were written up in English, reinforcing a dependency on English rather than local languages (Hountondji 1990). Therefore, this can be understood as a pragmatic response to fill gaps in teaching needs, rather than an overt gesture towards disciplinary decolonisation. The outcome of such efforts was a geography of African regions based on epistemic values bought to African geography departments by British geographers and emphasising, at least in the early years, the logic of land use surveys and regional geography. However, at the same time it also reveals a history of students contributing to the creation of foundational teaching materials and of the valuing and incorporation of local knowledge and experience within the geographical curricula.

Following the completion of his undergraduate degree at Makerere in 1961 Adolfo Mascarenhas returned to Tanganyika to take up a teaching post in a secondary school, which made him reflect on both his peculiar geographical expertise, shaped by his colonial education, and the knowledge of his students, at least in comparison to published work by the British 'great men' of geography:

> I knew more about the physical Geography of Britain than they [the students] knew about Tanganyika or, for that matter, about East Africa. Yet I knew that the knowledge that all students had about their social and physical environment was far greater than all text book information, especially Dudley Stamp.[47]

Here we see how one of the first generation of Tanzanian geographers Adolfo Mascarenhas's education at Makerere was steeped in the British geographical tradition. However, he was also very aware of the need to overhaul the geographical canon and take an approach that saw not just a *deficit* of prescribed geographical knowledge in his students, but rather an untapped *abundance* of local knowledge.

However, there were challenges for those open to change. For Len Berry, producing the curriculum for the first time at University College Dar es Salaam:

> the titles of the courses were more or less provided, but the content of the courses we did a lot of work on. The meteorologists set up a rain recording and hydrological network on campus. We really worked on getting a perspective, an East African or Tanzanian perspective on the courses.[48]

In this case, local data and local examples were the vehicle for establishing a Tanzanian perspective.

> We did the best we could. But the intent was not to create a UK syllabus, it was to create a Tanzanian syllabus ... it was very difficult. It was not difficult to get the concept, it was difficult to get the materials. And this was true I think right through the university ... It's shocking to think how far we had to go to get African materials.[49]

This kind of work took time, and to some extent explains the lack of research publications by early faculty members. The challenging conditions in university departments, sometimes with very little support for developing new courses, also often lead to the reproduction of British-centred content, by both British and African geographers. For example, for the Ghanaian George Benneh, on his appointment to the University of Ghana in 1964 shortly after he completed his PhD in London:

> what I found difficult was, you're just left to do your own thing, no, no guidelines, nothing. Since I'd done economic geography at LSE, precisely agricultural geog-

> raphy, the lecturer in economic geography who had lectured me before just called me and said I should take the class. So, that's that. Fortunately I'd bought some economic geography books at Dillons Bookshop [in London]. I prepared my notes and started lecturing.[50]

An important (and time consuming) contribution of African geography departments was producing the first textbooks on African countries and regions for schools and universities (Baker, 2015).[51] Moreover, the lines between secondary and higher education were often more blurred in these early departments than we see today: several early staff members were school teachers with degrees working locally (e.g. Ian Thomas, a demographer working in a school who joined the geography department in Dar es Salaam in 1965 when it opened). New teaching materials produced on African geography were utilised not only in university teaching but also in schools, and indeed, some were specially designed for this purpose. For example, the Dar es Salaam Geography Department produced a pamphlet *Geographical Fieldwork, a Handbook for Teachers* (Berry and Battersby 1963) to replace 'one of the books that was used in Tanzanian high schools' that said, Berry remembered, 'to visit Tanzania, you first had to take the train to Southampton. It was bizarre'.[52] Elsewhere, at Ibadan geographers regularly contributed short case studies for school teaching which were published in the *Nigerian Geographical Journal* in the 1950s and 1960s. The research of geographers at Makerere 'enabled the redirection of the secondary-school syllabus toward local concerns', whilst the same department, under both Baker and later Langlands, cooperated with the Faculty of Education 'in training secondary-school teachers and running refresher courses for teachers' (Sicherman 2005, p. 88).

The Kenyan geographer Simeon Ominde was also involved in school-level educational policy change in Kenya, alongside his research, on demography and development, and work towards Africanisation of staff. In 1964 he was asked to chair the Kenya Education Commission – which became known as the Ominde Commission – charged with restructuring the whole education sector after independence (Sifuna 1990; Ojiambo 2009). Part One of the Ominde Commission Report (1964) recommended, amongst a whole raft of measures, abolishing the Cambridge certificates in favour of localised curricula, advocating non-racial schools, and renaming schools that had western names. Writing at the time of this Commission, Ominde (1965) explored the role of teachers in 'revolutionary Africa'. Reflecting the concerns he had raised in his undergraduate essay on *The Luo Girl* (1952) about the loss of African culture and knowledge in the hands of colonial modernisation, Ominde (1965, p. 7) highlighted the need for teachers to contribute to national development not only economically through producing productive citizens but also to inform new generations about, and safeguard, 'cultural values in Africa' – which he termed African Socialism – 'in the face of change'.

One result of these Ominde Commission reforms, some of which anticipate still live debates about decolonising institutions, was the replacement of the

Cambridge Certificate syllabus in geography with a new one designed for East African needs. As Occiti (1971, p. 3) reported at the time:

> This new geography syllabus is but one aspect of the non-violent revolution which has been taking place in East Africa since Independence in the field of curriculum development ... The general feeling in East Africa at Independence was that the School Certificate geography syllabus, like other syllabi which were formulated under the umbrella of colonialism, was not in line with the needs and aspirations of a sovereign state, especially in terms of its contents, teaching approach and the way it was examined.

Whilst the radical geographer and writer Occiti cast the changes as revolutionary (Clayton 2020), the new curriculum focused on localised content, national development, and practical geographical skills (Occiti 1971), thus mirroring changes at university level, and looking to transform content rather than underlying epistemologies.

Finally, whilst syllabuses were an important part of the student experience of teaching, other aspects were also important, and here, the influence of key individuals could have a powerful role on the character of departments, including on the extent to which they were viewed as Africanising. At Ibadan, whilst Mabogunje didn't become Head of Department until 1972, he was a big presence in the department for a decade before that, drawing Nigerian and international students and staff members keen to work with him, as well as masterminding dramatic curriculum changes that are detailed in Chapter 4. At the University of Ghana from the early 1960s Ghanaians were leading the department and dominant in the lecturing ranks. At Makerere, however, Bryan Langlands wielded a substantial influence on the tone and content of the geography taught into the 1970s:

> He instituted in the department his own belief of what geography should be – a holistic discipline rooted in human ecology ... Unfortunately, Professor Langlands was a true product of the colonial university education system. He was always right. His word was nothing but the whole truth – the Gospels according to St. B. W. Langlands was the true word of geography. No student would dare debate Professor Langlands. (BakamaNume and Mutabuza 1997, p. 4)

Langlands, and the geography department as a whole, were notorious for strict, and overly harsh, marking in comparison to other departments, leading to an exodus of good undergraduate students leaving with average grades but later excelling in postgraduate study elsewhere. Very few East Africans gained PhDs at Makerere until the late 1970s and the wider departmental culture treated those studying for higher degrees – and working on teaching assistantships at the same time – with little respect. As BakamaNume and Mutabuza (1997, p. 5) record:

> The terms academic houseboy and housegirl were used to refer to graduate assistants in the university ... The Uganda government was an attractive employer of young graduates of Makerere. A civil servant position in the 1950s, 1960s, and early 1970s was more attractive than a teaching assistantship – an academic "houseboy" or "housegirl".

At Makerere, there was clearly no sense that the lecturer-student relationship might need to change as part of a wider process of educational decolonisation (Mazrui 1978; Ndlovu-Gatsheni 2018).

## Conclusion

Universities were not insulated from the politics of decolonisation, they were integral to them: to 'some extent, the politicians regarded the universities, dominated as they were by expatriate staff, as part of the apparatus of imperialism, comparable to multinational corporations, which had to be decolonised' (Ajayi et al. 1996, p. 95; Mazrui 1978). The 1960s saw an increased impetus for change at African higher education institutions, as countries became independent, and the university colleges set up in special relation with the University of London cut these links. These constitutional and institutional changes provided more freedom for modifications at departmental level, including in course structure, staffing, and curricula.

A huge growth in the number of universities in this era provided opportunities for African scholars, though these were uneven; post-colonial Nigeria saw many new positions available in new institutions but at Ibadan, founded as a colonial university college, change was sometimes slow to come. In South Africa new opportunities were almost entirely limited to white academics, and especially Afrikaans-speakers. The atmosphere and culture of departments was as important as the changes in curricula and staff, though all were interrelated. A combination of international, domestic and departmental politics conditioned the opportunities and experiences of African geography staff and students. And African geographers – in concert and sometimes against their British colleagues – played an active role in these decolonisation efforts.

A focus on the intersections between academic careers and different versions of decolonisation highlighted in the chapter demonstrates the labour of decolonising disciplines and universities in Africa in this period. Decolonising work could be career enhancing – for African and British geographers alike – whose work in these universities in this period earned them academic and political recognition. But it could also be career threatening – when it meant challenging the status quo in the colonial university.

This period saw vociferous debates across Africa, including at the Universities of Ghana, Ibadan, Dar es Salaam, and Makerere, about the need for the Africanisation of syllabuses – for example in English and history departments. Here, new methods and sources for recovering and centering African voices were proposed. But in geography departments Africanisation tended to mean localisation of content, rather than methodological or epistemological change: local case studies, data bases, and maps. However, in focusing on local places, both in staff and student research projects, researchers mapped local communities and collected the experiences of local people, often (in the case of student work) in local African languages.

Ndlovu-Gatsheni (2018, p. 175) has argued that, despite this kind of step, 'this type of Africanization and decolonization did not touch the structural European epistemological scaffolding on which the university in Africa is built'. It is hard to argue with this. Nevertheless, important changes did take place in African universities – and in African geography – in this period. Most importantly, a new generation of African geographers had at least some agency to study the places, people, and issues that they felt to be important, especially once postgraduate training began to take place in Africa. Many of this first generation of African geographers identified closely with the project of African nationalism (as well as sometimes pan-Africanism, see Mkandawire 1995) and shaped the discipline into one that could serve their independent nations. For them, decolonisation meant influencing the shape of the discipline – through pushing new research agendas, demanding new positions, building the reputation of national universities, introducing new locally relevant topics, and putting geography to work for their independent nations. Returning to work in increasingly vibrant African geography department in the 1950s and 1960s very much felt like success in what Ngũgĩ (1993) referred to as 'moving the centre'. It was *this* version of decolonisation that was primarily to shape geography in the 1960s and 1970s, as we shall see in the following two chapters. Yet, it also fell short of Ngũgĩ's demands in two ways: in sticking closely to Western geographical epistemologies, languages, and subjects, as well as notions of progress and development, it would be hard to see the changes made in the 1950s and 1960s as 'decolonising the mind' (1986). Despite the nuances of Ominde's writings about African culture and the impact of colonialism, his vision for geography remained tied to its role in development and state-building. Moreover, 'moving the centre' (Ngũgi 1993) meant more than an international shift towards Africa, it also meant challenging the elite, male, status of academic power within the nation. On that level, relationships changed very little.

In the next chapter, we further develop some of these arguments, highlighting the impact of the Cold War on the decolonisation and internationalisation of geography in Africa. As we will see, in the 1960s and 1970s, debates about disciplinary paradigms intersected with new forms and patterns of funding and mobility for African academics, leading to novel debates about theorising *from* as well as *about* Africa.

## Notes

1 Professor Simeon Ominde (deceased), Fellows, Kenyan National Academy of Sciences. https://www.knascience.or.ke/index.php/fellows/82-knas/125-prof-simeon-hongo-ominde-deceased [last accessed 3 December 2021].
2 Interview with George Benneh, Legon, Ghana, 17 January 2017.
3 Interview with Akin Mabogunje, London, UK, 7 July, 2016.
4 Interview with Olusegun Areola, Ibadan, Nigeria 9 July, 2017.
5 Interview with Akin Mabogunje.

6 Interview with Olusegun Areola.
7 Interview with Elizabeth Ardayfio-Schandorf, Legon, 20 January 2017.
8 Interview with George Benneh.
9 Interview with Elizabeth Ardayfio-Schandorf.
10 Interview with Johnson Ayoade, Ibadan, 9 July 2017.
11 Interview with Olusegun Areola.
12 Interview with Olusegun Areola.
13 Tony O'Connor, personal communication, March 2018.
14 Personal communication with Jennifer Bray, August 2018.
15 Professor B.W. Hodder memorandum, ODRP, MSS Afr. S. 1825 (51) p. 1.
16 Interview with Akin Mabogunje.
17 Interview with Ronnie Donaldson, Stellenbosch, South Africa, 21 October 2019.
18 Talbot, W.J. (1973) 'Lest we forget .... Annals of the University of Cape Town Department of Geography, 1936-1973' p. vii, ix. Professor William J. Talbot Papers, University of Cape Town Special Collections (Hereafter UCT) BC1079 D.5 Departmental Farewell Speech.
19 Adolfo Mascarenhas, Personal communication, April 2021.
20 Interview with Len Berry, via Zoom, 26 February 2021.
21 Interview with Len Berry.
22 Papers of Robert Steel, University of Liverpool (hereafter RS), Letter, Steel to R O Buchanan, 27 July 1960, RS D91/1/4/2iii (part 2 of 14).
23 Letter, Principal Stoughton to Robert Steel, 26 February 1960, University of Ghana Archives, UG/1/3/2/3/71.
24 Letter, Principal Stoughton to T.E. Hilton, 5 May 1960. University of Ghana Archives, UG/1/3/2/3/75.
25 Interview with George Benneh.
26 Kenneth (Michael) Barbour memo, Oxford Development Records Project here after ODRP, Bodleian Library, Oxford MSS Afr. S. 1825 Box III (4).
27 *Orbis Geographicus*, Weisbaden, 1966, p. 64.
28 E.g., Letter to the Vice-Chancellor, 29 April 1965; Letter to the Vice-Chancellor, 26 November, 1963, University of Ibadan Archives and Records Service Project, hereafter UI, University of Ibadan FASC.12.5 (Geography).
29 Kenneth (Michael) Barbour memo ODRP Oxford MSS Afr. S. 1825 Box III (4).
30 Interview with Reuban K. Udo, London, UK, 1 June 2018.
31 Further Particulars, Lecturer in geography, 1970. UI FASC.12.5.42.
32 Interview with Akin Mabogunje.
33 Interview with Akin Mabogunje.
34 Barbour memo; Barbour to Steel, 1962, D91/1/4/2iii (part 1 of 14).
35 Interview with Hugh Clout, London, UK, 30 November 2016; Interview with Reuban K. Udo.
36 Geography Department photograph albums, University of Ibadan, by kind permission of Chris Ikporukpo.
37 International Geographical Union Archive, Archiv für Geographie, Leibniz-Institut für Länderkunde, hereafter IGU: II-11–7 UGANDA Letter from Langlands to Boesch, 10 April 1967; Letter from Langlands to Chauncy Harris, 20 December 1968.
38 IGU II-11–7 UGANDA Letter from Langlands to IGU 19 August 1976.
39 Interview with Len Berry.

40 Interview with Elizabeth Ardayfio-Schandorf.
41 Interview with Elizabeth Ardayfio-Schandorf.
42 E.g. it was the subject of a panel discussion at a recent major conference in Lagos, Lagos Studies Association Conference, 2018.
43 For example interview with Len Berry.
44 Interview with Akin Mabogunje.
45 S.J.K. Baker 'Memorandum from S.J.K. Baker' Oxford Development Records Project (ODRP) MSS Afr. S. 1825 (3) BOX II, p. 8.
46 Undated speech discussing Baker's early experiences at Makerere, D402/1/2/1 from papers of S.J.K. Baker, University of Liverpool Special Collections and Archives (hereafter SJK) D402/1/2/1.
47 University of Dar es Salaam Alumni of the Month, Adolfo Caridade Mascarenhas https://udsm.ac.tz/web/index.php/alumni/name/7 [last accessed 13 December 2021].
48 Interview with Len Berry.
49 Interview with Len Berry.
50 Interview with George Benneh.
51 Interview with Len Berry.
52 Interview with Len Berry. The date given of this pamphlet seems to predate the department, and it hasn't been possible to find an original copy to confirm this. It is possible that it was published a little later in the 1960s.

## References


Adi, H. (1993). West African students in Britain, 1900–60: the politics of exile. *Immigrants & Minorities* 12 (3): 107–128.

Agbodeka, F. (1998). *A History of University of Ghana: Half A Century of Higher Education (1948–1998)*. Accra: Woeli Publishing Services.

Ahlman, J.S. (2017). *Living with Nkrumahism: Nation, State, and Pan-Africanism in Ghana*. Athens: Ohio University Press.

Ajayi, J.F.A., Lameck, G., and Johnson, G.A. (1996). *The African Experience with Higher Education*. Athens: Ohio University Press.

Areola, O. (1998). Postgraduate training at Ibadan. In: *50 Years of Geography in Nigeria: The Ibadan Story* (ed. O. Areola and S.I. Okafor), 47–74. Ibadan: Ibadan University Press.

Atieno Odhiambo, J. (2000). The worlds of Ojiji: luo women, 1900–1900. In: *Issues in Resource Management and Development in Kenya: Essays in Memory of Professor Simeon H. Ominde* (ed. R.A. Obudho and J.B. Ojwang), 8–16. Nairobi: East African Educational Publishers.

BakamaNume, B. and Mutabuza, M. (1997). Bryan Wooleston Langlands: Ugandan geography and the gospels of St. Langlands, 1953–1976. *East African Geographical Review* 19 (2): 1–12. https://doi.org/10.1080/00707961.1997.9756243.

Baker, R.A. (2015). Samuel John Kenneth Baker 1907–1992. In: *Geographers Biobibliographical Studies Volume 22* (ed. P.H. Armstrong and G.J. Martin), 1–11. London: Continuum.

Baker, R. and Polonskyi, M. (2015). *Cold War in a Hot Place*. Sofia: East-West Publishing.

Barnard, W.S. (2016). *Encountering Adamastor: South Africa's Founding Geographers in Time and Place*, vol. 1. Stellenbosch: African Sun Media.

Benneh, G. (2016). *My Time, My Nation*. Accra: Winmat Publishers Limited.

Berry, L. and Battersby, J. (1963). *Geographical Fieldwork, a Handbook for Teachers*. Dar es Salaam: Institute of Education.

Clayton, D. (2020). The passing of "geography's empire" and question of geography in decolonization, 1945–1980. *Annals of the American Association of Geographers* 110 (5): 1540–1558. https://doi.org/10.1080/24694452.2020.1715194.

Cooper, F. (2008). Possibility and Constraint: African independence in historical perspective. *The Journal of African History* 49 (2): 167–19. https://doi.org/10.1017/S0021853708003915.

Cooper, F. (2019). *Africa since 1940: The past of the Present*. Cambridge: Cambridge University Press.

Dubow, S. (2014). *Apartheid, 1948–1994*. Oxford: Oxford University Press.

Falola, T. (2004). *Nationalism and African Intellectuals*. Rochester: University Rochester Press.

Fortes, M., Steel, R.W., and Ady, P. (1947). Ashanti survey: 1945–46: an experiment in social research. *The Geographical Journal* 110 (4/6): 149–177.

Greenhalgh, S. (1996). The social construction of population science: an intellectual, institutional, and political history of twentieth-century demography. *Comparative Studies in Society and History* 38: 26–66. https://doi.org/10.1017/S0010417500020119.

Hountondji, P. (1990). Dependence in Africa today. *Research in African Literatures* 21 (3): 5–15.

Kithinji, M.M. (2012). An imperial enterprise: the making and breaking of the University of East Africa, 1949–1969. *Canadian Journal of African Studies / Revue Canadienne Des E´tudes Africaines* 46: 195–214.

Livsey, T. (2016). Imagining an imperial modernity: universities and the Nigerian roots of colonial development. *Journal of Imperial and Commonwealth History* 44: 952–275. https://doi.org/10.1080/03086534.2016.1210305.

Livsey, T. (2017). *Nigeria's University Age: Reframing Decolonisation and Development*. Basingstoke: Palgrave Macmillan.

Liyai, H.A. (2000). Appendix: Simeon Hongo Ominde a biographical note. In: *Issues in Resource Management and Development in Kenya: Essays in Memory of Professor Simeon H. Ominde* (ed. R.A. Obudho and J.B. Ojwang), 298–308. Nairobi: East African Educational Publishers.

Mabogunje, A.L. (1968). *Urbanization in Nigeria*. London: University of London Press.

Mabogunje, A.L. (1980). *The Development Process: A Spatial Perspective*. London: Hutchinson.

Mabogunje, A.L. (2011). *A Measure of Grace: The Autobiography of Akinlawon Ladipo Mabogunje*. Ibadan: Book Builders.

Mamdani, M. (1993). University crisis and reform: a reflection on the African Experience. *Review of African Political Economy* 58: 7–19. https://doi.org/10.1080/03056249308704016.

Mamdani, M. (2018). The African University. *London Review of Books* 40 (14). Retrieved from: https://www.lrb.co.uk/the-paper/v40/n14/mahmood-mamdani/the-african-university.

Mazrui, A.A. (1978). *Political Values and the Educated Class in Africa*. Berkeley: University of California Press.

Milford, I. (2020). Federation, partnership, and the chronologies of space in 1950s East and Central Africa. *The Historical Journal* 63 (5): 1325–1348. https://doi.org/10.1017/S0018246×19000712.

Mkandawire, T. (1995). Three generations of African academics: a note. *Transformation* 28: 75–83.

Mkandawire, T. (2005). African intellectuals and nationalism. In: *African Intellectuals: Rethinking Politics, Language, Gender and Development* (ed. T. Mkandawire), 10–55. London: Zed Books in association with CODESRIA Books.

Mngomezulu, B.R. (2004). *A Political History of Higher Education in East Africa: The Rise and Fall of the University of East Africa, 1937–1970*. PhD Thesis. Rice University.

Morgan, W.T.W. (1991). East Africa (Kenya, Tanzania and Uganda). In: *Modern Geography: An Encyclopaedic Survey* (ed. G.S. Dunbar), 46–47. Oxford: Routledge.

Ndlovu-Gatsheni, S.J. (2018). *Epistemic Freedom in Africa: Deprovincialization and Decolonization*. London: Routledge.

Ngũgĩ wa Thiong'o. (1986). *Decolonising the Mind: The Politics of Language in African Literature*. Portsmouth: Heinemann.

Ngũgĩ wa Thiong'o. (1993). *Moving the Centre: The Struggle for Cultural Freedoms*. Portsmouth: Heinemann.

Nkrumah, K. (1962). *Guide to Party Action: Address by Osagyefo Dr. Kwame Nkrumah at the First Seminar at the Winneba Ideological School on 3rd February, 1962*. Accra: Central Committee of the Convention People's Party.

Nkrumah, K. (2009 [1964]). *Consciencism: Philosophy and Ideology for De-Colonization and Development with Particular Reference to the African Revolution*. New York: Monthly Review Press.

Ocitti, J.P. (1971). A revolutionary school geography syllabus in East Africa. *East African Geographical Review* 9: 3–10.

Odingo, R.S. (2000). Professor Simeon. Hongo Ominde's contribution to geographical science. In: *Issues in Resource Management and Development in Kenya: Essays in Memory of Professor Simeon H. Ominde*. (ed. R.A. Obudho and J.B. Ojwang), 3–7. Nairobi: East African Educational Publishers.

Ogot, B.A. (1978). Three decades of historical studies in East Africa, 1949–1977. *Kenya Historical Review* 6 (1–2): 22–23.

Ogot, B.A. (2003). *My Footprints in the Sands of Time: An Autobiography*. Victoria: Trafford.

Ojiambo, P.C.O. (2009). Quality of education and its role in National Development: a case study of Kenya's educational reforms. *Kenya Studies Review* 1 (1): 133–149.

Ojwang, J.B. (2000). Scholarship in national development. In *Issues in Resource Management and Development in Kenya: Essays in Memory of Professor Simeon H. Ominde*. (ed. R.A. Obudho and J.B. Ojwang), 17–26. Nairobi: East African Educational Publishers.

Ominde, S.H. (1952). *The Luo Girl: From Infancy to Marriage*. London: Macmillan.

Ominde, S.H. (1964). *Kenya education commission report, Part 1*. Nairobi: Government Printer.

Ominde, S.H. (1965). Education in revolutionary Africa. *East Africa Journal* 2: 6–14.

Ominde, S.H. (1971a). Geography and African development. In: *Studies in East African Geography and Development* (ed. S.H. Ominde), 1–8. London: Heinemann Educational Books.

Ominde, S.H. (1971b). S.J.K. Baker: a biographical note. In: *Studies in East African Geography and Development* (ed. S.H. Ominde), xi–xiii. London: Heinemann Educational Books.

Ominde, S.K. (1981). Population and resource crisis: a Kenyan case study. *GeoJournal* 5: 539–556.

Pietsch, T. (2013). *Empire of Scholars Universities, Networks and the British Academic World, 1850–1939*. Manchester: Manchester University Press.

Rodney, W. (1990 [1975]). *Walter Rodney Speaks: The Making of an African Intellectual.* Trenton, NJ: Africa World Press.

Sicherman, C. (1998). Revolutionising the literature curriculum at the University of East Africa: literature and the soul of the nation. *Research in African Literatures* 29 (3): 129–148.

Sicherman, C. (2005). *Becoming and African University: Makerere 1922–2000.* Trenton: Africa World Press.

Sifuna, D. (1990). *Development of Education in Africa: The Kenyan Experience.* Nairobi: Initiatives Publishers.

Southall, R.J. and Kaufert, J.M. (1974). Converging models of university development: Ghana and East Africa. *Canadian Journal of African Studies / Revue Canadienne Des Études Africaines* 8 (3): 607–628.

Stockwell, A.J. (2008). Leaders, dissidents and the disappointed: colonial students in Britain as empire ended. *The Journal of Imperial and Commonwealth History* 36 (3): 487–507. https://doi.org/10.1080/03086530802318730.

Stockwell, S. (2018). *The British End to the British Empire.* Cambridge: Cambridge University Press.

Wesso, H. (1992). *The Colonisation of the Geographical Mind: A Critical Contextual Analysis of the Institutionalisation and Establishment of Geography as an Academic Discipline in South Africa.* PhD Thesis. University of the Western Cape.

Wolpe, H. (1995). The debate on university transformation in South Africa: the case of the University of the Western Cape. *Comparative Education* 31 (2): 275–292.

# Chapter Four
# International Networks, Decolonisation, and the Cold War

## Introduction

In 1960, the African American jazz artist Louis Armstrong embarked upon an African tour, playing in 13 countries (Monson 2007, p. 124). In Uganda, Armstrong played at Makerere College, and the young Tanganyikan (later Tanzanian) Adolfo Mascarenhas, a geography student at the College, found himself involved. 'There was a bit of a hitch' he recalled, 'and so I was asked to help in the lighting on the stage'.[1] Also invited on stage at the concert was Gordon Ominde, the four-year-old son of the geographer Simeon Ominde, then a lecturer at Makerere, for whom the opportunity to try his hand at conducting the band was an early memory.[2] After the concert Armstrong thanked Mascarenhas for his help and gave him an autograph. 'Shame it was eaten by white ants', the geographer concluded, when reminiscing about the occasion.[3] Armstrong's tour was sponsored by the US Department of State Cultural Presentation Programme, part of the cultural front in the Cold War (1946–1991). Concerned that 'domestic racial relations were having a negative impact' in the conflict, the State Department hoped the tours would help to dissipate negative associations with the US, and prevent newly independent African countries from seeing this as a reason to ally instead with the Soviet Union (Monson 2007, p. 108).

Alongside his brush with musical greatness, Mascarenhas's professional life would be touched in many ways by Cold War influences. Both the American State Department (and the American foundations that shared the State De-

*Decolonising Geography? Disciplinary Histories and the End of the British Empire in Africa, 1948–1998*, First Edition.
Ruth Craggs and Hannah Neate.

partment's goals), and the Soviet Union and its allies became increasingly interested in Africa in the 1950s and 1960s. They were particularly interested in its universities and their students who would become the new elites (Livsey 2017). At Makerere, Mascarenhas became friendly with the teenage children of the American geographer Ben Thomas, who was visiting Makerere on sabbatical and staying in flats sponsored by the Ford Foundation. Thomas arranged American funding for Mascarenhas to study for his PhD at the University of California Los Angeles after meeting him on campus. As Mascarenhas recalled: 'In the formal classroom atmosphere and the shared breaks at the Makerere University Canteen he awakened in me the first ambitions in post-graduate studies' (1971, p. x).

Mascarenhas's return to Tanzania was initially sponsored by the Rockefeller Foundation, and, with an American PhD under his belt, he became a rising star in the country.[4] Once installed as lecturer at the University of Dar es Salaam in the late 1960s, Mascarenhas was courted by visitors from both sides of the conflict. On one occasion he hosted 'five visitors who came to see me in my tiny office', he remembered, including the Head of the Rockefeller Foundation and the US Secretary of State. On another, Mascarenhas was visited by a delegation from North Korea: 'They wanted me to go for 6 weeks to their capital city Pyongyang via Moscow in a Russian Plane.' Mascarenhas was not keen, 'I *accidently* brought up the fact that I was at UCLA and what wonderful people they were and how I admired President Kennedy and President Johnson... Yes I never saw them again!'[5] But Mascarenhas was no American stooge; he was close to Tanzanian President Julius Nyerere who advocated a non-aligned position in world affairs, and to the Pugwash movement which sought to reduce the danger of nuclear war by stimulating discussions between scholars from across the Cold War divides.[6] Life at the University of Dar es Salaam, but also at many African universities was bound up in many ways with the direct and indirect influences of the global Cold War, as the superpowers and their allies sought to influence the new elites, and African politicians and academics sought to navigate the tense geopolitical terrain. These dynamics filtered into the careers of geographers, and the teaching and research of geography in Africa.

This chapter explores the intersections between decolonisation, the Cold War, and international mobility through the experiences of geographers like Mascarenhas. While it discusses a range of places and people, focus falls in particular on the Geography Departments of the University of Ibadan, Nigeria, and the University of Dar es Salaam, Tanzania, as well as on the academic mobility of the Nigerian Akin Mabogunje. It takes the imperial and national level debates explored in the previous chapters and examines other important transnational connections enabled by generous funding and scholarships that enabled travel and in turn shaped geography, and geographers. Mobility provided increasing opportunities for individual African geographers to develop their careers and their ideas, facilitating individual agency, and brought together networks of people, including in Africa, making Africa a centre for some of the most exciting scholarship of the 1960s and 1970s. However, whilst external funding provided individual opportunities, it also risked reinforcing economic and 'intellectual extraversion', and 'institutional nomadism, a restless going to and fro' of African academics primarily to the West (Hountondji 1990, p.10). Moreover, whilst mobility provided opportunities for some, these were primarily opportunities for the 'inclusion of Africans into a long-existing "European game" without changing the rules of the game' (Ndlovu-Gatsheni 2018, p. 175; Fanon 2019 [1961]).

The following account demonstrates the ways in which the history of geography in Africa was bound up not just with decolonisation, but also with the Cold War. The impact of the Cold War in Africa has only fairly recently been inserted into the historiography of the superpower rivalry (e.g. Onslow 2009). It shaped not only military interventions, but also relationships in the cultural and educational realms, particularly through development aid (Lorenzini 2019). Funding flowed into higher education in Africa primarily from the US, but also from the USSR and Eastern bloc. South Africa leveraged its position as a bulwark against communism on the continent to ensure continued relationships with the UK and US, including within higher education, despite criticism of its policy of apartheid. And countries like Ghana and Tanzania looked to non-alignment to carve an independent path between the two Cold War poles, with consequences for their university sectors. The chapter complements work on the history of geography that has highlighted the crucial role that the Cold War played in shaping research and its uptake, particularly in the United States and Soviet Union (Farish 2010; Oldfield 2018; Child and Barnes 2019), by demonstrating the global and transnational dimensions of its impact. A biographical approach, focusing on professional experiences far from the violence of the Cold War 'frontlines', provides a richer account of the social and cultural consequences of the superpower rivalry, as well as demonstrating how decolonisation and the Cold War were profoundly entangled in ways which impacted the experiences, opportunities, and challenges which geographers faced, and the research that they produced.

The chapter also demonstrates the value of more diverse histories of geography. These can reveal new dimensions to disciplinary histories which are often dominated by a narrow European and North American perspective (e.g. Livingstone 1992). Geographers travelling to the US from Africa were often immersed in the new techniques of spatial science, and viewing the Quantitative Revolution from this perspective provides new insights into this episode in geography's history. As geographers from within and beyond Africa were attracted to non-aligned Tanzania, they contributed to the elaboration of a radical development geography which was highly critical of quantitative techniques, their application to questions of modernisation, and of the underlying tenets of mainstream development ideas. This early history too, has often been overlooked in Anglo-American accounts of the rise of critical development geography.

The chapter begins by examining the shift away from the 'British world' of higher education in Africa, towards other nodes – most notably North America in the late 1950s. In this period, huge amounts of Rockefeller and Ford Foundation funding were available for staff in the universities of newly independent countries in Africa. Alongside on-going British funding, these grants facilitated huge mobility amongst academic geographers, not only taking US geographers to post-colonial universities, but also taking large numbers of staff from Africa to the US. The chapter considers the impact of this turn to the US for the direction of the discipline in Africa.

The second half of the chapter explores alternative networks, including those with Eastern Bloc and non-aligned countries, that took geographers to different locations. Attempting to forge an independent socialist path, Tanzania, and its new university in Dar es Salaam became a darling of many left-wing academics and a magnet for radical scholars globally, including geographers (Shivji 1993; Sharp 2019). This section highlights how networks including African universities and scholars benefitted not only Africans but also the academics from all over the world who were able to engage with them, shifting ideas around radical geography and postcolonial development.

## Diversifying Influences and Americanisation

### *Looking to America*

Students studying in African countries in the 1950s remember lots of academic visitors from the UK, alongside a preponderance of British titles amongst journal subscriptions.[7] This started to shift in the 1970s, and the South African Gordon Pirie remembered 'a sense of switch, British geography seeming a little old fashioned … Even the RGS [Royal Geographical Society], with its crest. The new generation journals, many of which were American, looked different.'[8] Whilst British connections continued to have an important impact on African geography departments after decolonisation, increasingly it was American influences that were most visible. Whilst this was true also in the UK and the rest of Europe in this period (Whitehand and Edmondson 1977), African geography was impacted more dramatically, and in particular ways, as a result of the geopolitics of the Cold War. Tim Livsey (2017) has argued that, in the late 1950s and 1960s, African development had become a new Cold War battleground and higher education was a substantial front in the battle.

A key outcome of the additional US funding available was an increase in American geographers visiting Africa, and often spending time in African geography departments. For example the Americans Malcolm Murray, Mike Mc-

Nulty, and Ed Soja each visited Ibadan in the 1960s, alongside ongoing British visitors such as Terry Coppock and Michael Chisholm, all travelling to Ibadan on a visiting lecturer scheme (Areola and Okafor 1998). Many faculty members and students at Ibadan remember the 1960s and 1970s as an exciting period. 'We had staff from all different parts of the world ... We had a vibrant academic community', remembered Albert Aweto.[9] Olusegun Areola agreed:

> That is one thing about Ibadan, up until the 1980s, there was this rich culture of having people from different parts of the world and I think that helped students from Ibadan ... The experience, the diversity of experiences. And it made us more cosmopolitan.[10]

Ghana also saw geographers based in America visiting for research: as noted by Ginelli (2018, n.p.) 'Brian Berry and Michael McNulty did research on Ghana, while Peter R. Gould ..., Edward Taaffe, and Richard L. Morrill constructed their much cited and debated ideal stage model of transportation system development partly based on their Ghanaian experiences.' (See for example Gould 1960; Berry 1962; Taafe et al., 1963; McNulty 1969). Tanzania was the site for Peter Gould's quantitative work on transport networks and Ed Soja's early modernisation surfaces research for his PhD focused on Kenya (Gould 1967; Soja 1968). Africa therefore became a key empirical site in the development of the new quantitative and theoretical geography.

The funding for visiting positions came largely from external sources in the United Kingdom and the United States. Rockefeller Foundation funding was particularly important for many Geography Departments in Africa. The Foundation funded US and UK based academics and students to research and study at African Universities: Peter Gould's research in Ghana in 1958–1959 was Rockefeller funded (Ginelli 2018), as was Randall Baker's study at the University of Makerere in 1965 (Baker and Polonskyi 2015, p. 15). According to Barbour, the Head of Department at the University of Ibadan, the Rockefeller money provided 'the air required to appoint a Visiting Lecturer for several years,' whilst several departments benefitted from funding for research projects.[11]

The American Foundations' priorities closely mirrored those of the US State Department. Ibadan was a particular beneficiary: Nigeria was close to the West and offered the opportunity to demonstrate to other African nations the benefits of aligning with the US in the Cold War (Livsey 2017). The growing interest in Ibadan was very clear to those who worked there, who noticed growing American influence at the university, as British curricula, faculty, and course structures began to change.[12] Barbour remembered 'parties with the American and French Ambassadors meeting [the] V[ice]/C[hancellor], Deans etc., and even Robert McNamara taking great pains to charm leading Nigerian academics and make them think better of the USA'.[13] Rockefeller and Ford funding allowed for creation of new departments, new buildings to house them, library acquisitions, and the rapid expansion of staff (Livsey 2017). The historian C.C. Wrigley, then at Ibadan, recalled that the Social Science Faculty was virtually 'taken over' by Rockefeller (Wrigley, cited in Livsey 2017, p. 139). And whilst Ibadan was a particular beneficiary, other university colleges were also successful in leveraging substantial funding. An exchange between the Principal of the University College of Rhodesia and Nyasaland, Walter Adams, and R. H. Stoughton, Principal of the University College of Ghana in early 1958 gives an insight into the availability of American foundation funding. Adams wrote to Stoughton about an upcoming visit by Ford Foundation staff. He signed off his letter, with humour, asking his colleague in Ghana 'How many more millions have your [*sic.*] been given in the past month?'[14] Stoughton replied a few days later: 'I am, as a matter of fact, trying for some more money on a fairly big scale at the moment.'[15] Whilst it is not clear what became of Stoughton's plan, shortly afterwards, the Ford Foundation funded a guesthouse to the tune of $112,000 (Agbodeka 1998, p. 154). Whilst Len Berry was a lecturer in the Geography Department at the University of Khartoum in Sudan in the early 1960s, he was visited by the Ford and Rockefeller Foundations and managed to get funding for an interdisciplinary Arid Zone Research Institute. The funding package included vehicles and he recalled 'I went there twenty years later and the Land Rover was still struggling on.' He noted that 'the link with the Rockefeller Foundation was important' throughout his career, and in the later 1960s, when he was the founding Head of Department at the University College of Dar es Salaam he again leveraged this connection to fund an interdisciplinary Bureau for Resource Allocation and Land-use Planning (see Chapter 5).[16]

## *American Orbits*

Alongside support for visiting staff, researchers, and the development of facilities, American funding supported the academic mobility of a generation of African scholars, taking them to the US for PhD training, conferences, and visiting positions. The archives of African geography departments covering the late 1950s and 1960s are full of requests for additional staffing to cover colleagues who were in the US, and advertisements for replacement positions.[17] Often cover was found courtesy of foreign staff paid for by their respective governments' grant aid schemes. Two of the first Ghanaian geographers appointed to the University of the Gold Coast/Ghana, Ernest Boateng and George Benneh followed a typical trajectory, studying first in Ghana at the University College (or its forerunner Achimota College) then for their PhDs in the UK courtesy of British and Commonwealth awards, and afterwards travelling to the US courtesy of Fulbright, US State Department, and individual US university support (Benneh 2016).[18] Whilst these trips could be academically productive, they were not always personally enjoyable. In his autobiography, Benneh recalls a 1964 visit to the US funded by a State Department Travelling Grant as 'the first time I was discriminated against on the basis of my race and colour', continuing 'it was very painful, and made me realise the kind of thing that the African Americans had to put up with all these years.' When he narrated his experiences of being ignored and refused service in a bar to his State Department attendant, the attendant went 'very quiet' (Benneh 2016). When Adolfo Mascarenhas was studying at the University of California Los Angeles (UCLA) he was often mistaken for a Mexican gardener.[19]

Despite these experiences, after the mid-1960s, it was increasingly common for African geographers to do their PhDs in the US, including those from South Africa, who benefitted from a desire to support a Cold War ally, despite concerns about Apartheid. Many Tanzanians went to America for postgraduate study by the early 1970s, following the early example of Adolfo Mascarenhas.[20] The experience below of another young Tanzanian geographer in the late 1960s is fairly typical, and illustrates the direct impact of the American Foundations on African scholars' careers. Having completed an undergraduate degree at the University of Dar es Salaam, and working as a tutorial assistant in the Department, this geographer received a visitor in his office on campus:

> A Professor from the Rockefeller Foundation came from America and he was going [around] Africa looking for "bright" [uses fingers to add scare quotes] people to sponsor. … he introduced himself and told me I was going to be interviewed. The man who was going to be head of the Faculty at the University of Minnesota was here also. So, he interviewed me. Afterwards, for 15 mins I found I was the one interviewing him, about the level of funding, about sponsorship. "Who is interviewing whom here?": I told him I wanted to know everything. So at the end of that he stood up and said I could have the scholarship … So, the same year 1969 I went to the US, and started studying there. I discovered that in Geography there had been a revolution, the "quantitative revolution", and a lot of other things.[21]

This quote highlights the impact that funding associated with the Cold War might have on geographers' careers far beyond the borders of the superpowers. Just as in the US (Barnes and Farish 2006), Cold War funding helped to found the Quantitative Revolution in Africa. However, this was not a passive process: geographers negotiated the terms of their travel, and, in the case of the Nigerian Akin Mabogunje (pictured in Figure 4.1), requested access to specific theoretical specialisms:

> with independence, Ford Foundation, Rockefeller Foundation, then started sending teams of their respected academics to come out to countries in Africa to find out in what way they can help. And so, in 1961, a team came from Rockefeller, it comprised… Professor [Gilbert] White of Chicago… And so they came to Ibadan…. they were asking what else can we do to help. And of course I was saying I wanted to go to a university in the U.S. where I can supplement my quantitative and theoretical capabilities. And of course Gilbert White comes from the University of Chicago, but he thought the best teacher at that point in time was at Northwestern, that was Bill Garrison… I had nine months.[22]

**Figure 4.1** Professor Akin Mabogunje at the University of Ibadan geography seminar, mid-1960s. Ford Foundation Photographs S3.10 B95 F1663. Courtesy of the Rockefeller Archive Center.

Though US intervention in African universities could feel imposed, it could also be negotiated and offer opportunities (Livsey 2017). Certainly, for Mabogunje, travelling to the United States was transformative: 'I would say what Northwestern did was throw me into orbit, you know. That anytime you then had something quantitative, and you were looking for a spread round the world my name always came up'.[23]

## *The Quantitative Revolution in Africa*

When in 1968 Mabogunje published his book *Urbanization in Nigeria*, which drew substantially on quantitative factor analysis of census data, the influence of his visit to the US and training was clearly articulated in the preface:

> This book derives in part from my doctoral dissertation. However, much of my thinking that has influenced its present form began only during my nine-month stay as Visiting Scholar in the United States of America in 1962. Before then, my insight into the urban situation in Nigeria was handicapped by my lack of acquaintance with sophisticated analytical methods. This deficiency was corrected during those pleasant months I spent with the staff at the Department of Geography and Economics at Northwestern University, Evanston, Illinois. (Mabogunje 1968, p. 6)

William Garrison and Duane Marble get particular mentions in the acknowledgements, as does 'the Computing Centre of Northwestern University ... for the use of their IBM 7090 Computer' (Mabogunje 1968, p. 7). Northwestern provided, then, both intellectual and technological resources crucial to developing a quantitative analysis of urban development in Nigeria (this was Mabogunje's first contact with a computer). Reviews from the time focus on the value of the new quantitative methods applied. Paul Wheatley, writing in *Economic Geography* noted that:

> No analysis of this degree of sophistication has, to my knowledge, been carried out anywhere else in the non-Western world .... All this adds up to a scholarly contribution of first rate importance to the study of urbanism and urbanization not only in Nigeria but in the non-western world at large: with its publication urban geography in Africa has come of age. (Wheatley 1970, p. 103)

Correspondence between various key protagonists in the Quantitative Revolution in the United States and Britain and the Department at Ibadan, discussing external examining, references, or research methods highlight the extent to which Ibadan was a part of these networks, not just observing them at a distance.[24] Mabogunje described this recently as being part of 'this fraternity of quantitative geographers' (and it was overwhelmingly male), and was in demand as part of this international community.[25] Mabogunje remembers 'a wonderful six weeks in Berkeley' in 1964 where funding from the US National Science Foundation brought together scholars from Nigeria, India, Sweden and the United States.[26] This included friendships as well as professional relationships. Brian Berry, who first met Mabogunje when they were both undergraduates at University College London (UCL), remembers that in the late 1960s Mabogunje

> came to Northwestern University as a visitor and we saw a lot of each other ... One memorable Thanksgiving Akin and family came to dine with Jan and I in a home in Park Forest at a time when America's suburbs were still highly segregated. I recall the neighbors' consternation when the Mabogunje and Berry children spilled out of the house to play after dinner. By then we were good friends and spoke frequently about the changes that our Washington group had initiated in American geography.[27]

Based on his quantitative urban work and broader esteem, over the course of the 1960s and 1970s, Mabogunje also held visiting professorships in several places central to the Quantitative Revolution, such as Lund, Northwestern, and Cambridge (Mabogunje 2011). Recent work has suggested that histories of the Quantitative Revolution should stretch further than a few institutions in the United States and United Kingdom (Barnes and Abrahamson 2017; Ginelli 2018) and this account suggests that Nigeria, and other locations in Africa too, could be stop offs on the flight map of quantitative disciplinary development (Taylor 1977). Mabogunje's academic status was significantly boosted by American funding, though it meant his primary academic audience was also in Anglo-American geography. Zoltan Ginelli has suggested we might also view Mabogunje's reliance on American computers as 'technological dependency',[28] with the combination risking reinforcing the intellectual extraversion of the colonial academic world (Hountondji 1990).

When Mabogunje returned to Ibadan from Northwestern, he, alongside Barbour, the British geographer who remained head of department until 1972, set about reshaping the curriculum to suit this new quantitative and theoretical era, aiming to share and reproduce the new insights he gained from travel. By 1970, Barbour noted in a letter to a colleague at University College London that they had fully embraced these new approaches: 'In recent years the department has veered quite heavily towards quantification and theoretical studies, and any visitor engaged in model-building, physical or intellectual, simulation, etc. could be sure of finding a congenial milieu.'[29] Syllabus changes and the everyday labour of teaching are an important, though often overlooked, part of broader disciplinary change. In this relatively young department, it was possible for dramatic changes to be made quickly, especially when led by a rising Nigerian star. 'Bola Ayeni remembers approvingly that when Mabogunje 'came [back], he made courses in statistics compulsory, courses in mathematics compulsory, courses in Research Design compulsory'.[30] Others were less keen:

> Mabogunje went to Northwestern and got so taken in by all this quantification, statistical revolution and so. We were still undergraduates then and so we were guinea pigs for his ideas. Nobody liked statistics but we had to go through it.[31]

Quantitative geography became a strong area of the discipline across African geography departments, but the extent of its take-up depended both on the opportunities for training in these new theories and methods available, through young visiting staff from the US and UK keen on these issues, through opportunities to study in the US, and then in the curriculum changes pioneered by returning African academics. It also depended on the attitudes of key faculty in what were often still small university departments: whilst Mabogunje, a rising international star, was able to transform the curriculum at Ibadan, Nigeria, in other places there was a varied uptake. Whilst in Kenya Simeon Ominde 'did not employ abstract quantitative methods' himself, he 'did not feel threatened by the quantitative revolution in geography. In fact, he encouraged its arrival and made appropriate arrangement for the appointment of staff who had statistical and mathematical skills' (Odingo 2000, p. 7). On the other hand, the first African Head of Department in Ghana, Ernest Boateng, who took charge in 1961, was not particularly interested in quantification, so any substantial changes in this direction had to wait until 1969 when the department got new leadership. Ghanaian student Stephen Kwami Awunyo-Akaba criticised this lack of innovation in the appendix of his undergraduate dissertation (1966):

> To those geographers who believe that geography is principally a descriptive science there would obviously not be much scope for participation [in development]… The department of geography here can really build itself up and contribute to the development of the resources of the country by strengthening the scientific bases of many of its courses, especially through the use of mathematical and statistical methods and by establishing a research side of the department to provide the theoretical and scientific basis for the applied aspects of the discipline. This would demand money, and geographers who would like to see the development of their subject in a tropical environment.[32]

South African geographers embraced quantification, though this took longer than in Nigeria: the university sector was larger and longer established, meaning that it was less easy for new approaches to sweep in unchallenged. In addition, the influence of universities dominated by Afrikaner academics diluted the initial influence of the growing Anglo-American dominance in geography – which favoured quantification. However, by the early 1970s, spatial science had been fully integrated in South Africa. By 1974, the Urban Geography reading list for a course at the University of Cape Town was divided into themes including 'concepts of distance', 'external' and 'internal relationships of cities', 'rank size', 'location analysis', 'models', and dominated by Anglo-American literature by the likes of Brian Berry, William Garrison and Peter Haggett alongside Walter Christaller and Anders Lösch (e.g. Lösch 1954; Berry and Garrison 1958; Bunge 1962; Haggett 1965; Christaller 1966). The 67-item list included only a small handful of references to work by South Africa based scholars or about South African cities.[33] Alan Mabin attended an intensive school given by Brian Berry at Witwatersrand in 1972 as a geography student and reflected that for career development it was at that time more important to keep up with conceptual trends in the US/Anglophone world rather than engaging with forced removals in South Africa (as he was outside of his studies).[34]

Mabogunuje (2004) has argued that the 1960s saw Nigerian geography shift from a discipline primarily influenced by British geography to one increasingly impacted by developments in the United States, including those of quantification (see also Sada 1982; Okafor 1989); and we can see similar trends in other countries in Africa. However, to view geography in African countries as solely derivative of other national traditions, and to understand the role of African geographers within this as merely passive is to overlook the extent to which individuals actively shaped their own academic careers and became important contributors to these very paradigms. Chapter 5 picks up on the story of quantitative geography in Africa to show how it became entwined in projects of national development – of different political hues – following a different trajectory to that of these news methods in Europe and the US. The next section shifts attention away from US influences towards others from the East and South.

## Other Eastern and Non-aligned Networks

### *Eastern Bloc Connections*

In Ghana from the 1950s, there were increasing numbers of international experts in the country, and at the University, not only from the US, but also from Eastern Europe and the Soviet Union, as well as from India (Stanek 2019). This included many in cognate fields to geography, for example architects, economists, and regional planners. Zoltan Ginelli (2018) has highlighted the role of Hungarian development economists in Ghana advising on the government's five-year plans in the late 1950s and early 1960s, as well as the presence of Soviet geologists prospecting for minerals in the country in the same period.

Staff at the University of Dar es Salaam in Tanzania remember Poles, East Germans, and other Eastern Europeans in other departments and in the Bureau of Resource Allocation and Land-use Planning (BRALUP) run by the Geography Department.[35] In the late 1960s, 'BRALUP had one East German woman who was gifted by the East German Government … she was a regional planner [with] … a degree in demography and geography. She was an integral part of the group', recalled Len Berry, who also remembered: 'We had a Russian geophysicist group visit the university and I got very friendly with that group.'[36] Whilst Makerere and Nairobi continued to use established Commonwealth networks for recruitment into the 1960s, 'Dar es Salaam set up its own recruitment scheme to attract expatriates from socialist countries', reflecting the different political orientations of Uganda, Kenya, and Tanzania after independence (Southall and Kaufert 1974, p. 626). However in comparison with the ongoing connections with the UK, and increasing links to the US and Western Europe, there remained fewer geographers from the Eastern bloc in Ghana, Tanzania or other African universities in this period.

Nevertheless, the Eastern bloc was increasingly influential in higher education in Africa from the early 1960s onwards. This was especially true in relation to student scholarships: in 1960, only three Tanganyikan (later Tanzanian) students studied abroad in Eastern Europe – in Hungary and Czechoslovakia – but by 1973–1974, 251 studied in eight different countries in Eastern Europe, whilst many more went to the USSR. As a comparison, 38 studied in Western Europe in 1960, and 79 in 1973–1974 (compiled from Burton 2020, p. 170–171). One British academic remembered that a few Tanzanian colleagues studied for their PhDs in Halle University in what was then East Germany, and Lumumba University in Moscow in the 1970s, whilst others in Tanzania and other parts of East Africa studied in Yugoslavia.[37]

These overseas experiences were sometimes blighted by racism: 'you talked to people who'd been in Moscow and they found it quite a difficult experience … not just because of the weather but there was a very overt racism in

Moscow at that time. So they found it a less than enervating experience'.[38] The Kenyan hydrologist George Stephensons Obegi Ongweny, who worked at the University of Nairobi in the 1970s, had been an undergraduate at Moscow State University in the 1960s but left after three years (of a five-year course), remarking that 'life has become intolerable in the Soviet Union for Kenyan students because of the atmosphere of intimidation and the constant propaganda to which they were subjected' (No Author 1966, p. 587). Ogweny, active in the Kenya Students' Union whilst in the USSR had found Russian students in his room photographing union documents and when he protested about KGB spies on campus he was imprisoned and beaten, as well as subjected to racist abuse (No Author, 1966).

## *Moving the Centre*

Whilst Tanzanian students travelled abroad for study across the globe in the 1960s, increasingly there was also substantial traffic in the opposite direction. Just as Ibadan had been a draw in the 1950s and early 1960s, as one of the most vibrant geography departments (and wider universities) on the continent, by the later 1960s, University College Dar es Salaam was building a substantial reputation amongst radical scholars from across the globe. Attracted by President Julius Nyerere's vision for a non-aligned, socialist, independent Tanzania, the university was at the heart of 'intense debates around the meaning of African knowledge, the role of the postcolonial university, and the most appropriate future for Tanzania, Africa, and the Third World.' (Sharp 2019, p. 75). Jo Sharp (2019, p. 83) has argued that 'Tanzania in the 1970s presented a material provincializing of Europe as intellectuals from the north moved to participate in and learn from Nyerere's Tanzania'. In this way, it can be seen as a manifestation of decolonisation by 'Moving the Centre' to Africa (Ngũgĩ 1993). The experiences of British geographers who worked at Dar in the 1960s and 1970s illustrate this pull: 'I was very much focused on the way in which this African country was trying to deal with the consequences of independence and the way it was trying in an apparently innovative way to reshape the way society was actually organised', remembered one British geographer.[39] Another British academic, offered a job in Swaziland and another at Dar at the same time, reflected:

> in many ways it was a no-brainer because at that time Tanzania was a rather more interesting country, it was a front-line state, it was a country that had been trying to build, ah, a socialist response to African development but not by falling into the Eastern bloc or indeed the Western bloc for that matter … What Nyerere talked about as African Socialism.[40]

The British joined a truly international community, as the British Head of the Geography Department recalled:

> because of Nyerere, because of Tanzania, because it was new, the University attracted an amazing range of talent, some of it radical, but not all of it … I ended up being Dean and there were 120 Faculty in Arts and Social Science, 30-40 nationalities. And including as Chairs of departments, Swedish, Hungarian, English, … Finland, and East and West Germany, and most importantly a few people from the Caribbean, including Walter Rodney. He and I played cricket together. He was a really good intellectual … John Saul and Walter Rodney were the most radical of the group, along with a guy called Arrighi, an Italian, who had been in Zimbabwe …. Really a very interesting dynamic time. Intellectually I think the most diverse and enlightening group I've ever been with and you wouldn't get the circumstances again.[41]

The Guyanese scholar and activist Walter Rodney taught at the University of Dar es Salaam 1966–1967 before being appointed Lecturer of African History at the University of the West Indies (UWE) in Jamaica, until he was barred from re-entering Jamaica by the government a few months later. This banning, on spurious grounds, lead to mass protests and uprisings both on campus and beyond it. The government was alarmed by Rodney's speeches in the poor black communities of Kingston highlighting the continuity between colonial and post-colonial Jamaica, connecting race and class based critiques and articulating a renewed black power for the Caribbean (Ledgister 2008). Unable to return to his UWE position, Rodney was welcomed back to Dar – a potent symbol of the difference between the Jamaican government – then allied closely with the US – and that of non-aligned Tanzania under Nyerere. Rodney was joined by Saul from Canada, and the Italian Giovanni Arrighi, and 'lots of very, very engaged and active Tanzanians' such as Issa Shivji as well as Mahmood Mamdani from Uganda in becoming prominent figures in the discussions at Dar. The radical Brazilian geographer Milton Santos was also at Dar from 1974–1976, researching socialist development projects and refining his ideas about underdevelopment. This group of scholars became leading intellectuals in the debates about decolonisation, development, class and race in Africa, and connected Tanzanian and African struggles with those in the Caribbean, South America, and beyond (see also Al-Bulushi 2022a).[42]

British geographers joining the department in the 1970s were drawn by this intellectual ferment, which was matched by the broader geopolitics of southern Africa in these years: Tanzania was a front-line state in the struggle against white minority rule in what was then Rhodesia, as well as Mozambique, Namibia, Angola, and South Africa. One British academic remembered:

> there were some exciting times. That was '78, that was the height of the liberation struggle for Zimbabwe, Tanzania was very much, always had been a front-line state, … that was really the denouement of the Rhodesia Zimbabwe [liberation struggle] and here it was a very exciting time.[43]

Being at the university provided the opportunity to develop political ideas and scholarship against this backdrop. A Tanzanian geographer reflected that as an undergraduate student between 1965 and 1969, the experience was

> Interactive, very challenging. We had professors here who came from all over the world to support this idea, we had professors from neighbouring Kenya and Uganda, we had Jaramogi Oginga Odinga who was the vice president who used to come here in his pantaloons! He never wore long trousers. There was Ali Mazrui [visiting from Makerere] ... It was very exciting and we participated in a lot of debate so I think intellectually it was very awakening. Some of the professors became government advisors ...[44]

Another Tanzanian remembered:

> It was a very, very special time for us in Africa, I would say, not only in Tanzania. The kind of lecturers also who came, ... they were from the West, but the West which believed in the liberation of South Africa... In the seventies, Tanzania was at the centre of the decolonization movement. (quoted in Sharp 2019, p. 84)

These experiences had a significant impact on the research and broader political outlook of western scholars, many of whom acknowledged this opportunity and their position as learners rather than teachers. One British geographer, at Dar in the early 1970s reflected: 'I didn't think I had very much to say. I was there to learn. I wasn't there to give views. I used to go along and lap the stuff up, it was an amazing education experience.'[45] A focus on Dar es Salaam, and the contributions of some of the scholars that taught there, demonstrates that geography in the global south was not a passive receptor of northern theory but rather a key site for learning for Northern colleagues who were 'taught' by scholars from Africa and South America at Dar 'nourishing their ideas on development and underdevelopment from original perspectives' (Ferretti and Viotta Pedrosa, 2018, p. 2; Ferretti 2020).

Geographers (alongside colleagues in other disciplines) working at Dar developed several key insights important for the future development of their own work, and the wider discipline of geography. These insights, around interdisciplinarity, ethnocentrism and the lack of voices from the South, and the coloniality of development, were not only important and often overlooked in geography in the 1970s, but are absolutely central to debates about decolonising the discipline today.

One British scholar explained:

> The first thing I learned was loosening the disciplinary cage ... Because of the interdisciplinary staff and exposure to all these cosmopolitan views in terms of the geography of where people actually came from ... but also the cosmopolitan nature of the perspectives ... both in political theory but also in terms of their own disciplinary views.[46]

Interdisciplinarity was central to debates at Dar about the place of university learning for the development of Tanzania. Following Nyerere's Arusha Declaration 'which promoted equality, self-reliance, "traditional" African communal values, and the virtues of education and hard work' alongside a model for development that would follow a 'non-aligned' path independent of the US or USSR (Sharp 2019, p. 78), proposals were developed for a new compulsory interdisciplinary course that aimed to fit students for the task of understanding the needs of Tanzanian development. According to Sharp (2019, p. 87–88):

> This new curriculum represented a conscious decentering of colonial knowledge in two ways. First, instead of starting with Europe and its history and experience, Africa and Tanzania were to be put on a central stage. This was not in any way parochial, however, as this understanding of Africa and Tanzania was firmly related to an examination of the colonial experience and the neocolonial capitalist world order with which the new states were struggling. Second, there was a resolutely interdisciplinary way in which issues were to be addressed.

The view from Dar also highlighted more clearly the Anglo-American focus of geographical scholarship and publications:

> I guess it's sort of solidified my feelings about the ethnocentrism in geography, working in a Third World country, and I sort of wrote some polemic pieces about the poverty of modern geographical enquiry which came out of being in that peripheral location which in many ways was an extremely positive experience.[47]

## *Radical Geography and Underdevelopment*

The British geographer David Slater's time Dar es Salaam in the early 1970s was crucial in the development of his concern 'with where knowledge was made and who had the power to articulate it. His reference would be to "the persistence of absence", where he would wonder how academics could talk of globalisation without mentioning the global South' (O'Keefe 2020, p.157). Following from this, O'Keefe (2020, p. 156) argues, was the question of academic practice: 'Radical thought, for Slater ... carried a moral imperative suggesting how geographers should practice their craft which, all too often for Slater, had been complicit with imperialism.' These concerns underpin many of the critical questions that postcolonial theory has asked of academic work in geography more widely, and Slater was to go on to write about postcolonial theory and development in the early 1990s (e.g. Slater 1993).

Following on from the need to learn from the South, and from Southern scholars, and the ethnocentrism of geography, came imperatives to publish and engage with scholars from the South in northern 'international' journals. Whilst in Tanzania, Milton Santos completed much of the work putting together the first of two special issues of *Antipode* which appeared in 1977, together with Richard Peet and David Slater. The special issues include contributions from Slater (2) and Santos (3), along with Bashir Datoo, Michael Keith McCall (2), and Joe Doherty, all who had been a part of the geography department at Dar, plus an article from Ben Wisner who was also at the University of Dar es Salaam in Community Health (Figure 4.2). This line-up demonstrates the vibrancy of radical geography scholarship there, as well as the extent to which Santos valued and utilised these Tanzanian links. Santos's explicit aim was to publish scholars from the global south in his special issues (Ferretti and Viotto Pedrosa 2018). These issues also contained contributions from academics based in France (Anouar Abdel-Malek), Venezuela (Sonia Barros), Argentina (Alejandro Forman, Ruiz Alberto Romero) and Mexico (Jose Coraggio) 'showing the effort to involve different continents and linguistic areas in the exchange' (Ferretti and Viotto Pedrosa 2018, p. 8). Santos made this point explicitly in his introduction to the first special issue (co-authored with Peet) where he wrote:

> what is new in this project is that specialists of the Third World were invited to discuss the subjects they consider the most relevant. This is rather unusual in the Anglo-Saxon world where debates about poor countries have been for long a "de facto" monopoly of western scholars. (Santos and Peet 1977, p. 1)

Elsewhere, Santos elaborated on this point: 'why should we not instead rally expertise from the underdeveloped countries themselves: to develop theories which would make sense to them both as geographers and citizens? At the moment, "official" geography operates as though the West had the monopoly of ideas' (1974, p. 4). In the same article he noted though, that 'there are, of course, many Third-World geographers who prefer to remain silent either through timidity or careerism.' He did however highlight in a footnote the bind in which many African academics found themselves: 'the university system in many underdeveloped countries (especially in Africa) is still a relic of colonialism: it is still organised in such a fashion that decisions concerning the future prospects of young professors rest in the hands of western, or Westernized "patrons"' (Santos 1974, Footnote 7). In this he echoed the criticisms of many contemporary politicians and scholars that despite independence, African universities remained colonial spaces in need of decolonisation. Many African geographers in this situation faced a dilemma: build a successful and fulfilling career working for decolonisation in some areas only, or push more radical research and institutional agendas but risk damaging your career prospects and livelihood. In many ways, international scholars working in African institutions had more freedom to innovate and critique than their African colleagues.

Alongside this critique of the dominance of Northern voices and theories in the discipline, geographers who worked at Dar es Salaam (and other universities in Africa in the 1970s) were also central to the critique of modernisation and growth of ideas about underdevelopment and dependency in geography. Ferretti

ANTIPODE

VOLUME 9, NUMBER 1 (FEBRUARY, 1977)

UNDERDEVELOPMENT IN THE THIRD WORLD

I: Socio-Economic Formation and Spatial Organization

CONTENTS



ii

**Figure 4.2** Contents page of the first of two special issues of *Antipode* co-edited by Milton Santos (Santos and Peet 1977). Courtesy of the Antipode Foundation.

and Viotta Pedrosa (2018, p. 7) argue that the *Antipode* special issues edited by Santos, and key contributions from Slater, 'inaugurated a systematic interest on the part of Anglophone radical geographers in matters of underdevelopment' (see also Davies 2019; Al-Bulushi 2022b on the role of Dar in the development of World-Systems Analysis). Richard Peet (no date), the *Antipode* editor in 1977, remembered that 'Many people have subsequently told me that these two issues deeply affected their thinking on the geography of development'. The Tanzanian Bashir Datoo's contribution to the 1977 special issue demonstrated his shift in research from historical geography towards a radical focus on underdevelopment, dependence and regional planning, reflecting the ideological currents at the university and in Tanzania more widely at this time.[48] Also published in the special issue was Michael Watts, whose article with Alasdair Drysdale questioned the 'shortcoming and ethnocentrism of the "geography of modernization"' (Drysdale and Watts 1977, p. 40). Watts, writing from the perspective of two years researching and teaching at Ahmadu Bello University in Northern Nigeria, was intensely critical of geographers' work on modernisation as ahistorical, obsessed with refining 'technique' over understanding, and both uninterested in and 'insensitive to upheavals and traumas associated with the modernization process' (Drysdale and Watts 1977, p. 40). Drysdale and Watts picked out Peter Gould's paper on transport networks in Tanzania published in 1967, entitled 'On the geographical interpretation of eigen values' as an exemplar of these dehumanising trends: the title providing an indication of the importance of the experiences of Tanzanians and the specificity of Tanzania to the author (something noted by several commentators at the time, see Brookfield 1975).

In an *Antipode* article published in 1973 whilst he was at Dar es Salaam, Slater argued that some of the key concepts in new geographical studies of development: spatial differentiation, spatial diffusion, and spatial integration, and the closely connected ideology of modernisation, were ahistorical, ethnocentric, and neglectful of the role of colonialism and capitalism in producing and perpetuating inequalities. Slater (1973, p. 26) highlighted how the data utilised in new quantitative models of modernisation contained colonial biases and criticised an inherent assumption that modernisation 'leads to improvement and progress' for poorer areas. He concluded by noting that regional planning strategies take for granted a capitalist economic structure, rather than evaluating the possibilities of alternatives. In the paper, he drew heavily on his colleagues at Dar: including Giovanni Arrighi, and the historians John Iliffe and Walter Rodney. Part II of this paper, published in the second special issue of *Antipode* edited by Santos in 1977 drew heavily on the work of Mahmood Mamdani and Issa Shivji, two further intellectual giants of the University of Dar es Salaam, in its outline of the role of imperialism in producing underdeveloped capitalism in East Africa. The paper was critical of the abject failure of development geography to take into account the histories of capitalist penetration through colonialism in their studies of development, and to consider revolutionary socialist alternatives to the capitalist status quo. Slater named the new Development Geography Research Group of the Institute of British Geographers as a particularly good example of the 'irrelevancies and inadequacies of development geography' (Slater 1977a, p. 22, and footnote 14), but also highlighted the 'African offshoots' of this 'Anglo-Saxon literature' which follows a similar path and namechecked the work of Akin Mabogunje (1973) in this context.

Milton Santos's intellectual contributions are only now being recognised in Anglophone geography (Melgaço 2017; Davies 2019; Neate 2022), but his work has provided key theoretical contributions in the fields of 'Third World urbanization, globalization, and Brazilian society and political economy' (Davies 2019, p. 584). Writing from Dar es Salaam, his research focused on radical understandings of underdevelopment, arguing that much development geography, and wider social science, failed to focus on 'world space as a whole', instead focusing on the flows of 'giant firms' and 'persons of leisure' (Santos 1974, p. 3–4). A more holistic approach, he argued, 'would produce a true geography of poverty, i.e., a world geography where wealth and poverty are not treated as separate entities, but as complementary parts of a single reality' (p.4). Moreover, 'underdeveloped space' was different to that in the North, he argued, returning to his point about ethnocentrism in geography: 'this is something that Western Geographers have had great difficulty in understanding' (p.4).

Whilst O'Keefe (2020, p. 156) argues that 'Dar es Salaam was a missed opportunity to send the history of radical geography southwards', we can see a shift in radical scholarship from the early 1970s onwards towards questions of development in what was then known as the 'Third World', and the bifurcation of development geography as more radical critiques challenged established theories of development as modernisation (Ferretti and Viotto Pedrosa 2018). Santos, Slater, Wisner and others working in and around Dar es Salaam were deeply influenced by their experiences and discussions there, and in turn were central in shaping radical development geography, and radical geography more broadly (Sidaway et al. 2020). Wisner and Joe Docherty (at Dar from 1971–1975) both acted as *Antipode* editors for a time. When Richard Peet published his edited collection *Radical Geography* in 1977, it reprinted an article, 'The Poverty of Modern Geographical Enquiry', by David Slater (1975), initially prepared for a staff-student seminar on 'Trends in Modern Geography' at Dar es Salaam in August 1974. It also included contributions from several others who were influenced by their experiences working in colonial and post-colonial universities in Africa, Asia, and the Caribbean, including James Blaut who worked at the University of Malaya (now the National University of Singapore) in the early 1950s, and Keith Buchanan, who worked at the University of Natal (Durban) and the University of Ibadan in the late 1940s and early 1950s (See Power and Sidaway 2004; Bunnell et al. 2013; Kobayashi 2019).

Ongoing discussions between Milton Santos, Ben Wisner, and Joe Doherty after each had left Tanzania – evidenced by correspondence in Santos's archives –

highlight the longstanding connections between radical scholars forged through their time in Dar es Salaam (Ferretti 2020). Santos and Slater left Dar es Salaam increasingly critical not only of modernisation but also of the model of socialist development through modernisation practiced in Tanzania (Ferretti and Viotto Pedrosa 2018, p. 7) and their work (independently) in the 1970s contributed to the wing of development geography that critiqued modernisation theory and turned to the radical geography of dependency. In this way, they elaborated a critique of neo-colonialism and articulated the need for economic as well as political decolonisation. However, their work went beyond the economic, beginning to touch upon some of the demands more closely linked to questions of decolonisation of the mind (Ngũgĩ 1986).

Beyond critiques of modernisation, Tanzanian geographers and colleagues at Dar also lead the way in widening radical and development geography to consider gender, the environment, and decolonisation, often drawing on historical perspectives. For example, Adolfo Mascarenhas' PhD at the University of California (UCLA, 1971) focused on Sisal plantations in Tanzania (Mascarenhas 1971). It highlighted in particular how the introduction of the plantations to the country and ongoing organisational structures reflected colonial power, benefitting the foreign owners and producing poor pay and labour conditions, with these uneven relationships facilitating the suppression of locally owned sisal growing and labour unions, even close to independence. Featuring a chapter entitled 'Decolonizing Sisal' and discussing the transformation of the Tanzanian landscape and environment through the plantation, this thesis brought together environmental, political, and economic concerns through the lens of colonialism and its aftermath. Mascarenhas' later work remained attentive to the politics of human-environment relations and focused on the every-day experiences and expertise of rural people caught up in large-scale and rapid changes to their living and livelihoods (Veit et al. 1995; Mascarenhas 2000, 2004). An example published in *Economic Geography* from 1977 – addressing experiences of desertification – provides evidence of this innovative focus:

> To Chimwenda, a Wagogo peasant in the district, there are two seasons, *cifuku* (the rainy season) and *cilimila* (the dry season). When the *cilimila* (the constellation Pleiades) appears in the sky in an early November evening, Chimwenda, like his grandfather before him, knows that it is time to hoe… Chimwenda can remember the onslaught of the *Mau Mau* famine (1953–1954), when only scattered clouds covered the sky, sterile thunder rumbled across the earth, and people patiently waited for rain … (Mascarenhas 1977, p. 376)

The story-telling style is extraordinary, particularly when contrasted with other articles published in the journal that year, which typically focused on quantitative and modelling work on topics such as industrial location.

Ophelia Mascarenhas also made important contributions to questions of development, primarily through the lens of gender. She travelled to California alongside her husband Adolfo and gained a Masters in Library Science, later getting her PhD in geography from Clark University in the US. Working with Marjorie Mbilinyi, she published:

> a definitive account of writing on the question of women and development. This volume was expanded into an edited collection that mirrored Rodney's *How Europe Underdeveloped Africa* in its sweeping historical account focused on the struggles of Tanzanian women during the pre-colonial, colonial and post-colonial periods. (Al-Bulushi 2022a, p. 12)

According to Mbilinyi, this collection (Mascarenhas and Mbilinyi 1983)

> deliberately challenged the usual western feminist view of African women as being powerless victims, or the bourgeois nationalist view that the concept of gender equity was a foreign importation, or the Marxist view that it was possible to separate gender and class struggles in the world of marginalised women.[49]

Whilst never a member of the Geography Department, in her role as Director of the Library at Dar es Salaam, and the first Tanzanian Professor of Library Studies, Ophelia Mascarenhas played a key role. Her obituary records that

> Throughout her life, as a teacher in Zanzibar and later librarian/researcher at UDSM, Ophelia fought all types of discrimination and infringement of the rights of workers, women, rights of people to information on resources, including land.[50]

Dar es Salaam provided then, an important influence, and meeting point, for radical thinkers interested in colonialism, rights to land and resources, (under) development, and in challenging perspectives produced by Western scholars. The next chapter returns to the work of Adolfo Mascarenhas on Tanzanian development, further elucidating the complex position of many African geographers working between applied and critical research.

## Conclusion

### *Mobility – for Some*

The Cold War impacted upon everything from academic mobility to research trajectories in African geographers' careers from the late 1950s to the 1970s and beyond. Funded academic travel – for sabbaticals, study, networking and conferencing – was common for African geographers in the 1960s and 1970s who were able to take advantage of scholarship and mobility schemes primarily from North America, the Soviet Union, and Eastern and Western Europe to develop their careers. Travel for PhD study, further training, or intellectual exchange boosted the careers of individual geographers as they became more prominent domestically and internationally. They were not mere pawns in a Cold War game, however. African geographers utilised and created opportunities for themselves and used their agency to shape their careers in ways that suited them. Whilst these opportunities were often provided as part of a Cold War strategy on the part of the superpowers and their allies, as Burton argues,

> from the perspective of travellers, it often mattered much more to go "North" rather than going "East" or "West": system competition was a minor concern to many nationalist activists, students or politicians from Africa who tried to use, rather than be subject to, the dichotomies surrounding the mental maps of the Cold War. (2018, p. 8–9)

African geographers often took up opportunities from wherever they came: the Kenyan geographer Simeon Ominde was awarded a 'United States Government award to tour several universities in the States' making 'contacts with departments of geography and institutions having population programmes' in 1973 and soon after was on sabbatical 'touring planning institutes in Yugoslavia'.[51]

Whilst many benefitted from travel opportunities, they were not equally open to all: trade unionists from Zanzibar were prevented from leaving for study in the early 1960s, whilst in Uganda, Baganda students dominated the scholarships to study at Makerere and abroad (Burton 2020). Regionally and ethnically differentiated schooling advantaged some groups, for example in Nigeria where the system in the South was much closer to western educational styles (Livsey 2017), or in East Africa where more was spent by the colonial authorities on segregated Asian, rather than African, schools.[52]

Gender and caring responsibilities also influenced the ability to travel for study, further reinforcing the male dominance of academic staff. This was not always true of course: in 1961 Akin Mabogunje's wife Titilola left Nigeria for legal training at Grays Inn, London, leaving the children in Nigeria, where their father would oversee baths and breakfasts before work, returning twice throughout the day to check up on them before returning to the office (Mabogunje 2011). Mabogunje could, though, also draw upon the paid and unpaid female labour of a housemaid and his mother-in-law. Such examples are relatively unusual though, and gendered expectations often impacted the possibilities for travel for career development.

## *Decolonisation or Incorporation?*

The American representatives of the foundations travelling around Africa recruiting the 'brightest' students and staff for funded training in the US wielded enormous power, with the competition and decision-making process unclear. Given the context, it is likely that the successful candidates reflected Western ideas of the modern academic and were politically friendly to the United States, or at least seen as convertible to an American political ideology. The most 'successful' geographers, such as the Nigerian Akin Mabogunje, often cited, internationally mobile and highly decorated, were successful primarily because they were able to fit into western networks and practices, rather than because they challenged them or were able to fundamentally alter the shape of the discipline. In this way their careers perhaps succeeded in playing the 'European game' but failed to challenge the ongoing coloniality of the academic world, despite their individual success (Ndlovu-Gatsheni 2018; Fanon 2019 [1961]). As the Nigerian scholar Claude Ake, also at Dar es Salaam in the 1970s argued (1982, p. 136),

> many, indeed most Third World social scientists have become propagandists of capitalism ... They have accepted too uncritically the social science which they learned from their Western universities, books and journals and they are busy reproducing and spreading this social science, often with unshakeable conviction.

As much as through grants and scholarships, knowledge and theories could reproduce political and economic ideologies in the context of the global Cold War:

> The battle is on for the control of the Third World and one critical area in which this battle will be fought out is the mind. Western social science scholarship on Africa and other Third World regions is an important weapon in this battle ... this scholarship is an important tool for controlling Third World perceptions of their world and their problems and eventually Third World behaviour. (Ake 1982, p. 139)

For Ake, the rise of quantification exacerbated these issues, reinforcing 'the conservatism of Western social science; some fundamental issues are glossed over, presumably because they are not amenable to quantification, other questions are trivialized in the course of quantification' (1982, p. 133). The rise of quantitative techniques in African geography can be traced to the international training that African academics received (and sometimes demanded), often in the US, and to their subsequent work reforming curricula that had been previously shaped on the model of British regional description. The technical nature of these methods, relying on learning complex statistics and modelling techniques which few of the British founding members of African geography departments were familiar with, as well as expensive hardware such as new computing facilities, meant that initially American funding – tied to Cold War – provided the opportunities to make these changes.

The Cold War also exacerbated some of the issues of 'intellectual extraversion' produced by the colonial university system. It promoted 'vertical exchange and dialogue with scientists from the North [rather] than in any horizontal exchange with their fellow scholars from the South', and provided funding which reproduced northern theories, methods, and knowledge, as well as replicating an imagined Western audience for research (Hountondji 1990, p. 13). As Hountondji (1990, p. 11) argued:

> This is one of the most pernicious forms of extraversion: theoretical, or socio-theoretical extraversion, the fact that we allow the content of our scientific production, the questions we pose, and the way we deal with them to be preoriented, predetermined by the expectations of our potential readers.

The Cold War context, and the effort to prevent countries being torn apart by it, dominated the thinking of academics at African universities, especially in non-aligned countries like Tanzania (Leys 2018; Mamdani 2019; Al-Bulushi 2022a), and fundamentally undermined some of the most radical political and economic projects in independent Africa. It provided the rationale for covert intervention in the politics of post-colonial nations, as well as for on-going support for apartheid South Africa as a supposed bulwark against communism in the continent.

## *Innovation*

Despite this pernicious impact of the Cold War, President Nyerere's attempts to articulate a non-Western non-aligned and independent path to development also provided fertile ground for intellectual innovation. The University of Dar es Salaam became a global hub for radical scholarship, 'moving the centre' of academic thinking in Ngũgĩ's (1993) terms in two ways: both to Tanzania from the West *and* questioning the place of the university in producing a neo-colonial elite within postcolonial African societies (Ngũgĩ 1993). Geographical research at Dar drew from the Tanzanian context, and from the diverse local and international networks of scholars working there, prompting radical critiques of modernisation, and the elaboration of theories of neocolonialism, underdevelopment, and dependence, as well as critiques alive to environmental and gendered perspectives. Geographers with connections to Tanzania were at the heart of some of the early critiques of the often quantitative studies of modernisation that dominated the emerging field of development geography, sowing the seeds of some of the post-colonial critiques of the development project. The Cold War thus had a complex and ambivalent impact on African geography, and the professional lives of geographers, though this was not something experienced equally by all. Radical scholars may have rallied intellectually against the 'fraternity of quantitative geographers' described by Mabogunje, but early radical geography too was often dominated by men (Jazeel et al. 2019; Chouinard 2020).

This chapter has shown how the Cold War provided mobility and career development for individual scholars to participate in international geography, though this international geography was primarily Anglo-American and the opportunities were on particular terms. Indirectly it produced the conditions for the intellectual innovation which developed critiques of neo-colonialism, and the need for economic decolonisation, and began to touch on some of the questions about the need for epistemological decolonisation too. In parallel, it supported the rise of quantitative geography and the linked projects of modernisation and spatial planning which were so criticized by academics at Dar. But even this was ambivalent; as we shall see in the next chapter, quantitative techniques and spatial science were also viewed as central to post-colonial development, itself crucial for securing political and economic decolonisation.

## Notes

1 Personal correspondence with Adolfo Mascarenhas, May 2021.
2 Gesthuizen, T. 'Here's some psychedelic rock from Kenya', *OkayAfrica*. Retrieved from: https://www.okayafrica.com/kenya-music-psychedelic-rock-black-savage [last accessed 15 December 2021].
3 Personal correspondence with Adolfo Mascarenhas.
4 University of Dar es Salaam Alumni of the Month, Adolfo Caridade Mascarenhas https://udsm.ac.tz/web/index.php/alumni/name/7 [last accessed 13 December 2021].
5 Personal correspondence with Adolfo Mascarenhas.
6 Personal correspondence with Adolfo Mascarenhas.
7 E.g. Interview with Olusegun Areola, Ibadan, Nigeria 9 July 2017; personal correspondence with Adolfo Mascarenhas.
8 Interview with Gordon Pirie, London, UK, 12 November 2019.
9 Interview with Albert Aweto, Ibadan, Nigeria, 19 June 2017.
10 Interview with Olusegun Areola.
11 Kenneth (Michael) Barbour memo, Oxford Development Records Project here after ODRP, Bodleian Library, Oxford MSS Afr. S. 1825 Box III (4) p. 10.
12 Interview with Akin Mabogunje, London, UK, 7 July 2016.
13 Kenneth (Michael) Barbour memo, p. 16.
14 University of Ghana Archives, UG/3/8/5/5/27 Letter from Walter Adams, UC Rhodesia to Principal of the University of Ghana R. Stoughton, 29 January 1958.
15 University of Ghana Archives, UG/3/8/5/5/28 Letter from R. Stoughton to W. Adams, 4 February 1958.
16 Interview with Len Berry, via Zoom, 26 February 2021.
17 E.g. Letter, T.E. Hilton to the Secretary of the Board of Faculties, 1 February 1960. University of Ghana Archives UG/3/2/3/56/61.
18 'Ernest Amano Boateng' *The International Who's Who 1991–2* International Press Publications p. 170.
19 Adolfo Mascarenhas, personal correspondence.
20 Interview with a British academic who worked at Dar es Salaam in the Faulty of Arts and Social Sciences in the 1970s, Dar es Salaam, Tanzania, 11 August 2011. Interview conducted by Professor Joanne Sharp.
21 Interview with a Tanzanian geographer who studied at Dar es Salaam in the 1960s, Dar es Salaam, Tanzania, 28 April 2014. Interview conducted by Professor Joanne Sharp.
22 Interview with Akin Mabogunje, London, UK, 7 July 2016.
23 Interview with Akin Mabogunje.

24 Interview with 'Bola Iyeni, Ibadan, Nigeria, 8 July 2017; Letter, Peter Haggett to Registrar, UI, 16 April 1972. University Archives and Records Project, University of Ibadan (Herefter UI). UI/FASC.12.5.42.
25 Interview with Akin Mabogunje.
26 Interview with Akin Mabogunje.
27 Personal correspondence with Brian Berry, August 2018.
28 Zoltan Ginelli, Discussion at the International Conference of Historical Geographers, Warsaw. In session: 'Globalising the histories of geography', July 2018.
29 Letter, Kenneth Barbour to Bill Mead, 1 July 1970, Geography Department Archive, University College London. With thanks to Hugh Clout.
30 Interview with 'Bola Ayeni, Ibadan, Nigeria, 8 July 2017.
31 Interview with Olusegun Areola.
32 Stephen Kwami Awunyo-Akaba, 'The Volta River Project: The agricultural programme of the Volta Resettlement Scheme 1965-66' Undergraduate dissertation. Department of Geography Archives, University of Ghana, Legon, Ghana.
33 'Geography Department 1974' Professor William J. Talbot Papers, University of Cape Town Special Collections (Hereafter UCT) BC 1079 D3.1 Salaries; teaching schedules, some course outlines, photographs.
34 Interview with Alan Mabin, via Zoom, 22 January 2021.
35 Interview with Len Berry.
36 Interview with Len Berry.
37 Interview with a British academic who worked at Dar es Salaam in the Faulty of Arts and Social Sciences in the 1970s, Dar es Salaam, Tanzania, 11 August, 2011. Interview conducted by Professor Joanne Sharp.
38 Interview with a British academic who worked at Dar es Salaam in the Faulty of Arts and Social Sciences in the 1970s, Dar es Salaam, Tanzania, 11 August 2011. Interview conducted by Professor Joanne Sharp.
39 Interview with a British geographer who worked at Dar es Salaam in the 1970s, University of Glasgow, UK, 17 August 2012. Interview conducted by Professor Joanne Sharp.
40 Interview with a British academic who worked at Dar es Salaam in the Faulty of Arts and Social Sciences in the 1970s, Dar es Salaam, Tanzania, 11 August, 2011. Interview conducted by Professor Joanne Sharp.
41 Interview with Len Berry.
42 Interview with a British academic who worked at Dar es Salaam in the Faulty of Arts and Social Sciences in the 1970s, Dar es Salaam, Tanzania, 11 August 2011. Interview conducted by Professor Joanne Sharp.
43 Interview with a British academic who worked at Dar es Salaam in the Faulty of Arts and Social Sciences in the 1970s, Dar es Salaam, Tanzania, 11 August 2011. Interview conducted by Professor Joanne Sharp.
44 Interview with a Tanzanian geographer who studied at Dar es Salaam in the 1960s, Dar es Salaam, Tanzania, 28 April 2014. Interview conducted by Professor Joanne Sharp.
45 Interview with a British geographer who worked at Dar es Salaam in the 1970s, University of Glasgow, UK, 17 August 2012. Interview conducted by Professor Joanne Sharp.

46 Interview with a British geographer who worked at Dar es Salaam in the 1970s, University of Glasgow, UK, 17 August 2012. Interview conducted by Professor Joanne Sharp.
47 Interview with a British geographer who worked at Dar es Salaam in the 1970s, London, UK, 28 March 2013. Interview conducted by Professor Joanne Sharp.
48 Personal correspondence with Bashir Datoo, May 2021.
49 Gender and Politics in Africa: An Interview with Marjorie Mbilinyi. *Review of African Political Economy*. Retrieved from: https://roape.net/2017/08/24/gender-politics-change-africa-interview-marjorie-mbilinyi.
50 Ophelia Mascarenhas, *Tanzanian Affairs*. Retrieved from: https://www.tzaffairs.org/2018/05/obituaries-73.
51 *University of Nairobi News & Views* No.6 (September 1975), p. 37.
52 Interview with Len Berry.

## References


Agbodeka, F. (1998). *A History of University of Ghana: Half A Century of Higher Education (1948–1998)*. Accra: Woeli Publishing Services.

Ake, C. (1982). *Social Science as Imperialism: The Theory of Political Development (Second Edition)*. Ibadan: Ibadan University Press.

Al-Bulushi, Y. (2022a). Dar es Salaam on the frontline: red and black internationalisms. *Third World Thematics: A TWQ Journal*. https://doi.org/10.1080/23802014.2022.2141851.

Al-Bulushi, Y. (2022b). Thinking racial capitalism and black radicalism from Africa: an intellectual geography of Cedric Robinson's world-system. *Geoforum* 132: 252–262. https://doi.org/10.1016/j.geoforum.2020.01.018.

Areola, O. and Okafor, S.I. (eds.) (1998). *50 Years of Geography in Nigeria: The Ibadan Story*. Ibadan: Ibadan University Press.

Baker, R. and Polonskyi, M. (2015). *Cold War in a Hot Place*. Sofia: East-West Publishing.

Barnes, T. and Abrahamson, C.C. (2017). The imprecise wanderings of a precise idea: the travels of spatial analysis. In: *Mobilities of Knowledge: Knowledge and Space* (ed. H. Jöns, P. Meusburger, and M. Heffernan). Cham: Springer.

Barnes, T. and Farish, M. (2006). Between regions: science, militarism, and American geography from World War to Cold War. *Annals of the Association of American Geographers* 96 (4): 807–826. https://doi.org/10.1111/j.1467-8306.2006.00516.x.

Benneh, G. (2016). *My Time, My Nation*. Accra: Winmat Publishers Limited.

Berry, B.J.L. (1962). Urban growth and the economic development of Ashanti. In: *Urban Systems and Economic Development* (ed. R.P. Forrest). Eugene: University of Oregon Press.

Berry, B.J.L. and Garrison, W.L. (1958). The functional bases of the central place hierarchy. *Economic Geography* 34 (2): 145–154. https://doi.org/10.2307/142299.

Brookfield, H. (1975). *Interdependent Development*. London: Methuen.

Bunge, W. (1962). *Theoretical Geography*. Lund: Royal University of Lund.

Bunnell, T., Ong, C.E., and Sidaway, J.D. (2013). Jim Blaut and the trajectories of tropical geography. *Singapore Journal of Tropical Geography* 34: 285–291. https://doi.org/10.1111/sjtg.12035.

Burton, E. (2018). Introduction Journeys of education and struggle: African mobility in times of decolonization and the cold war. *Vienna Journal of African Studies* 34 (18): 1–17. https://doi.org/10.25365/phaidra.262_01.

Burton, E. (2020). Decolonization, the Cold War, and Africans' routes to higher education overseas, 1957–65. *Journal of Global History* 15 (1): 169–191. https://doi.org:doi:10.1017/S174002281900038X.

Child, E. and Barnes, T.J. (2019). Area studies and American imperial expansion without geography. *Journal of Historical Geography* 66: 43–54. https://doi.org/10.1016/j.jhg.2018.08.001.

Chouinard, V. (2020). Spatial histories of radical geography: North America and beyond. *The AAG Review of Books* 8 (4): 236–258. http://doi.org/10.1080/2325548X.2020.1781463.

Christaller, W. (1966). *Central Place Studies in Urban Geography*. Englewood Cliffs: Prentice Hall.

Davies, A. (2019). Milton Santos: the conceptual geographer and the philosophy of technics. *Progress in Human Geography* 43 (3): 584–591. https://doi.org/10.1177/0309132517753809.

Drysdale, A. and Watts, M. (1977). Modernization and social protest movements. *Antipode* 9 (1): 40–55.

Fanon, F. (2019 [1961]). *The Wretched of the Earth*. Milton Keynes: Diana Publishing.

Farish, M. (2010). *The Contours of America's Cold War*. Minnesota: University of Minnesota Press.

Ferretti, F. (2020). Subaltern connections: Brazilian critical geographers, development and African decolonisation. *Third World Quarterly* 41 (5): 822–841. https://doi.org/10.1080/01436597.2020.1722095.

Ferretti, F. and Viotto Pedrosa, B. (2018). Inventing critical development: a Brazilian geographer and his Northern networks. *Transactions of the Institute of British Geographers* 43: 703–717. https://doi.org/10.1111/tran.12241.

Ginelli, Z. (2018). Historical geographies of the "quantitative revolution": towards a transnational history of central place theory. *International Conference of Historical Geographers Conference*, Warsaw.

Gould, P.R. (1960). The development of the transportation pattern in Ghana. *Northwestern University Studies in Geography* (number 5): Evanston: Northwestern University.

Gould, P.R. (1967). On the geographical interpretation of eigenvalues. *Transactions of the Institute of British Geographers* 42: 53–86.

Haggett, P. (1965). *Locational Analysis in Human Geography*. London: Edward Arnold.

Hountondji, P. (1990). Dependence in Africa today. *Research in African Literatures* 21 (3): 5–15.

Jazeel, T., Kent, A., McKittrick, K. et al. (2019). *Keywords in radical geography: antipode at 50*. Wiley. https://doi.org/10.1002/9781119558071 (accessed 5 July 2023).

Kobayashi, A. (2019). Issues of "race" and early radical geography: our invisible proponents. In: *Spatial Histories of Radical Geography: North America and Beyond* (ed. T.J. Barnes and T. Sheppard), 39–58. Oxford: Wiley.

Ledgister, F.S.J. (2008). 'Intellectual Murder': Walter Rodney's *Groundings* in the Jamaican Context. *Commonwealth & Comparative Politics* 46 (1): 79–100. https://doi.org/10.1080/14662040701838043.

Leys, C. (2018). Letters: the African University. *London Review of Books* 40: 15. Retrieved from: https://www.lrb.co.uk/the-paper/v40/n15/letters.

Livingstone, D. (1992). *The Geographical Tradition*. Oxford: John Wiley and Sons.

Livsey, T. (2017). *Nigeria's University Age: Reframing Decolonisation and Development*. Basingstoke: Palgrave Macmillan.

Lorenzini, S. (2019). *Global Development: A Cold War History*. Princeton: Princeton University Press.

Lösch, A. (1954). *The Economic Location*. New Haven: Yale University Press.

Mabogunje, A.L. (1968). *Urbanization in Nigeria*. London: University of London Press.

Mabogunje, A.L. (1973). Role of the city in the modernization of developing countries. *Canadian Geographer* 17 (1): 68–76.

Mabogunuje, A.L. (2004). Geography in the Nigerian public policy domain: the impact and influence of American geographers. *GeoJournal* 59 (1): 63–67.

Mabogunje, A.L. (2011). *A Measure of Grace: The Autobiography of Akinlawon Ladipo Mabogunje*. Ibadan: Book Builders.

Mamdani, M. (2019). Decolonising universities. In: *Sharing Knowledge Transforming Societies* (ed. T. Halvorsen, K. Skare Orgeret, and R. Krøvel), 48–67. Cape Town: African Minds.

Mascarenhas, A.C. (1971). *Resistance and Change in the Sisal Plantation System of Tanzania*. Los Angeles: University of California. Ph.D., 1970.

Mascarenhas, A.C. (1977). Resettlement and desertification: the Wagogo of Dodoma District, Tanzania. *Economic Geography* 53: 376–380.

Mascarenhas, A.C. (2000). *Poverty, Environment and Livelihood along the Gradients of the Usumbaras in Tanzania*. Dar es Salaam: Research on Poverty Alleviation, Tanzania Printers Ltd.

Mascarenhas, A.C. (2004). Knowledge, indigenous knowledge, peace and development. *Indilinga African Journal of Indigenous Knowledge Systems* 3: 1–15.

Mascarenhas, O. and Mbilinyi, M. (1983). *Women in Tanzania: An Analytical Bibliography*. Uppsala: Nordiska Afrikainstitutet.

McNulty, M.L. (1969 January). Urban structure and development: the urban system of Ghana. *Journal of Developing Areas* 3 (2): 159–176.

Melgaço, L. (2017). Thinking outside the bubble of the global North: introducing Milton Santos and "the active role of geography". *Antipode* 49: 946–951. https://doi.org/10.1111/anti.12319.

Monson, J. (2007). *Freedom Sounds: Civil Rights and Call Out to Jazz and Africa*. Oxford: Oxford University Press.

Ndlovu-Gatsheni, S.J. (2018). *Epistemic Freedom in Africa: Deprovincialization and Decolonization*. London: Routledge.

Neate, H. (2022). Finding 'new' geographies in dedications, acknowledgements, and citations. *Dialogues in Human Geography* 12 (3): 467–469. https://doi.org/10.1177/20438206221088386.

Ngũgĩ wa Thiong'o. (1986). *Decolonising the Mind: The Politics of Language in African Literature*. Portsmouth: Heinemann.

Ngũgĩ wa Thiong'o. (1993). *Moving the Centre: The Struggle for Cultural Freedoms*. Portsmouth: Heinemann.

No Author. (1966). Violence at the University of Rome. *Minerva* 4: 585–588. https://doi.org/10.1007/BF02207991.

O'Keefe, P. (2020). David Slater: a leading geographical theorist. *Human Geography* 13 (2): 154–159. https://doi.org/10.1177/1942778620944562.

Odingo, R.S. (2000). Professor Simeon Hongo Ominde's contribution to geographical science. In: *Issues in Resource Management and Development in Kenya: Essays in Memory of Professor Simeon H. Ominde* (ed. R.A. Obudho and J.B. Ojwang), 3–7. Nairobi: East African Educational Publishers.

Okafor, S.I. (1989). Research trends in Nigerian Human Geography. *Professional Geographer* 41 (2): 208–214.

Oldfield, J. (2018). Imagining climates past, present and future: Soviet contributions to the science of anthropogenic climate change, 1952–1991. *Journal of Historical Geography* 60: 41–51. https://doi.org/10.1016/j.jhg.2017.12.004.

Onslow, S., (ed.) (2009). *Cold War in Southern Africa: White Power, Black Liberation*. Abingdon: Routledge.

Peet, R. (no date). Past editors' reflections: reminiscing on the early antipode: Richard Peet (1970–85). *Antipode*. Retrieved from: https://onlinelibrary.wiley.com/page/journal/14678330/homepage/editor_s_past_reflections.htm#Peet (accessed 5 July 2023).

Power, M. and Sidaway, J.D. (2004). The degeneration of tropical geography. *Annals of the Association of American Geographers* 94: 585–601. https://doi.org/10.1111/j.1467-8306.2004.00415.x.

Sada, P.O. (1982). Geography in Nigeria: perspectives and prospects. *Presidential Address to the 25th Annual Conference of the Nigerian Geographical Association*, Ibadan.

Santos, M. (1974). Geography, marxism and underdevelopment. *Antipode* 6 (3): 1–9.

Santos, M. and Peet, R. (1977). Introduction. *Antipode* 9 (1): 1–3.

Sharp, J. (2019). Practising subalternity: postcolonial Tanzania, the Dar School and pan-African geopolitical imaginations. In: *Subaltern Geographies* (ed. T. Jazeel and S. Legg), 74–93. Athens: University of Georgia Press.

Shivji, I.G. (1993). *Intellectuals at the Hill*. Dar es Salaam: Dar es Salaam University Press.

Sidaway, J., Lin, S., Chouinard, V. et al. (2020). Spatial histories of radical geography: North America and beyond. *The AAG Review of Books* 8 (4): 236–258. https://doi.org/10.1080/2325548X.2020.1781463.

Slater, D. (1973). Geography and underdevelopment – 1. *Antipode* 5: 21–32. https://doi.org/10.1111/j.1467-8330.1973.tb00568.x.

Slater, D. (1975). The poverty of modern geographical enquiry. *Pacific Viewpoint* 16: 159–176. https://doi.org/10.1111/apv.162003.

Slater, D. (1977a). Geography and underdevelopment – Part II. *Antipode* 9: 1–31. https://doi.org/10.1111/j.1467-8330.1977.tb00089.x.

Slater, D. (1993). The geopolitical imagination and the enframing of development theory. *Transactions of the Institute of British Geographers* 18 (4): 419–437. https://doi.org/10.2307/622559.

Soja, E.W. (1968). *The Geography of Modernization in Kenya: A Spatial Analysis of Social, Economic and Political Change*. Syracuse: Syracuse University Press.

Southall, R.J. and Kaufert, J.M. (1974). Converging models of University development: Ghana and East Africa. *Canadian Journal of African Studies / Revue Canadienne Des Études Africaines* 8 (3): 607–628.

Stanek, M.B. (2019). Decolonial education and geography: beyond the 2017 Royal Geographical Society with the Institute of British Geographers Annual Conference. *Geography Compass* 13: e12472. https://doi.org/10.1111/gec3.12472.

Taaffe, E.J., Morrill, R., and Gould, P.R. (1963). Transport expansion in underdeveloped countries: a comparative analysis. *The Geographical Review* 53: 503–529.

Taylor, P. (1977). *Quantitative Methods in Geography: Introduction to Spatial Analysis*. Boston: Houghton Mifflin.

Veit, P.G., Mascarenhas, A., and Ampadu Agyei, O. (1995). *Lessons from the Ground Up: African Development that Works*. Washington: World Resources Institute.

Wheatley, P. (1970). Urbanization in Nigeria. *Economic Geography* 46 (1): 102–104. https://doi.org/10.2307/142689.

Whitehand, J.W.R. and Edmondson, P.M. (1977). Europe and America: the reorientation in geographical communication in the post-war period. *The Professional Geographer* 29 (3): 278–282. https://doi.org/10.1111/j.0033-0124.1977.00278.x.

# Chapter Five
# Geography and National Development: Knowing, Planning, and Exploiting Resources for Independent Africa

## Introduction

The Nigerian geographer Akin Mabogunje was only an undergraduate student at the University of Ibadan when he was first asked to assist the government in (then still colonial) Nigeria. Working as an enumerator for the census in 1952, Mabogunje's role was to collect data on the population characteristics of the dense urban centre of Ibadan, for use in planning and resource allocation. When he returned to Nigeria following postgraduate study in London, Mabogunje was straightaway back in the fray as part of the leadership team for the 1962 census in the Western Region. After that, in his words, 'for nearly fifty years of my life, there was no year I wasn't serving government in an advisory capacity … urban development, rural development, population …'.[1] This chapter focuses on a key preoccupation of the decolonising era: national development, and on the ways that African geographers worked with governments and other agencies on development projects in the 1960s and 1970s.

Debates about the role of geography in Africa, and its potential contributions to national development reflected wider debates about the role of the university, and the scholar, in postcolonial Africa. To what extent should the idea and practice of the academic change in independent countries? How could and should

*Decolonising Geography? Disciplinary Histories and the End of the British Empire in Africa, 1948–1998*, First Edition.
Ruth Craggs and Hannah Neate.

the university contribute to national development as part of a wider project of state-building? According to Mamdani, in empire,

> the university was in the frontline of the colonial 'civilising mission'... Its aim was to create universal scholars, men and women who stood for excellence, regardless of context ... The first critical reflection on this colonial project took place in the nationalist movement. From the ranks of the nationalist movement emerged a different kind of intellectual, the public intellectual. If the hallmark of the global scholar was *excellence*, that of the public intellectual was *relevance*. Excellence was said to be universal, measured without regard to context; relevance, however, was necessarily contextual, place-specific. (Mamdani 2019, p. 51, emphasis in original)

In setting out the 'trajectories of struggles for an "African University"', Ndlovu-Gatsheni (2017, p. 61) highlights the emergence of the 'developmental' university in the 1960s, with the University of Dar es Salaam 'soon known as the prototype ... truly responsive to its society' (Ajayi et al. 1996, p. 119). The government of independent Tanzania required that The University of Dar es Salaam be 'relevant', to contribute to national development, whilst at the same time avoiding the 'dangers of looking "inward" at the expense of the universal aspect of knowledge' (Ndlovu-Gatsheni 2017, p. 61). Despite these dangers, Tanzanian President Julius Nyerere was clear, demanding that the university, of which he was Chancellor (Figure 5.1), took 'an active part in the social revolution' (Nyerere 1966, quoted in Ndlovu-Gatsheni 2017, p. 63). This social revolution was 'predicated on nation-building, economic development and promoting African consciousness' (Ndlovu-Gatsheni 2017, p. 63). Whilst Nyerere, Kwame Nkrumah, and other African leaders embraced various pan-African visions, much of their thinking on black emancipation remained tied to the nation-state as the vehicle for delivering social, economic, and cultural uplift.

The report of an influential 1972 workshop on 'Creating the African University' highlighted the extent to which a vision of the university as central to national development came to dominate the continent in the 1970s. The workshop proceedings recorded:

> an emergent African university must, henceforth, be much more than an institution for teaching, research, and dissemination of higher learning. It must be accountable to, and serve, the vast majority of people who live in rural areas. The African university must be committed to active participation in social transformation, economic modernization, and the training and upgrading of the total human resources of the nation, not just of a small elite. (Yesufu 1973, p. 41)

To what extent could geography be relevant in these terms? Drawing attention to the intersections of debates about decolonisation and the practice of urban, rural, and regional planning, resource evaluation and census taking, this chapter demonstrates how geographers contributed to the multifaceted project of knowing, planning and shaping the postcolonial state through their contributions to

**Figure 5.1** President Julius Nyerere of Tanzania being installed as Chancellor of the University of Dar-Es-Salaam, Tanzania, 29 August 1970. Popperfoto/Getty Images.

'national development'. Building on the themes of the last chapter, this exploration also provides further insight into the role of Cold War dynamics within the development projects of postcolonial Africa (Lorenzini 2019). Whilst we draw on a range of examples, the focus falls substantially on the contributions of the Geography Departments at Dar es Salaam, Tanzania, and Ibadan, Nigeria.

The chapter begins with a discussion of the rise of applied geography for development before moving on to discuss how geography departments were tasked with training development experts through their teaching programmes. Here Ghanaian and Tanzanian changes to teaching are explored, with the latter providing an example of potentially radical curriculum transformation later being domesticated into a less critical, more technocratic ones aimed at producing development experts rather than critiquing the ideology, policy, and practice of development. The rest of the chapter explores how geographers proactively set out to support new national governments with development issues, aiming to produce policy-relevant research. This work, sometimes as government consultants, contributed to national censuses, spatial planning, land-use and rural development, large infrastructure projects, and the development of new capital cities. In this work, geographers were involved with some of the most ambitious, but also controversial, development projects of the period in Africa: such as building Nigeria's new capital in Abuja and 'villagisation' in Tanzania. As the next chapter highlights, they were also involved in planning the new 'homelands' – known critically as bantustans – created for Black South Africans under apartheid.

## Geography and African Development

What sort of geography was required for development? Speaking in 1968 on the theme of 'Geography and African Development', Simeon Ominde described the evolution of the new geography departments of East Africa as they attempted to produce the knowledge and skills relevant for new nations. The Department of Geography at Dar es Salaam, founded in 1965 had set out:

> an outline of research topics in both physical and human fields that underline a clear awareness of the development role of the discipline. In the expanding programme of research covering land use, land classification and regional planning, population problems, problems of water resources, political geography, biogeography and geomorphology, the department is stretching its resources to realize its role as regards the urgent practical needs of the nation. (1971a, p. 5)

Geography was envisaged as of use in two main ways: providing data to inform government policy making, and providing students trained in planning and resource questions to fulfil these roles in the postcolonial states (Darkoh 1981). In this way it could contribute directly to the first of the United Nation's key proposals for action in the so-called development decade of the 1960s: '(i) The more systematic survey, development and utilization of physical and human resources in underdeveloped countries.'[2] According to the Ghanaian geographer Elizabeth Ardayfio-Schandorf, it was 'human geography appraisals and that applied aspect of geography that relate[d] to development'.[3]

Whilst on the one hand applied research could be aligned with independence, this focus has often lead to the exclusion of basic research lead solely by curiosity from African universities. This has been criticised as part and parcel of a wider 'intellectual extraversion' (Hountondji 1990). In 1990, Hountondji (p. 11) argued that in the global south:

> There is still a collective prejudice against basic research, which is thought to be too costly, quite useless for developing countries and, therefore, appropriate only to the North. It even happens that, although adequate facilities sometimes exist, we still refrain from doing basic research out of conviction. The prevailing ideology in our countries is utilitarianism or pragmatism. Obviously inherited from the colonial period, this ideology illustrates the degree to which we have internalized the norms of the colonizer.

Geography departments certainly prioritised applied research in the 1960s and 1970s, in ways that in some respects echoed the types of research conducted in the name of colonial development (and exploitation). However, 'applied geography', which in the mid twentieth century consisted primarily of land-use and population surveys and mapping, had grown in popularity globally from the Second World War, and was therefore not solely an African colonial or post-colonial phenomenon. Championed by the British geographer Dudley Stamp, gearing the discipline 'towards the solution of some of the great world problems' (1960, p.10), applied geography dominated British geography into the 1970s and beyond. It nevertheless had a particular and more longstanding purchase in Africa than in Britain or other parts of the West and applied approaches remain dominant in many African departments today.

Akin Mabogunje was very clear about the types of theoretical and methodological approaches required to understand and contribute to development, and for him, this involved more than just the mapping and cataloguing of land use past and present. Reflecting on his undergraduate and postgraduate training in the UK at University College, University of London under the historical geographer H.C. Darby, he noted: 'what I got in London was … really the substantive development process of a people over time in space'.[4] After his PhD, whilst on his sabbatical trip to North Western University in the US,

> What I got … was a better appreciation of the process … Which allows me therefore in looking at my country to see how the two [historical and quantitative geography] can become focused on to how you do development … So, the ability to become more theoretical than just being historical allowed me to start interrogating the development process in Nigeria in a way geographers weren't used to doing.[5]

This had consequences for the discipline back at the University of Ibadan in Nigeria:

> Part of the consequence of that was that the relationship of geography with social sciences became a lot more obvious. And so in fact that's how geography came to move from the arts faculty to the social sciences [in 1967] and the expectation is that a lot of what we are doing will now become more relevant to development … people started to see geography in a slightly different light, that it's not just a teaching subject, but it's a subject that really affects the day to day development.[6]

By the middle of the 1960s then, geography in Africa was being cast as central to the project of 'National Development'.

## Producing Development Experts

In 1962, the first Ghanaian Head of Department at the University of Ghana, Ernest Boateng wrote to the Pro-Vice Chancellor Wright to explain his vision for the department: 'it is my aim to make geography in this university a practical subject, which is closely geared through the courses in applied geography to the contemporary problems of Ghana.'[7] The focus of the curriculum and student interests can be seen by the topics of undergraduate dissertations from the 1960s, which focused primarily on industrial and agricultural development, as well as specific development projects such as the Akosombo Dam on the Volta River. Student Stephen Kwami Awunyo-Akaba completed a dissertation 'The Volta River Project: The agricultural programme of the Volta Resettlement Scheme 1965–66' (1966) and later became a development planner.[8] Awunyo-Akaba was clear about the relevance of his study to the development of the country, recording in the preface that

> this particular topic for the dissertation – the agricultural programme of the V[olta] R[iver]P[roject] – was chosen because it represents one of the three important facets of the economic drive which is sweeping through Africa (agricultural improvement). The others are the drive to industrialise and to increase both intra-continental and foreign trade. (n.p.)

And in the Appendix, the question of geography's role in contributing to development was addressed head on:

> the multipurpose development of the basin raises the pertinent question of what role the geographer can play in such an exercise … for those geographers who believe that the main task of geography in the twentieth century is to help in "the transformation of nature" for the benefit of mankind, there is great scope to make contributions to multi-purpose development projects, such as of the Volta basin.

The phrase the 'transformation of nature' was popular in the mid-twentieth century, coming to prominence in Soviet plans for largescale development – most notably 'The Great Stalin Plan for the Transformation of Nature' (1948) – which aimed to utilise scientific expertise to bring about rapid progress through the stabilisation of soils, new agricultural techniques, afforestation, dams, and hydropower (Shaw 2015). Geographers, experts in synthesising different kinds of knowledge, played a substantial role in these plans in the Soviet Union (Shaw 2015; e.g. Geramisov 1962). Awunyo-Akaba, studying in a Ghana which under Nkrumah was close to the Soviet Union and replete with scientific experts from the Eastern Bloc (Ginelli 2018; Stanek 2019) drew on these visions of geography as central to a project of development driven by mega-projects, large-scale changes, and modern technical solutions. This undergraduate dissertation therefore reflected wider ideologies of development as modernisation dominant in Nkrumah's vision for Ghana (Ahlman 2017).

Tanzania under Nyerere also aimed to follow a socialist development path, but focused primarily on rural development and decentralisation, rather than on the more urban vision of modernity favoured by Nkrumah (Lal 2015; Ahlman 2017). At the University of Dar es Salaam, Tanzania, from the late 1960s, teaching was reformulated explicitly to produce graduates who could support this version of socialist development. Following the Arusha Declaration in 1967, a new three-year interdisciplinary course was introduced which was compulsory for most students: East African Society and Environment (EASE) (Doherty 1982). This had its roots in the debates between radical scholars such as Walter Rodney, John Saul, Issa Shivji, and Giovanni Arrighi (Sharp 2019), and geography contributed to this interdisciplinary effort. EASE focused on providing students with the tools to think critically about Tanzanian development in the context of postcolonial Africa and the wider world. Reflecting the late 1960s debates, the Faculty of Arts and Social Sciences at Dar es Salaam, of which geography was a part, also restructured its degrees in the early 1970s (though after vociferous debate disciplinary departments remained).

Opinion on the extent to which geography should embrace the most radical ideas and curriculum changes had often been split. Whilst radical thought had put the University of Dar es Salaam on the international map, it was not unanimously supported even within the university. A leading Tanzanian geographer Adolfo Mascarenhas had weighed in on these debates, allying with several liberal voices in the university, against concerns of 'political indoctrination' in the proposals to further overhaul teaching, much to the chagrin of 'the strong left-wing section of the university, both black and white'.[9] In a demonstration of the level of public interest in the debate about the role of the university in radical development, the controversy surrounding the issue made headline news in the *Tanzanian Standard*. This gave Mascarenhas a nasty shock when he emerged from a late-night showing of the film *Tora! Tora! Tora!* at the cinema with external examiner Akin Mabogunje, and saw the newspaper stands.[10]

The interdisciplinary course EASE was discontinued after a few years, something which has been understood as an attempt to

> reassert the dominance in university education of the official ideology of the Tanzanian state. EASE created among students a critical awareness of the social structures of East African society which threatened one of the principal aims of university education in Tanzania, namely the production of graduates who would 'adhere to and promote the ideology and political attitudes of government and party' (Briggs and Gray 1982, p. 40). (Doherty 1982, p.174)

One British geographer who was at Dar in the mid-1970s reflected that

> My reading of it … was that the university authorisers had given the task of actually disbanding that course and actually re-charting it in a way which would make it more technocratic, I suppose is the best word. So it would be about resource development, for example, rather than political development.[11]

In the wake of intense debates in Tanzania about the role of the university, and individual disciplines, for national development (in which some Tanzanian and foreign staff had not had their contracts renewed or had been forced to leave the country), Joe Doherty, a British geographer at Dar in the 1970s believed that 'Geography had survived because it had "kept its head down"' and had 'successfully presented itself as a "technical" subject and thus avoided controversy' (1982, p. 175). The curriculum was, according to one British geographer at Dar in the 1970s, 'very safe and very conservative … it really was outside the geography department that more exciting things were happening'.[12] Courses that were taught by the geography department 'retained a strong empiricist orientation with the emphasis on data collection and analysis and … differ little from those on offer in many Western universities. As taught, geography fits well into the mould of a technocratic institution' reported Doherty (1982, p.175).

In the 1970s and 1980s Geography contributed to a range of interdisciplinary vocational degree streams in planning, industrial development, agricultural and rural development, statistics, management and administration, resource assessment, international relations, and social development. It specialised in resource assessment in particular, as well as in producing teachers, and its contributions to other courses consisted of issues such as regional planning, location, urban and rural systems, physical resource development, and transportation (Briggs and Gray 1982).

By 1982, the Ministry of High-Level Manpower Planning required the Faculty (led by the Department of Geography) to produce 14 Resource Assessors a year, but Resource Assessment and Land Use Planning students also took up a variety of roles in the Ministry of Lands, the Capital Development Authority (responsible for the development of the new national capital at Dodoma), in the Prime Minister's Office, the National Transport Corporation, 'District Development Directors' Offices (which plan and implement development projects at the district level), the survey department, posts in social service planning, agricultural marketing and various types of rural and agricultural development agencies' (Briggs and Gray 1982, p. 45; Mashalla, 1988).

John Briggs and Alex Grey (1982), who were members of the Geography Department in the late 1970s, were positive overall about the changes in university structures and curriculum. They argued that 'the move to "relevance" within university teaching ha[d] … reinforced the notion of nation building and self-reliance and created the feeling of contributing actively towards Tanzania's development, a goal for which geography ha[d] … an important role to play.' (p. 45). Others were more critical. Briggs and Gray (1982, p. 45) themselves admitted that, by the late 1970s, 'for a department to offer esoteric courses of little relevance to contemporary Tanzanian and African society would be to jeopardise its position in the University'. And many feared that more radical visions for and critiques of development had been replaced by more technocratic training at the university by the middle of the 1970s.

Whilst Tanzania under Julius Nyerere had become a hub for radical development thinking, technical approaches were central to post-colonial development plans across the political spectrum, including in Tanzania (Ake 1982). For the Nigerian scholar Claude Ake, who worked at the University of Dar es Salaam in the 1970s, technicism encouraged the view that all countries were on the same path to linear development and concealed 'the culpability of the imperialist powers for the present plight of the underdeveloped countries' (1982, p. 177). According to Ake, social science disciplines had

> been promoting … the technicist view of development … the reduction of development to a series of simple problems requiring essentially technical solutions … to think of development in these terms is to implicitly take an apolitical view of development … of course, this apoliticism only conceals a conservative political and ideological stance, namely the presumption that the revolutionizing of the present international order and the economic and political order of the Third World countries is not pertinent. The truth is otherwise. (Ake 1982, pp. 171–173; pp. 193–194)

Whilst the last chapter highlighted the place of the University of Dar es Salaam in the history of radical geography, and especially in radical critiques of development and theories of modernisation, a focus on the teaching in the department suggests that despite attempts to reform the curriculum, geography teaching remained fairly conventional, even if there was a greater focus on specific roles in the country after graduation. Geography was able to secure itself a central role within the fulfilment of the 'manpower' [*sic.*] needs of socialist Tanzania, and the particular skill sets of students taught by the geography department suited the requirements of a country attempting to reconfigure its spatial and social development. However, this central role relied precisely on the exclusion of some of the more critical perspectives on development policies domestically which emerged in geography, influenced by a radical interdisciplinary group of scholars at the university in the late 1960s and early 1970s. The rest of this chapter explores the ways that geographers were more directly involved in development work, if not often directly critiquing it.

## Knowing New Nations (and Resources) through the Census

> Under powerful Fluorescent lights
> Young Ghanaians write in a book of gold […]
> With smiles on their lips they ask:
> Where did you sleep the night before?
> How old are you?
> Answer! […]
> And save Ghana from Economic Want
> And kill poverty and illiteracy
> Hail the Census Night.[13]

As the poem above, published in the *Evening News* in Ghana on the eve of the first census after independence demonstrates, these projects of national accounting were closely linked, in the political and popular imagination, with questions of national development. And geographers were instrumental in many of these post-colonial censuses.

As Kertzer and Arel (2002, p. 2) argue, 'the census does much more than simply reflect social reality; rather it plays a key role in the construction of that reality'. Censuses are part of what James Scott suggests is the 'state's attempt to make a society legible' (1998, p. 2). In post-colonial Africa, they allowed for the boundaries of a state to be drawn, for populations to be incorporated into a nation, and also contributed to the imaginative construction of newly independent states for their governments and citizens. They were also understood as crucial for the allocation of resources for equitable development, and 'an important precondition of economic and social modernization' (Serra 2018, p. 659). As such, censuses were a crucial step for national development in newly independent African states, both symbolically and practically (Mbalisi et al. 2019). They also connected to the so-called national question in post-colonial Africa: 'how to order the relations between the different ethnic, linguistic and cultural groupings so that they have the same rights and privileges, access to power and equitable share of national resources' (Ajayi 2000, p. 218). As a result, censuses were sometimes contested as they became bound up with regional struggles for power and resources within newly independent states (Ahonsi 1988).

The American geographer John Hunter 'was planning officer and adviser for the preparation and enumeration of area maps and census reports' for the 1960 census in Ghana, whilst working in the Geography Department at the University of Ghana.[14] According to Hunter (1961, p. 1), before implementing its five-year development plan, Ghana needed an 'accurate stocktaking in all sectors of the economy', including 'an inventory of the human resources.' In an article entitled 'An exercise in applied geography' (1961), Hunter set out the role of the geographer in the census 'in providing maps on which the enumeration will be based', a role, he noted, 'of considerable importance' (p. 3). Hunter oversaw the mapping of 23 towns; his colleagues at the University of Ghana T.E. Hilton, from the UK, and the Ghanaian E.A. Boateng, were involved in the mapping of several rural areas (1961, p. 8 footnote 14). Hunter described the importance of the census not only in providing detail about populations levels for development planning, but also in capturing, for researchers, the dramatic changes taking place in West Africa.

> [O]ld order is giving way to the new. Traditional ways of life are fast disappearing; and among the many agencies at work affecting this transition that of urbanization is of paramount importance. Towns are meeting points, melting pots, always in the vanguard of change. Their mushroom growth in recent decades, and the changes of which that growth is both symptomatic and instrumental in causing, have attracted the attention of sociologists, demographers and students of government. But with few exceptions in the past they have been poorly served with statistical information. Perhaps it is not too much to hope that, given inter alia adequate geographical planning within the framework of a carefully conceived and executed

census operation, firm statistical basis will be provided for future investigations. (Hunter 1961, p. 8)

Hunter's comments reflected an increasing interest in urbanisation in Africa, and a growing frustration in the lack of adequate data on which to base research (see for example Mabogunje 1968 on this issue). Urbanisation was a focus for African researchers, most notably in this period Akin Mabogunje (1968) whose work fused careful archival work reconstructing histories of towns and cities in Nigeria with new statistical techniques. It was also of interest to those from the West, including the British colonial anthropologists who had from the 1940s been interested in cultural change amongst urbanising populations (Wilson and Wilson, 1945; Schumaker 2001), and post-war interest in urbanisation and modernisation, primarily from American geographers and other social scientists at the vanguard of the quantitative revolution (e.g. Gould 1960; Taaffe et al. 1963; Soja 1968). Serra (2018, p. 685) records that in Ghana, the census, coming so soon after independence, 'entered the literary imagination, informing visions of a utopian future where poverty and illiteracy would be eradicated.' Whilst Hunter mostly discussed the census in much drier technical terms, his words quoted above betray some of the intellectual excitement felt by international researchers about the rapid transformation of African countries they were witnessing. Hunter shared the optimism of the anonymous poet at the start of this section, about the power of the census to bring about positive change in the new nation.

Boateng, Hunter, and Hilton in Ghana were not the only geographers involved in African census projects. In Nigeria Akin Mabogunje, Reuban K. Udo, and E.O. Adeniyi of the University of Ibadan, were all involved (Udo 1981; Mabogunje 2011). In Nigeria, censuses were politically controversial. As Ahonsi (1988, p.554) notes, 'all the colonial censuses were technically deficient, while the 1962 and 1973 counts were dropped due to the intense political tensions they provoked', as there were fears of inflation in counts in different regions for the purpose of securing increased revenue and political representation.

Mabogunje was involved in several difficult counts over decades from the 1950s– 1970s. The Department of Geography at Ibadan was asked to provide students as enumerators for the 1952 colonial census and Mabogunje was one of those enrolled. In his autobiography, Mabogunje recounted the challenges of counting children in polygamous families where this was seen as 'tempting the gods for their bountiful mercies', in a dense old part of the city of Ibadan. In this part of the city, large compounds had begun to be broken up into smaller units leaving 'a confused jumble of houses'. As a result, Mabogunje reflected, he realised, the results of the census 'could only be an approximation.' (2011, pp. 371–372).

In preparation for the census in 1962, Mabogunje was approached, alongside his friend the eminent historian Jacob Ade-Ajayi 'to give critical leadership' to the preparations in the Western Region (Mabogunje 2011, p. 374). Ajayi was to build tables of local historical events in different areas to help determine people's ages; Mabogunje to demarcate the enumeration areas, no easy task when the Survey Office had no appropriately scaled maps for the process and were not willing to produce them. In the end, field assistants produced sketch maps linked to the available maps which showed the boundaries of each area. Mabogunje was relatively happy with the census collection in his region, though the unreliability of the data overall and the fact that Northern Nigeria was shown to have a larger population than the South lead to a political crisis. As a result, the census was held again in 1963, showing ten million more people than the year before, and leading to accusations of population inflation from Southern politicians, though the results were allowed to stand (Ahonsi 1988). A decade later, Mabogunje was Special Adviser to the National Census Board for the 1973 count, with responsibility for 'the demarcation of the whole country into enumeration areas' (Udo 1998, p. 358); Reuban Udo and E.O. Adeniyi were also on the board. Again, however, the census was to collapse into political controversy and was eventually annulled. Reflecting on his 'census years' Mabogunje was sanguine, he had done his best but in challenging circumstances. However, when the opportunity arose to take charge again in the early 1990s, Mabogunje remembered, smiling, 'my wife said enough, … they can find someone else.'[15]

The first postcolonial census in Tanzania was less controversial. Taking place in August 1968, preparations began almost immediately after the Department of Geography was founded, and geographers were heavily involved. Len Berry, the first Head of Department remembers:

> the whole department was enlisted to help with the training of the census takers, and the programme for the census, so the department right from step one became heavily engaged in this practical thing. And we ended up with the department staff, and in many cases their wives, who had been trained, going to different parts of the country and training the census takers, talking about the techniques and how you save the data and so on. So that was the first practical thing we did. My wife and I went up to Lake Victoria, spent some time with the Commissioner, calling meetings literally under mango trees, with translation into the local language. It sounded very haphazard, it was, but the census did take place.[16]

Reflecting longstanding trends across the foreign lecturing staff in African geography departments, Berry's wife Eileen, herself a University College London geography graduate, was heavily involved with the work of the department at Dar es Salaam. Crucially for Berry, this involvement with the census demonstrated the value of geography to national development. His memories again reflect the atmosphere of excitement in African universities in the years immediately following independence:

> [The census] was the first indicator of the way that geographers could be practical, practically involved, and there was no way at that time that you could be in Tanzania at the time and not feel that you had to be involved in the problems of the country and the dynamism that was going on.[17]

## Research for Rural Development in Tanzania

### *BRALUP and Applied Research*

Alongside its work for the census, and following from the department's experience of it, the Geography Department at Dar es Salaam aimed to enhance its relevance to national development through a new institute for applied research. The Bureau for Resource Assessment and Land-use Planning (BRALUP), set up in 1967, was the brainchild of Berry, who remembered:

> The university had already begun to establish an Economic Research Bureau, which was standard practice in African countries … I felt that there should be a counter punching geography or resource based centre, because the economists were dealing with macroeconomics at the national level, and in that particular period, were not looking at rural development, so I had this concept … of setting up a parallel institute geographically based, but not geographically constrained in terms of discipline, to look at water, rural and resource issues.[18]

The Institute was founded at a time when 'resources were available' for African universities, reflecting both the Cold War funding dynamics outlined in the last chapter and the wider emphasis on higher education in development policy.[19] As Berry recounted:

> The amazing thing is we got money for it very quickly, including a building … it was international, the building came from NORAD [The Norwegian Agency for Development Cooperation], I remember going to a donors' meeting in Nairobi, and they were asking what did we need, and I said well we need a building …. We got 84,000 dollars … which was enough to build the four storey building that is still there. And the other resources were people resources from different countries, you could get a visiting person for two years from whatever country you wanted to name at that time.[20]

Rockefeller also contributed the salary of the Director for the first six years, meaning that the set up and initial staffing costs of BRALUP did not come out of the Tanzanian government budget.[21] Through the late 1960s and 1970s, BRALUP was staffed by a cosmopolitan mix of temporary staff funded by international government aid budgets and permanent Tanzanian researchers, reflecting the huge amount of foreign aid pouring into Tanzania in this era as decolonisation, international development and Cold War priorities intersected (Rugumamu 1997; Lal 2015).

Berry, who had met the American geographer Gilbert White in Dar es Salaam when he was in the country setting up a water resources research project (utilising Dar geography undergraduates as field researchers), asked the American for advice on a first Director for the new Institute. Berry did so because he 'felt it needed an American approach not a British one. I hadn't been to America but I had this image of people being much more involved in practical things'.[22] White suggested the American Robert Kates, a geographer who worked at Clark University on applied development research. Betraying his views of the relative innovation of US and British geography (Berry would move to Clark in the US from Tanzania in 1971), Berry's request also highlighted the fact that until the late 1960s, even in stridently independent Tanzania, it was still possible to appoint foreign staff to important roles. An external review in 1979 confirmed that 'During its first four years of operation BRALUP was essentially an expatriate institution', and whilst 'a number of competent Tanzanian researchers' had joined by 1979, of the nine Tanzanian staff, still only three were at a senior level.[23] In this the Institute reflected wider patterns: writing about Kenya, Tanzania, and Uganda, Cook (1979, p. 97) reported that 'Much of the resource assessment in all three countries has been carried out by expatriate scientists both in the pre- and post-independence eras.' It was not until 1972, when several Tanzanians had returned from PhD study abroad that BRALUP got its first Tanzanian Director, Adolfo Mascarenhas. Revealing the closeness of the Bureau to government and the more technocratic nature of its research and consultancy work, Mascarenhas was appointed directly on the instruction of President Nyerere.[24]

Describing the rationale for the Bureau in 1967, an early publication recorded:

> the research program should be directly related to national goals, particularly as affirmed in the Arusha Declaration. "All citizens together possess all the natural resources of the country in trust for their decedents..." The Declaration urges that emphasis in developments should be on land and agriculture, the self-reliance of the people, good policies and good leadership. Specifically this suggests our research program be in the rural sector, geared to providing knowledge with which to help implement the goals now being developed for the 1969–74 plan. (BRALUP Research perspectives, 1967, quoted in Berry and Kates 1973, p. 268)

The Arusha Declaration proclaimed a radical rethinking of natural resources as jointly owned and held in trust, in marked contrast to more bombastic accounts of the 'transformation of nature' envisaged in visions of development as modernisation dominating both the West and the Eastern bloc in this period. However, BRALUP's practical policy work did not depart substantially from that of many others across the political spectrum at the time. A brief review in 1970 set out the Bureau's work as follows:

> It is a multi-disciplinary body, orientated towards rural areas ... Current research covers (1) basic data service; (2) land use and land potential; (3) rural water supply and rural development; (4) river basin studies; (5) roads and rural development; (6) population; (7) regional planning and urban studies. (Berry 1970, pp. 67–74, 67)

In its first six months the Bureau staff (along with Eileen Berry in a voluntary capacity) produced a series of summaries of the latest research available in forestry, soil erosion, water supply and similar areas, which were sent to government ministries.[25] Later, they inaugurated a series of Service Papers building on this and aiming to alert the government directly of problems the country would face.[26] The Director of BRALUP also sat on a series of government committees on Water, and Regional Planning, for example. In the early years, BRALUP worked closely with the Tanzanian government, as well as with international development bodies, and their research fed into policy on the ground. Ajayi et al. (1996, p. 118) pick out BRALUP as one of several 'service institutions' located on the University of Dar es Salaam campus that illustrated the university's development orientation, and its closeness to the government.

This worked well in the early years, and both Berry and his successor as Head of Department Paul Temple sat on government committees for hydrology and minerals, providing further connections between the geography department and Nyerere's government (Dorrnkamp, 1986).[27] Whilst an external review in 1979 noted that this governance structure had been 'dormant for several years', lessening the Bureau's impact, it still reported that overall 'BRALUP has been engaged in policy-relevant work, carried out in a competent manner and had a definite impact on both official thinking and policy-making in Tanzania'.[28]

## *Research for Ujamaa*

In the late 1960s and 1970s BRALUP was involved in research to evaluate and support the Tanzanian government's vision for development, *ujamaa*, which, alongside the nationalisation of key institutions such as banks and major industries, focused on decentralisation of power and 'reorganizing the Tanzanian countryside into Socialist villages' (Lal 2015, p.3). This resettlement project began as a voluntary one following the articulation of *ujamaa* in the Arusha Declaration of 1967, but from 1973–1975, under Operation Vijiji 'morphed into a compulsory drive in which millions of peasants were forcibly relocated into concentrated settlements' (Lal 2015, p. 3). The vision for what became known internationally as 'villagisation' drew on nostalgic notions of the village as a centre of African community and solidarity, plans for a radical decentralisation of power, and the cultivation of self-reliance. But it also reflected seemingly contradictory impulses towards modernisation and state control. Welfare services could be more easily delivered to rural populations living in more concentrated communities, villages could be sites for the modernisation of agricultural production through the application of science and cooperative farming, and these communities were more easily monitored than dispersed populations (Lal 2015). These rural interventions were not entirely new: Tanganyikan villages had already been the site for colonial era development projects as well as post-colonial interventions aimed at bringing about agricultural 'transformation' supported by organisations such as the World Bank prior to the implementation of *ujamaa* policies (Kates and Berry 1970).

How exactly was BRALUP involved in this huge experiment in geographical reorganising? Early on in the process, around 1969, Len Berry, then the Bureau's Director, received a call from the President's Office 'saying they had a plan to bring villages together and would we be willing to do some background work on it?' There was no follow up, which Berry felt was because 'the guy I was speaking to realised I was an expatriate'.[29] Despite this, BRALUP was involved in the policies of *ujamaa* in a number of ways, in the following decade. One of its six major areas of inquiry was listed in 1973 as 'Land use and village level planning' focusing on 'developmental planning of villages' and devising 'criteria for success or failure in new village settlements'; their studies in the West Lake region were used directly in 'ujamaa (collective) village planning in the area' (Berry and Kates 1973, p. 268; see also Berry and Kates 1970).[30] One BRALUP staff member 'worked very closely with officials in his region and played an unobstructive but significant part in the approach to village problems in that area' (Berry and Kates 1973, p. 270). Bureau staff were seconded to the Ministry of Planning and Development to try to coordinate the various regional plans being produced by a range of international consultants and BRALUP also won a contract to produce a regional development plan for the Tanzanian government in Rukwa, 1975–1979 (Belshaw 1982).[31] Mascarenhas recalls that BRALUP was lent the Presidential plane for the Rukwa survey, leading Tanzanian security services to report with alarm that BRALUP 'would know where all the bases of the Freedom Fighters for Southern Africa were located' (Tanzania was at that time hosting military cadres from Mozambique, Rhodesia, and South Africa).[32]

Despite *ujamaa* placing self-reliance central, implementation of the new villages policy involved not only the Tanzanian government, but also experts from international organisations like the World Bank as well as the development programmes of individual governments, consultancy firms, and charities such as Oxfam. There was increasing external funding and influences as the period progressed and Tanzania became increasingly reliant on foreign donors (Rugumamu 1997). BRALUP hosted development researchers funded by Northern governments' development aid, who generally worked at the Bureau on short two-to-three-year contracts and made up the bulk of its staff. It also collaborated with development agencies, most notably the Swedish SIDA, with whom BRALUP worked to develop and evaluate the provision of water in rural settlements, as part of the wider resettlement project.[33]

An early evaluation of village resettlement from BRALUP: 'Twelve new settlements in Tanzania, a comparative study of success' was presented in 1969 (published 1970), examining a range of villages formed in response to colonial development schemes, early capital intensive transformation schemes lead by the World Bank, and the initiation of *ujamaa* villages (though the research was conducted before Operation *Vijiji* which introduced compulsion into the policy) (Kates et al. 1970, p. 73–75). Drawing on an intensive household survey (conducted by undergraduate geography students, 'an essential part of the research effort'), interviews with village leaders, minutes of meetings and government records, the study looked to assess quantitatively the success of different types of village settlement and draw lessons for future policy. The report responded in part to 'specific governmental requests', for example for 'a base-line survey' of one of the new *ujamaa* villages recently founded, so can be seen to be directly feeding into policy making (Kates et al. 1970, p. 64, footnote 1). The authors suggested that the settlement schemes had had some limited success, when evaluated in relation to the government's stated aims of increasing standards of living, opportunities for communal life, increasing production, and therefore contributing to national wealth, practising the values of 'hardwork, self-reliance, socialist cooperation, democratic participation', and providing an example to the surrounding areas (Kates et al. 1970, p. 63).

The study showed that the capital-intensive schemes, primarily those initiated by colonial development teams in the late 1950s and the World Bank around the time of independence were the least successful, and noted that their conclusions suggested that 'the prescription for success sounds remarkably like the prescription for Ujamaa Vijijini … – encourage, more or less spontaneously, communal, low capital, settlement on fertile land with youthful enthusiastic members' (Kates et al. 1970, p. 91). They tempered their conclusions however, reiterating that the villages' 'record of achievement by their own objectives (40%) is not impressive', and suggesting that the qualities that were most likely to lead to successful settlements – including 'local initiative' and 'self-direction' – were 'easily destroyed by institutionalising them' (Kates et al. 1970, pp. 93, 95). The paper also cautioned that 'Settlements where exhortation or deportation has been a major factor have a half-life measured in days', suggesting that not all the lessons of the study were taken on board by policy makers who introduced coercion into the scheme a few years later (Kates et al. 1970, p. 83).

BRALUP was also involved, along with the Geography Department as a whole, with the cross-department Decentralisation research project at the University of Dar es Salaam (Baker 1980). This aimed to evaluate the impact of 1972 reforms of planning and regional development which devolved manpower, decision-making, and budgets to a local level (Belshaw 1982). Maro (1990, pp. 680, 690) noted that there was a 'mutual interrelationship between villagization and decentralisation': the reforms allowed for the acceleration of the resettlement project, increased availability of social services and contributed to the aims of 'spatial equity' that resettlement hoped to achieve. Research conducted in 1976 by Dar geographers Paul Maro and Wilfred Mlay (1979) aimed to assess this decentralisation development policy, contrasting it with the 'capitalist models of private enterprise and "growth pole" theories' which had, on top of colonial era legacies, 'led to lop-sided development' with dominant primate cities and poor rural areas (p. 291, see also Santos 1977). Their quantitative analysis, which drew upon field surveys and government statistics suggested that the provision of piped water, health facilities and primary education had expanded substantially between 1970 and 1976 (though government statistics were notoriously prone to inflation, see Lal 2015). Research published a decade later and drawing on the same project, plus a follow up study from 1984–1985 (Maro 1990, p. 690) drew more nuanced conclusions; decentralisation and villagisation had increased the availability and access to services, and contributed to reducing regional inequality, but when moving beyond the numbers, 'the narrative reports indicated that most villagers were dissatisfied because social services suffered from inadequate supplies of essential materials, shortage of staff, and lack of maintenance.'

### *Between Commitment and Critique*

The creation of *ujamaa* villages has most often been narrated as a catastrophic failure, leading to disruption to agricultural production, food shortages, and famine, and contributing to the economic dependency and indebtedness which the country faced from the mid-1970s onwards (e.g. Scott 1998). However, Lal (2015, p. 11) reminds us that 'at the level of individual lives and livelihoods, *ujamaa* bought successes as well as failures, opportunities as well as hardships'. 'Articulated as a utopian vision, *ujamaa* staged a sharp break with existing approaches to national development by positioning itself as open-ended, improvisational, and experimental' (Lal 2015, p. 164). Geographers, both Tanzanian and foreign, were excited by this vision, and keen to contribute to it. However,

> when filtered through the matrix of Tanzanian government and translated into policy … *ujamaa* accrued blueprints, deadlines, and sometimes rigid developmental scripts. Official policy relied on planning at both the national level, in the form of the cyclical generation of five-year plans, and the local level, in the form of detailed maps for *ujamaa* villages among other things. (Lal 2015, p. 164)

Geographers' contributions sometimes amounted to these mundane planning documents and evaluations which both tied down the improvisational and open-ended visions of *ujamaa* and tried to animate them with useful evidence to help them succeed. These tensions, alongside those of supporting and collaborating with the government whilst also providing meaningful and critical evaluations of government policy, can be seen in the work of BRALUP in its first decade and in the research of its Director, Adolfo Mascarenhas. As the last chapter highlighted, Mascarenhas produced nuanced research exploring the tensions of rural development throughout his career. Yet he was also committed to the vision of development outlined by Nyerere, and close to him personally. It is worth returning to one article by Mascarenhas first discussed in the last chapter, as it encapsulates these tensions between commitment and critique. The article, from 1977, focuses on the Ugogo region in central Tanzania and the experiences of one man, Chimwenda and his family:

> Following the drought of 1969, the government announced that a major resettlement operation would take place and people would have to live in planned villages. One morning in 1970 a lorry stopped by the roadside about a kilometer away from Chimwenda's home and he and his family were moved to an Ujamaa Village. As days passed many other families joined Chimwenda, and a series of major changes in Wagogo life and livelihood were initiated ... (Mascarenhas 1977 p, 9. 378)

Whilst there were improvements – especially in the provision of medical care and wells for water – there were major issues which the article outlines in detail. For example, new housing structures didn't allow for growing melons or gourds or provide separate quarters for each wife. There was no space for cattle, new crops were introduced that didn't grow well, and the conditions for socialising that 'reaffirmed kith and kin' were no longer present. Soil erosion and desertification were substantial problems and women faced long walks to collect wood (Mascarenhas 1977, p. 379). The narrative and focus on one family leave little space for doubt as to the concerns of the author. Nevertheless, whilst the situation for the rural inhabitants of Ugogo was clearly devastating, the article reiterated that 'these problems are real, but they arise from a program undertaken with the highest and best of motivations and with much support from the populace.' (Mascarenhas, 1977, p. 379). Moreover, the government was hemmed in by its 'dependent position in the global economic system' (Mascarenhas, 1977, p. 380). The article's nuanced and ambivalent conclusions capture the position of Mascarenhas and many of his colleagues researching rural development in Tanzania:

> Integration of the perceptions of both decision makers and traditional peoples is clearly called for, but this takes time and is often a tension-filled process .... Much of the social change of the world gives only lip service to the fate of those affected. In the Tanzanian case, change in Ugogo is truly directed towards improving the lot of its rural inhabitants. Nevertheless, change has its price and great care and sensitivity is needed ... (Mascarenhas, 1977, p. 380)

The Geography Department at Dar es Salaam, and its research institute BRALUP were designed from the outset to support development in independent Tanzania. Staff members were compelled, emotionally, ideologically and by the prevailing political climate to produce research that would inform and assist in Nyerere's *ujamaa* development plans. These plans aimed to secure Tanzania's political and economic independence through self-reliance, and to lift the population out of the poverty and inequality left by colonialism. In this way, the applied work of geographers at Dar could be understood as undergirding a holistic project of decolonisation. However, at the same time, the methods used and the projects proposed reinforced visions of development and resource utilisation that have been criticised as western and extractive, and often drew on western funding and expertise, on which BRALUP was completely reliant in the early years.

## Regional Development Planning and New Urban Spaces in Nigeria

### *Rebalancing After War*

Nigeria, like Tanzania, was, from the late 1960s attempting dramatic transformations of the country. Nigeria looked to spatial planning in order to reduce regional inequalities, balance development and importantly, lead reconstruction and enhance national unity after the devastating civil war (1967–1970). The war wrought huge damage to some universities, like The University of Nigeria, Nsukka, where university buildings were damaged beyond repair by fire and the library 'ransacked' (Livsey 2017, p. 170). At Ibadan, whilst there was no physical damage, across the university many Igbo staff and students left, and some students and faculty members, including in the geography department, suffered huge personal losses in the ensuing violence (Bird 2011). In 1970, as the war drew to a close, the Federal Government produced its second national development plan, and, as in Tanzania, Akin Mabogunje remembers that 'the government was asking universities that every subject showed its relevance to national development'.[34] Mabogunje was critical of the 1970 plan 'because of its non-regional orientation', and, through connections with the Economics Department, was able to get into 'the circles of the people concerned with the planning of the country' to argue for a more spatial focus for development policy.[35] This was part of a wider trend, what Hougham (1981, p. 527) called a 'dramatic surge in spatial planning' in national development plans across Africa in the 1960s and 1970s. This replaced the solely sectoral planning which located projects only on the basis of what was best for that specific industry. Instead, spatial planning focused on projects within a wider spatial system, and in Africa, often aimed for 'the furthering of social justice (meaning, usually, a more egalitarian distribution of the benefits of modernization) and the fostering of national unity' (Hougham 1981, p. 528).

Whilst Nigeria was not following the alternative socialist development path of Tanzania, a focus on spatial equity through regional planning was still central, and this gave geography an obvious role. The British geographer Debby Potts, whose PhD focused on African spatial planning reflected:

> regional planning, it was all the rage in independent Africa, before the neo-liberal era, it was all the rage, a way of redistributing resources and wealth and opportunity across space, it was great stuff for geographers, I mean you really felt that you had a role in development.[36]

At Ibadan, the Faculty of Social Sciences had been founded in 1960 (the year of Nigerian independence) 'to provide high level manpower for the needs of the various governments of the federation' (Areola 1998, p. 55); and in 1967 geography moved from the Faculty of Arts to join it, a reflection of Mabogunje's vision for the discipline. In the late 1960s PhDs awarded at the University of Ibadan increasingly moved from those of regional geography (e.g. Agboola, 1963), towards quantitative studies dominated by the language of growth poles, diffusion, and central places (e.g. Akinbode 1973; Kumuyi 1973). Mabogunje published a report in 1971 for the UN Institute for Social Development entitled *Growth poles and growth centres in regional development in Nigeria*. Physical geography also placed 'a new emphasis on development potential' stressing areas such as 'water resources, river basin development, urban climate and land capability studies' (Mabogunje 1998, p. 15).

Geography at the University of Ibadan, working with the Economics Department, established a Planning Studies Programme in response to the demands for relevance from the National Universities Commission in the early 1970s (Mabogunje 1998). 'The basic idea behind the establishment of the programme was the need to ensure that the two categories of planners, that is macro-economic development planners (mainly economists) and the urban and regional development planners (mainly geographers) had a degree of common core training' (Mabogunje 1998, p. 13). The programme also had an 'outreach and consultancy section' and was contracted by the Federal Government in 1972 to provide short courses for planning officials from the Federal ministries and the regional states, a programme supported by the World Bank (Mabogunje 1998). Whilst the Programme was never very successful at securing students (Areola 1998), it highlights the vision for geography as a driver of national development held by Mabogunje and other senior department members.

### Akin Mabogunje and the Geographer as Consultant

The Planning Studies Programme at Ibadan was 'a prototype of attempts to encourage university institutions to earn additional resources through consultancy' and reflected Mabogunje's wider interest in working beyond the university (Mabogunje 1998, p. 14). An early scheme he was involved with was the Kainji dam, initially planned under the colonial government, but built 1964–1969. The dam was a large modernisation project involving huge infrastructural developments and top-down planning, providing opportunities to develop the nation through electrification. At a regional level, it was meant to provide more 'basic public utilities', in newly planned, central places amalgamating smaller villages into larger settlements (Adeniyi 1976, p. 237). Mabogunje, working with the Nigerian Institute for Social and Economic Research (N.I.S.E.R.), was tasked with a pre-inundation survey 'of socio-economic conditions with a view to identifying possible bottlenecks which might hamper the process of rehabilitation and development in the area after inundation.' (Terms of reference, N.I.S.E.R., quoted in Mabogunje 2011, p. 390). Whilst the idea was to carefully plan the resettlement in consultation with the local communities and after the report of the research team, the pressure to build the dam meant that 'it was not possible to delay resettlement until the necessary research activities were concluded.' (Mabogunje 2011, p. 392). An initial cash compensation scheme proved disastrous, and the British modernist architects Fry, Drew, and Atkinson (who had designed the University of Ibadan, including the N.I.S.E.R. building located on the University of Ibadan campus, Figure 5.2) were called in to provide replacement housing. Atkinson collaborated with Mabogunje's research team in designing the four housing types to be provided: 'urban', 'semi-urban', 'semi-rural' and 'rural'.

Mabogunje's autobiography gives a vivid description of what he describes as the 'bizarre' habits of one of the groups living in the area (the Kamberi, who were at that time pagan, and rubbed charcoal onto their skin) (2011, p. 392), and the shock of a Nigerian colleague in encountering them during reconnaissance. This demonstrates the educational and cultural distance between Mabogunje and the communities he was working in, despite them sharing a nationality. He also records that the research team, present in the area as the dam was constructed, couldn't help but notice the 'destitution', 'dislocation and disruption' caused in the lives of many of the villagers (Mabogunje 2011,

**Figure 5.2** N.I.S.E.R. building, designed by Fry, Drew, and Partners. RIBA Collections.

pp. 394–350). Mabogunje edited a volume exploring the project's socio-economic impacts published by N.I.S.E.R. in 1973. It highlighted many of these problems, including large-scale displacement and inundation of agricultural lands, and especially a large impact on fishing communities as their boats and transport networks would no longer be effective on the large lake that replaced the river (see also Adeniyi 1976).

Mabogunje's experiences with the Kainji Dam, and his later reflections on this work, highlight a number of pertinent themes running through this chapter. First, they demonstrate a fundamental belief in the necessity of modernisation, to bring electricity to the modern nation. The dam's modern design graced one of the first banknotes issued after the introduction of the naira in 1973, highlighting its centrality to the iconography as well as infrastructure of the new nation. Mabogunje's work with the famous modernist architects of Fry, Drew, and Atkinson in designing new homes for these communities reflected in architectural style this wider drive. Second, and related, they demonstrate a view of the local communities affected as shockingly different to the modern world of the Nigerian university. This perceived gulf was of great concern to many critiquing African universities in this period as an ivory tower, divorced from the cultures and needs of society, and especially from the rural areas (Mazrui 1978). Mabogunje himself was urban through and through, born in Kano and moving to Ibadan for schooling (Mabogunje 2011). Third, they highlight the priorities of government and development agencies in pushing forward with development schemes regardless of researcher and community concerns, and the relative powerlessness of even well-connected high-profile researchers in these contexts. Fourth, Mabogunje's writings about the dam – both his academic evaluations published at the time and his autobiographical reflections published half a century later – highlight a sensitivity to processes of social change and the devastation that could take place at a local level as a result of development projects.

Mabogunje drew on his early experience at Kainji when he became involved in the project to establish a new national capital in 1976 (Figure 5.3). He was tasked with bringing together a team to report on the physical geography of the area, and 'to provide a census of human settlement, their population, buildings, economic trees, farm lands and other immovable assets' in preparation for resettlement and compensation.[37] The new capital Abuja was to act as a growth pole for a poorer and less developed area of Nigeria, drawing on modern regional planning ideas as well as providing a geographically central capital replacing Lagos on the south coast (Darwent 1969; Salau 1977). Sylvester Abumere, another Ibadan geographer also heavily involved in the project underscored the poverty of the region prior to the development, noting that the 8000 square kilometres of the Federal Capital Territory had no secondary school or tarred road in 1976 (1998, p. 398). Abumere described the project in the language of regional development planning:

**Figure 5.3** Akin Mabogunje preparing for fieldwork, 1977. Courtesy of Akin Mabogunje.

> concepts of centrality, neighbourhoods, accessibility and linkages were important in the planning of the Federal Capital City… [in the wider region], central place concepts of lower and higher order centres, growth poles and growth centres, diffusion and linkages, have played roles in the planning. So also has been the geographers' concern for the distribution of public facilities that is both optimal and equitable. (Abumere 1998, p. 387)

Mabogunje and his colleagues were working on the plan for Abuja in the same year that Milton Santos (1977, p. 86), drawing on his experiences at Dar es Salaam, was publishing his critique of theories of central places and growth poles as worsening under-development and inequality by allowing the 'rapid and brutal intrusion of capital in underdeveloped countries' in *Antipode* (see Chapter 4). Geography for development was a contested enterprise in 1970s Africa.

The development work at Abuja involved Mabogunje and Abumere alongside many other department members, providing an important and enjoyable professional opportunity (Abumere 1998). According to Areola, 'everybody was excited, and it was virgin territory at that time'.[38] Involvement could also be financially rewarding, and for several of the human geographers, involved detailed work over more than two years which frequently took team members out of the classroom at Ibadan, much to the frustration of some students (Mabogunje 2011).

The demands of balancing this project with academic work, combined with the problems of the university, lead Mabogunje to leave academia in 1981 and set up his own consultancy business, alongside his former colleague, the economist Ojetunji Aboyade. Pai Associates International carried out various contracts including for the Federal Government, state governments, and the Nigerian Industrial Development Bank (Mabogunje 2011). They were involved in various projects, including preparing a geo-coding system for the major urban areas of Oyo state (Mabogunje 2011, p. 221). Aboyade and Mabogunje also worked closely with the military government of Babangida in the mid-1980s, as the country braced for the impact of structural adjustment. Mabogunje became Deputy Chair of the new Directorate of Food, Roads and Rural Infrastructure established in 1986, which aimed to bring about community led rural development to mitigate the worst impacts of currency devaluation on food prices and supplement the more top-down World Bank led Agricultural Development Programme.

Mabogunje was far from alone in holding consultancy roles, with many African geographers doing work for national and local governments, as well as international organisations such as UNESCO, the World Bank, UNEP, UNDP, and Habitat International alongside their academic positions. The Ghanaian geographer George Benneh acted as a consultant for the Food and Agriculture Organization providing expertise on agriculture and population. This was a well-paid role which allowed him to build his new house on a plot of land close to the university. 'Money was still a problem' in the early 1980s, he recalled in his autobiography, so the building required a bank loan. When asked if he did any consultancies by the bank chairman, Benneh was able answer in the affirmative and was therefore eligible for a loan, despite having no collateral. Each time he did some consultancy work he would pay back the loan and take out another (Benneh 2016, p. 54–55). The fact that the bank manager asked about consultancies highlights their ubiquity amongst Ghanaian academics at the time.

Western academics with experience working in Africa were also able to secure sometimes lucrative consultancy work with similar organisations. The British geographer Mansell Prothero who had worked at the University of Ibadan 1950–1954, was seconded to the World Health Organization as a consultant in 1960, and continued to work as a consultant for them and other international organisations over the course of his career.[39] When Len Berry left Tanzania in 1971 for a new role at Clark University in the United States, it didn't take long for him to be enrolled in the networks of USAID, UNDP, and the World Bank, providing expertise in desertification, environmental issues, and East African development. 'I at one point was a consultant to 8 or 9 of those agencies, which was kind of crazy' he remembered.[40] The next two chapters explore some of the consequences of this consultancy trend in Africa.

## Conclusion

Between 1930 and 1970, 'African agrarian societies became objects of both state intervention and expert knowledge' in the name of development. These schemes also 'played an important part in the construction of the African state' in both the colonial and postcolonial periods (Bonneuil 2000, p. 258). Geographers – as academics and consultants – were heavily involved in the project of national development in postcolonial Africa, alongside colleagues from a whole range of disciplines. They worked within and for a range of different political ideologies. In each context these development projects had different, complex and often ambiguous outcomes, with legacies that have outlasted both the political regimes that implemented them and the wider visions for development that dominated the 1950s–1970s.

Geographers sometimes utilised models, theories and techniques imported from the West, but developed them within the diverse frameworks of individual African states and their particular political systems. The transformations sought by postcolonial states produced unique opportunities for the utilisation of geographical knowledge and African geographers took these ideas in new directions, just as the development policies of African states were more than just the importation of western modernisation projects or the continuation of colonial-era policies (Lal 2015). African geographers became involved in domestic policy making and development projects much more commonly than did their counterparts in Europe or North America. In 1973 an American and British geographer working in Tanzania reflected together on 'the vast potential of the now simmering applied revolution in geography' that had become clear during their time in Africa (Berry and Kates 1973). Applied geography had a larger and longer lasting impact and legacy in Africa than elsewhere as a consequence.

The chapter has highlighted the excitement and optimism of the 1960s period, with these affective atmospheres galvanising the research, teaching, consultancy, and other work of both African scholars and many of their foreign counterparts. The intersections between the independence of great swathes of Africa between 1957–1965 and the rise of modern urban and regional planning provided a potent mix, and additional impetus – ethical, political, and financial – for involvement in development projects. Geography as a discipline was very much enrolled in projects of black freedom, but aligned to what Al-Bulushi (2022b, p. 259) describes as 'statist conceptions of liberation', rather than the non-sovereignty based visions of freedom imagined by many in the black radical tradition.

Despite the excitement of working in post-colonial state-building, the experiences of geographers in development projects demonstrate the ambivalence, challenges, and complexity of the work on the ground. The often top-down technocratic visions for development that dominated in the 1950s and 1960s, and which were accompanied by a shift to quantitative model-based geography, were often blind to the human and environmental impacts of development. Even when geographical assessments provided more nuanced or critical evaluations of development projects, such as in Nigeria and Tanzania, they were often ignored, or overtaken by events, such as political instability or changing international priorities.

African geographers should not be seen as any more responsible for the deficits in the development project than their counterparts globally. Indeed, the failures of these projects were in many ways driven by the demands and political economic structures through which the West remained dominant even after constitutional independence. African geographers were no less critically engaged than their counterparts elsewhere. Nevertheless, the political and social contexts – both nationally and internationally – also narrowed the opportunities for more critical scholarship, in ways that would have lasting consequences for the discipline. The continued requirement to contribute to development and be relevant left little space for geographers to engage with the more critical development work which emerged, primarily in Tanzania, in this period, exacerbating problems of intellectual extraversion (Hountondji 1990) and stifling critical African geographical accounts of development policy. Writing in 1977 soon after returning from Tanzania, the British geographer David Slater was to reflect on 'resource utilization' – a key area in which geographers working in Africa, including his colleagues in Tanzania at BRALUP, were contributing:

> when analysing resources, I would argue that we need to ask the questions: resource utilization for what and for who? The politics of resource questions are often ignored, but in fact they are of overriding significance. (Slater 1977b, p. 70)

These questions were at the heart of project of equitable development but were sometimes squeezed out in the technical applied work which came to dominate geography curricula, methods, techniques and consultancy work in Africa, and in much of development geography and regional planning globally. Ake (1982) argues that the fundamentally Western and capitalist social science dominant in Africa was conservative – looking to uphold the status quo over revolutionary change – and confident – that technical interventions could deal with the problems of development. African geography reflected these wider trends.

Fred Cooper (2008) has described the experiences of the postcolonial African state as a mix of 'possibility and constraint' and it is useful to consider development projects, and geographers' roles in these in similar terms. Rather than narrating a failure of development in this era, it is more useful to explore processes, opportunities and closures:

> What gets lost in narrating history as the triumph of freedom followed by failure to use that freedom is a sense of process. If we can, from our present-day vantage point, put ourselves in the position of different historical actors, in 1958 - or 1945,

1966 or 1994 - we see moments of divergent possibilities, or different configurations of power, that open up and shut down. (Cooper 2008, p. 169)

The economic crashes of the early 1970s, followed by structural adjustment and neoliberal reforms of the 1980s, brought a huge reduction in national budgets and the international development aid available to African countries for largescale state-led development projects. The geography of economic investment was increasingly led by the demands of capital rather than any concern to address uneven spatial development. Reflecting on the demise of regional planning since its 1970s heyday, Debby Potts noted: 'it all went down the [pan], I mean it was completely destroyed [with structural adjustment and neoliberal reforms].'[41] More broadly, the 1980s brought an end to the optimism of the early postcolonial period, that the future would bring social development and prosperity to African citizens in newly independent countries (Ferguson 1999).

Lal argues that the challenges and inconsistencies of Tanzanian development were the result of 'the dialectical friction at the heart of processes of state formation, socialism, and national development across postcolonial contexts – what we might call the tensions of national development' (2015, p.11). We can usefully use this phrase to think about the role of geographers contributing to the project of national development across Africa: navigating the relationships between scholarship, consultancy, career development, domestic political ideologies, and international schemes. Academic lives provide a useful lens through which to explore these tensions of decolonisation and development. Cooper asks: 'Can we build networks of piped water and electricity and work out an epistemological critique of "development" at the same time?' (2008, p. 186) and in doing so encapsulates much of the challenge facing African geographers in this period, attempting applied work which would both assist and critically assess development.

## Notes

1 Interview with Akin Mabogunje, London, UK, 7 July 2016.
2 United Nations Department of Economic and Social Affairs (1962) *The United Nations Development Decade: Proposal for Action*. Report of the Secretary General. United Nations.
3 Interview with Elizabeth Ardayfio-Schandorf, Legon, 20 January 2017.
4 Interview with Akin Mabogunje.
5 Interview with Akin Mabogunje.
6 Interview with Akin Mabogunje.
7 E. Boateng to Pro-Vice-Chancellor Wright, 26 February 1962, University of Ghana Archives, UGI/3/2/3/56/89.
8 Unpublished dissertation accessed in the Department of Geography Library and Resources Room at the University of Ghana. 'Establish fund for childhood immunisation – Dr Joan Awunyo-Akaba', Modern Ghana, 26 October 2013. https://www.

modernghana.com/lifestyle/5183/establish-fund-for-childhood-immunisation-dr.html [last accessed 4 January 2022].

9 John Iliffe, personal correspondence, May 2021. Iliffe notes that 50 years have passed since these events so his memories may not be entirely accurate.

10 Personal correspondence with Adolfo Mascarenhas, May 2021.

11 Interview with a British geographer who worked at Dar es Salaam in the 1970s, University of Glasgow, UK, 17 August 2012. Interview conducted by Professor Joanne Sharp.

12 Interview with a British geographer who worked at Dar es Salaam in the 1970s, University of Glasgow. Similar points were made in a further interview with a British geographer who worked at Dar es Salaam in the 1970s, London, UK, 28 March 2013. Both interviews conducted by Professor Joanne Sharp.

13 Census night, anonymous poem in the *Evening News*, 20 March 1960 quoted in Serra 2018 p. 659.

14 American Association of Geographers Memorials and Tributes: Distinguished Professor Emeritus John M. Hunter. Available at: http://www.aag.org/cs/membership/tributes_memorials/gl/hunter_john_m [last accessed 6 January 2022].

15 Interview with Akin Mabogunje, Ibadan, 16 July 2017.

16 Interview with Len Berry, via Zoom, 26 February 2021.

17 Interview with Len Berry.

18 Interview with Len Berry.

19 Interview with Len Berry.

20 Interview with Len Berry.

21 Interview with Len Berry, via Zoom, 5 March 2021.

22 Interview with Len Berry.

23 BRALUP Review, 1979, courtesy of Len Berry.

24 Personal correspondence with Adolfo Mascarenhas.

25 Interview with Len Berry.

26 Personal correspondence with Adolfo Mascarenhas.

27 Interview with Len Berry via Zoom, 26 February 2021.

28 BRALUP Review, 1979, pp. 3–5. Courtesy of Len Berry.

29 Interview with Len Berry.

30 See for example: Research Report 10 Ujamaa: problems of implementation, with special reference to the West Lake Region.

31 Personal correspondence with Adolfo Mascarenhas.

32 Personal correspondence with Adolfo Mascarenhas.

33 Interview with Len Berry, via Zoom, 26 February 2021.

34 Interview with Akin Mabogunje, Ibadan, 16 July 2017.

35 Interview with Akin Mabogunje.

36 Interview with Debby Potts, London, October 2018.

37 Commissioning letter, 21 April 1976, quoted in Mabogunje 2011 p. 451.

38 Interview with Olusegun Areola, Ibadan, Nigeria 9 July 2017.

39 Obituary, Mansell Prothero, 1924–2013. Royal Geographical Society. Available at: https://www.rgs.org/geography/news/r-mansell-prothero-(1924–2013) [last accessed 6 January 2022].

40 Interview with Len Berry, via Zoom, 12 March 2021.

41 Interview with Debby Potts.

## References


Abumere, S.I. (1998). Abuja: planning and development of the new federal capital territory. In: *50 Years of Geography in Nigeria: The Ibadan Story* (ed. O. Areola and S.I. Okafor), 387–410. Ibadan: Ibadan University Press.

Adeniyi, E.O. (1976). The Kainji Dam: an exercise in regional development planning. *Regional Studies* 10 (2): 233–243.

Agboola, S.A (1963). *Some Geographical Influences upon the Population and Economy of the Middle Belt West of the Niger*. Unpublished PhD Thesis. University of Ibadan.

Ahlman, J.S. (2017). *Living with Nkrumahism: Nation, State, and Pan-Africanism in Ghana*. Athens, Ohio: Ohio University Press.

Ahonsi, B.A. (1988). Deliberate falsification and census data in Nigeria. *African Affairs* 87 (349): 553–562.

Ajayi, J.F.A. (2000). The national question in Nigeria in historical perspective. In: *Tradition and Change in Africa: The Essays of JF Ade Ajayi* (ed. T. Falola), 217–241. Trenton: African World Press.

Ajayi, J.F.A., Lameck, G., and Johnson, G.A. (1996). *The African Experience with Higher Education*. Athens: Ohio University Press.

Ake, C. (1982). *Social Science as Imperialism: The Theory of Political Development*, 2e. Ibadan: Ibadan University Press.

Akinbode, A. (1973). *The location of kolanut production in Southwestern Nigeria: A case study of the spatial diffusion of an agricultural innovation*. Unpublished PhD Thesis. University of Ibadan.

Al-Bulushi, Y. (2022b). Thinking racial capitalism and black radicalism from Africa: an intellectual geography of Cedric Robinson's world-system. *Geoforum* 132: 252–262. https://doi.org/10.1016/j.geoforum.2020.01.018.

Areola, O. (1998). Postgraduate training at Ibadan. In: *50 Years of Geography in Nigeria: The Ibadan Story* (ed. O. Areola and S.I. Okafor), 47–74. Ibadan: Ibadan University Press.

Baker, S.J.K. (1980). Geography in Tanzania. *The Geographical Journal* 149 (2): 470–473.

Belshaw, D.G.R. (1982). An evaluation of foreign planning assistance to Tanzania's decentralized Regional Planning Programme, 1972–81. *Applied Geography* 2: 291–302.

Benneh, G. (2016). *My Time, My Nation*. Accra: Winmat Publishers Limited.

Berry, L. (1970). Research in BRALUP [Bureau of resource assessment and land use planning], University of Dar-es-Salaam. *Rural Africana* 13: 67–74.

Berry, L. and Kates, R.W. (1970). *A study of four villages in dodoma and Singida regions*. University College of Dar es Salaam. Bureau of Resource Development and Land Use Planning Research Paper No. 10. University of Dar es Salaam.

Berry, L. and Kates, R.W. (1973). Applied geographical research in East Africa. *The Professional Geographer* 25: 267–271.

Bird, L. (2011, October 17). *A final interview and plans for next year*. Retrieved from: https://asabamemorial.wordpress.com/2011/10/17/a-final-interview-and-plans-for-next-year (last accessed 6 January 2022).

Bonneuil, C. (2000). Development as experiment: science and state building in late colonial and postcolonial Africa, 1930–1970. *Osiris* 15 (1): 258–281. https://doi.org/10.1086/649330.

Briggs, J. and Gray, A. (1982). Geography in a developing country: the case of the University of Dar es Salaam, Tanzania. *Journal of Geography in Higher Education* 6 (1): 39–46. https://doi.org/10.1080/03098268208708832.

Cook, A. (1979). Resource assessment in East Africa. *Geography* 64 (2): 96–103.

Cooper, F. (2008). Possibility and constraint: African independence in historical perspective. *The Journal of African History* 49 (2): 167–19. https://doi.org/10.1017/S0021853708003915.

Darkoh, M.B.K. (1981). Geography and society. *Journal of Eastern African Research & Development* 11: 32–61.

Darwent, D.F. (1969). Growth poles and growth centers in regional planning—a review. *Environment and Planning A* 1 (1): 5–31.

Doherty, J. (1982). Geography, education and ideology in Tanzania. *Journal of Geography in Higher Education* 6 (2): 173–176. https://doi.org/10.1080/03098268208708858.

Dorrnkamp, J.C. (1986). Professor Paul Harrison Temple, 1936–1985. *Transactions of the Institute of British Geographers* 11 (2): 243–246.

Ferguson, J. (1999). *Expectations of Modernity*. Berkeley: University of California Press.

Geramisov, I.P. (1962). Soviet geographic science and problems of the transformation of nature. *Soviet Geography* 3 (1): 27–38. https://doi.org/10.1080/00385417.1962.10769900.

Ginelli, Z. (2018). Historical geographies of the "quantitative revolution": towards a transnational history of central place theory. *International Conference of Historical Geographers Conference*, Warsaw.

Gould, P.R. (1960). The development of the transportation pattern in Ghana. *Northwestern University Studies in Geography* (number 5): Evanston: Northwestern University.

Hougham, D. (1981). The Spatial Components of Anglo-African development plans, 1943–1956. *Canadian Journal of African Studies/La Revue Canadienne Des Études Africaines* 15 (3): 527–538.

Hountondji, P. (1990). Dependence in Africa today. *Research in African Literatures* 21 (3): 5–15.

Hunter, J.M. (1961). An exercise in applied geography. geographical planning in urban areas for the 1960 census of Ghana. *Geography* 46 (1): 1–8.

Kates, R.W.J., McKay, J., and Berry, L. (1970). Twelve new settlements in Tanzania: a comparative study of success. In: *Proceedings of the University of East Africa Social Science Conference, 1968–69* (ed. University of East Africa), 63–100. Kampala: University of East Africa.

Kertzer, D.I. and Arel, D. (eds.) (2002). *Census and Identity: The Politics of Race, Ethnicity and Language in National Censuses*. Cambridge: Cambridge University Press.

Kumuyi, A.J. (1973). *Spatial interaction of central places in Kano State, Nigeria*. Unpublished PhD Thesis. University of Ibadan.

Lal, P. (2015). *African Socialism in Postcolonial Tanzania*. Cambridge: Cambridge University Press.

Livsey, T. (2017). *Nigeria's University Age: Reframing Decolonisation and Development*. Basingstoke: Palgrave Macmillan.

Lorenzini, S. (2019). *Global Development: A Cold War History*. Princeton: Princeton University Press.

Mabogunje, A.L. (1968). *Urbanization in Nigeria*. London: University of London Press.

Mabogunje, A.L. (1971). *Growth poles and growth centres in the regional development of Nigeria*. United Nations Research Institute for Social Development (Report No. 17/3).

Mabogunje, A.L. (1998). The history of the department. In: *50 Years of Geography in Nigeria: The Ibadan Story* (ed. O. Areola and S.I. Okafor), 1–18. Ibadan: University of Ibadan Press.

Mabogunje, A.L. (2011). *A Measure of Grace: The Autobiography of Akinlawon Ladipo Mabogunje*. Ibadan: Book Builders.

Mamdani, M. (2019). Decolonising universities. In: *Sharing Knowledge Transforming Societies* (ed. T. Halvorsen, K.S. Orgeret, and R. Krøvel), 48–67. Cape Town: African Minds.

Maro, P.S. (1990). The impact of decentralization on spatial equity and rural development in Tanzania. *World Development* 18 (5): 673–693. https://doi.org/10.1016/0305-750X(90)90017-R.

Maro, P.S. and Mlay, W.F. (1979). Decentralization and the organization of space in Tanzania. *Africa* 49 (3): 291–301.

Mascarenhas, A.C. (1977). Resettlement and desertification: the Wagogo of Dodoma District, Tanzania. *Economic Geography* 53: 376–380.

Mashalla, S.K. (1988). Geography in Tanzania. *The Professional Geographer* 40 (2): 229–232. https://doi.org/10.1111/j.0033-0124.1988.00229.x.

Mazrui, A.A. (1978). *Political Values and the Educated Class in Africa*. Berkeley: University of California Press.

Mbalisi, C.N., Ekwueme, A., and Okwudili, V.-A.,P. (2019). The politics of population census: challenge to democracy and nation building in Nigeria since 1960. *Journal of Nation-Building and Policy Studies* 3 (2): 67–84.

Ndlovu-Gatsheni, S.J. (2017). The emergence and trajectories of struggles for an 'African university': the case of unfinished business of African epistemic decolonisation. *Kronos* 43 (1): 51–77. http://dx.doi.org/10.17159/2309-9585/2017/v43a4.

Nyerere, J. (1966). *Freedom and Unity: Uhuru na Umoja. A Selection from Writings and Speeches 1952–65*. Dar-es-Salaam: Oxford University Press.

Ominde, S.H. (1971a). Geography and African development. In: *Studies in East African Geography and Development* (ed. S.H. Ominde), 1–8. London: Heinemann Educational Books.

Rugumamu, S. (1997). *Lethal Aid: The Illusion of Socialism and Self-Reliance in Tanzania*. Trenton: Africa World Press.

Salau, A.T. (1977). A new capital for Nigeria: planning, problems and prospects. *Africa Today* 24 (4): 11–22.

Santos, M. (1977). Planning underdevelopment. *Antipode* 9: 86–98. https://doi.org/10.1111/j.1467-8330.1977.tb00095.x.

Schumaker, L. (2001). *Africanizing Anthropology*. Duke: Duke University Press.

Scott, J. (1998). *Seeing like a State*. New Haven: Yale University Press.

Serra, G. (2018). "Hail the census night": trust and political imagination in the 1960 population census of Ghana. *Comparative Studies in Society and History* 60 (3): 659–687. https://doi.org/10.1017/S0010417518000221.

Sharp, J. (2019). Practising subalternity: postcolonial Tanzania, the Dar School and pan-African geopolitical imaginations. In: *Subaltern Geographies* (ed. T. Jazeel and S. Legg), 74–93. Athens: University of Georgia Press.

Shaw, D.J. (2015). Mastering nature through science: soviet geographers and the Great Stalin Plan for the Transformation of Nature, 1948–53. *Slavonic & East European Review* 93 (1): 120–146. https://doi.org/10.5699/slaveasteurorev2.93.1.0120.

Slater, D. (1977b). Resources and class structure: notes on an alternative Marxist perspective. *Antipode* 9: 68–70. https://doi.org/10.1111/j.1467-8330.1977.tb00083.x.

Soja, E.W. (1968). *The Geography of Modernization in Kenya: A Spatial Analysis of Social, Economic and Political Change*. Syracuse, NY: Syracuse University Press.

Stamp, L.D. (1960). *Applied Geography: How the Geographer's Survey and Analysis Can Help in Understanding the Britain of Today and in Planning for Its Future*, Vol. 449. Penguin books.

Stanek, M.B. (2019). Decolonial education and geography: beyond the 2017 Royal Geographical Society with the Institute of British Geographers Annual Conference. *Geography Compass* 13: e12472. https://doi.org/10.1111/gec3.12472.

Taaffe, E.J., Morrill, R., and Gould, P.R. (1963). Transport expansion in underdeveloped countries: a comparative analysis. *The Geographical Review* 53: 503–529.

Udo, R.K. (1981). The 1983 population census of Nigeria; ideals and realities. *Nigerian Geographical Journal* 24: 175–185.

Udo, R.K. (1998). Geography and population censuses in Nigeria. In: *50 Years of Geography in Nigeria: The Ibadan Story* (ed. O. Areola and S.I. Okafor), 348–372. Ibadan: Ibadan University Press.

Wilson, G. and Wilson, M. (1945). *The Analysis of Social Change Based on Observations in Central Africa*. Cambridge: Cambridge University Press.

Yesufu, T.M. (ed.) (1973). *Creating the African University: Emerging Issues in the 1970s*. Ibadan: Oxford University Press.

# Chapter Six
# Geography, Apartheid and Anti-Apartheid Activism in South Africa

## Introduction

Calvyn Gilfellan studied geography at the University of the Western Cape (UWC), designated as a 'coloured' institution under apartheid. Whilst special permission could be sought by students of colour to study at a white university, many students and their families were not willing to make such a request. Gilfellan recalled that in the early 1980s 'I had no choice because the other choice was not one to contemplate … my father would never allow me to go and beg FW de Clerk [Education Minister] to study at a white university.'[1] By the time Gilfellan joined UWC as an undergraduate he was coming into a radical space, with a new mission statement issued in 1982, declaring the institution's intent 'to break from the apartheid project' (Wolpe 1995, p. 284). However whilst the university was increasingly home to anti-regime scholarship and activism, the geography department remained deeply conservative, still overseen by the two Afrikaans-speaking Stellenbosch graduates that had founded it in the early 1960s. Gilfellan's experience, as a student and later lecturer in the Geography Department at UWC, and as anti-apartheid activist and community educator, provides evidence for the ways in which geographers, and the discipline itself, were caught up in the struggle against apartheid and the broader colonial legacy in South Africa.

Geographers working in South African universities in the 1980s were part of a segregated system in which institutions were designated for each of the so-called 'racial groups': black (also divided further by language group), coloured, indian,

*Decolonising Geography? Disciplinary Histories and the End of the British Empire in Africa, 1948–1998*, First Edition.
Ruth Craggs and Hannah Neate.

and white. This chapter explores primarily the anti-apartheid and decolonising work of South African geographers like Gilfellan, focusing mainly on the 1980s and 1990s when an increasing number of Black geographers began to get jobs in South African universities, and when geographers began to engage more critically with apartheid within and outside the academy. In this period geographers experienced both the increasing violence of apartheid as it reached crisis point, and signs that transformation – or perhaps decolonisation – might be possible. Whilst the main approaches to apartheid within geography (active support through applied planning focused research, or alternatively ignoring apartheid as a topic for research or critique) are outlined, focus falls especially on anti-regime work. As a result, the chapter focuses for the main part on universities designated for 'non-white' students – institutions that have now become known in South Africa as Historically Black Universities (HBUs), especially UWC and the University of Durban-Westville (UDW) (Figure 6.1). Whilst these universities were

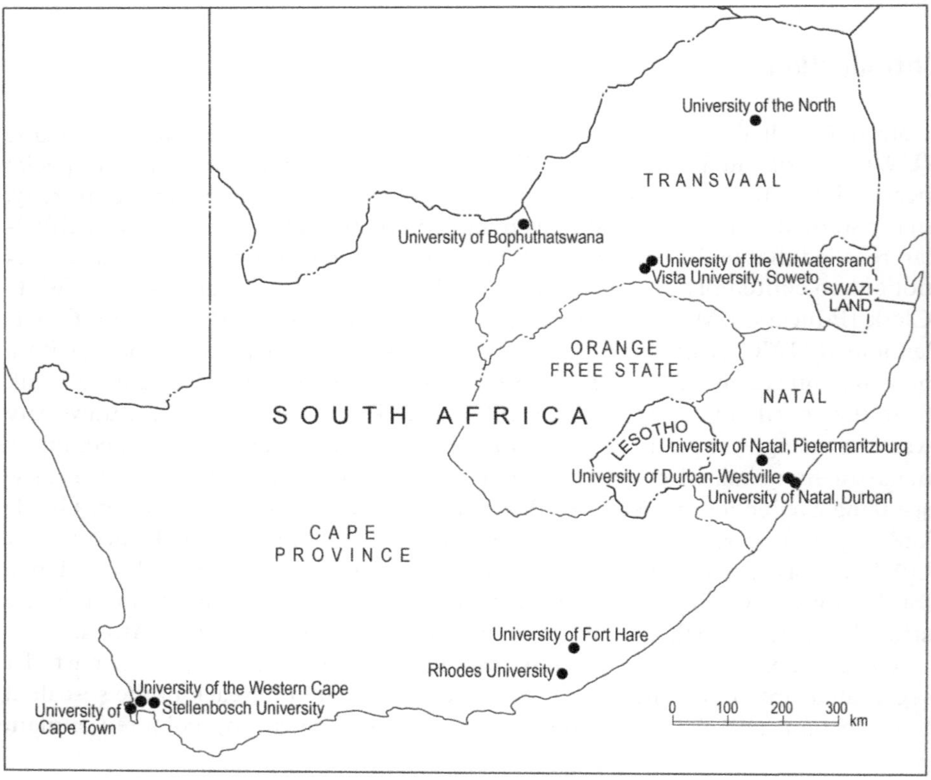


**Figure 6.1** South African universities discussed in Chapter 6 (pre-1994 names). Produced by Philip Stickler.

not representative of country as a whole, the chapter tells an important story of radical geographical work under apartheid.

In this chapter, we use the term Black in its political sense to encompass those classified as 'black' (black African), 'coloured' (mixed race), or 'indian' (of Indian descent) in the apartheid classifications, who were often at the forefront of the fight against apartheid. In places in the chapter, where discussing the particular experiences of certain 'groups', we will utilise apartheid categories. These categories, including 'white', 'are taken to refer to social constructs embedded in in the apartheid system, and not natural subdivisions of humankind in South Africa or anywhere else' (Smith 1992, p. 4).

The next section details the extent to which spaces of geography – from the tea-room to the classroom and the conference session – could act not only to reinforce apartheid but as sites for transgression of apartheid racial categories and the contestation of the regime. Thus the chapter begins with a shift in scale from the big infrastructure projects of the last chapter to the micro-spaces of disciplinary geography. Following this, the chapter explores geography student and staff anti-apartheid activism on and off campus. Then discussion turns to geography's problematic engagement with apartheid, before exploring the growing critiques of the discipline as white and the emergence of new critical geographies in the 1980s, as well as the first mention of the need for decolonisation of the discipline. The final section explores the transformation of geography in the late 1980s and 1990s, highlighting especially the role of Black academics in demanding change to the institutional structures of geography and its publication infrastructure. This section also examines the changing opportunities for geographers as South Africa begun to open up in the wake of the release of Nelson Mandela and the unbanning of the African National Congress (ANC) in 1990.

## Departmental Spaces, Geography and the Contestation of Apartheid

The structures of apartheid had a huge impact on geography staff and students, impacting their choices of subjects and institutions and their experiences when at university. At an institutional level, South African geography was represented by two bodies – The South African Society for Geography, and the South African Geographical Society. Both were overwhelmingly dominated by white academics (the former being Afrikaans speaking and the latter serving the English speaking community) despite being officially open to black scholars from the mid-1970s or earlier.[2]

Alongside separation and differentiation between university systems and in professional bodies, there was also segregation and racism within universities and individual geography departments. At universities with a mixed body of staff, such as at the University of the Western Cape, there were separate staff associations.[3] And beyond these formal divisions, micro-aggressions and explicit racism often characterised departmental working relationships. Interviewees recalled senior colleagues who would not greet them,[4] Black students left out of away days because a secretary 'didn't want to deal with catering for, quote, "them"',[5] and a head of department who was 'not very open to the idea of appointing black staff in his department' but who was protected in his position by white university management until 1999.[6] Brij Maharaj, the first non-white discipline chair at the University of Natal, Pietermaritzburg recalled that when he started in the role in 2000, he was a first mistaken for a student, and his friend commented that 'some of these people think you are here to deliver the tea'.[7] These moments could be degrading and depressing, but also a catalyst for change: In his autobiography, Nelson Mandela attributed his time at the University of Witwatersrand School of Law, where he experienced 'the steady accumulation of a thousand slights, a thousand indignities, a thousand unremembered moments' as producing in him 'an anger, a rebelliousness, a desire to fight the system that imprisoned my people' (quoted in Ajayi et al. 1996, p. 71). In the next section, taking the example of the University of the Western Cape, we explore how, within the formal and informal spaces of geography, apartheid and its colonial ideology was experienced and upheld, but also contested.

## *The Tearoom*

Universities were divided along racial lines, but they still represented spaces for interaction, contestation and occasionally conviviality across different communities. A key site within academic departments for this was the tearoom. Younger 'coloured' lecturers and technicians at the University of the Western Cape described often polite but superficial tea-room chat which underlined the different lives led by those privileged and discriminated against by apartheid. In the context of differentiated wages and property prices, having to 'talk about lawns' sometimes stuck in the throat.[8] Reflecting on this from the present day, Calvyn Gilfellan, by the mid-1980s a lecturer at UWC noted that when it is said that 'people are unreasonable when they articulate demands about apartheid and colonialism, it's the real thing, it beats you permanently behind … that's the tragedy.' He continued:

> I didn't have a car, so I couldn't speak about cars, I didn't have a house so I couldn't speak about houses, I don't drink wine so I couldn't speak about wine … all of those things I couldn't relate to … nobody is conversing with me about my political [activism].

However, this also provided space for such activism alongside an academic career. It meant that

> you could always play a double life. That you could listen to all this shit of these professors about students and whatever, and then go into your community and live out your aspirations for a better South Africa. So you 'eat' a lot of what your colleagues say.[9]

The tearoom was not always a place of polite conversation. Gilfellan recalled that in the UWC tearoom his colleague, the radical geographer Harold Wesso would often 'jump up ... and run out after a heated argument.' At the time, Gilfellan felt that 'Frankly speaking for me at that stage those arguments were not important to me ... Because ... we were too busy [with other colleagues in the struggle].' However, in retrospect he felt that these more mundane spaces could be seen as 'important sites of contestation, negotiation, and mediation. And we giving in and they [giving in]'.[10]

## *Conference Spaces*

Whilst senior academic conferences continued to be very white spaces until the late 1980s, student geography conferences were more mixed from a much earlier period. Keyan Tomaselli, then studying at the 'white' University of Witwatersrand, remembered these meetings provided the opportunity to interact with 'black students from the homeland universities' and learn 'to appreciate their own experiences'.[11] Student conferences also provided the opportunity to challenge conservative and racist scholarship through questioning and debate, particularly as the 1980s progressed. Students and junior staff members (there was often a crossover as those studying for higher degrees also often worked as technicians and junior lecturers) from UWC in particular got a reputation for such challenges, alongside those from Durban-Westville. Calvyn Gilfellan remembers

> they knew our reputation and our philosophies ... the anti-apartheid would always come in. And when we listened to other people's work they would be very nervous if we were in the audience. There was always the expectation that UWC would make ructions at the conferences.[12]

Gilfellan's UWC colleague Michael Dyssel elaborated:

> a typical student conference at Wits [Witwatersrand] would see UWC students coming in, and not sitting together as a clique but positioning ourselves all over the audience floor. And then start interrogating ... it was deliberate ... But it was also fun, just to see to what extent we can rattle some of the students from Stellenbosch University with a collar and a tie when they know that the professor is sitting there ...[13]

Gilfellan remembers one specific event in the early 1990s where he challenged Professor Norman Tait, Head of Department at the University of the North, on his paper about squatter camps and black home ownership. In Gilfellan's words, 'he was basically saying poor people want to live like that. And it can't go unchallenged ... I asked him a question and he was so furious he left the conference ...!'[14] Also in the audience was Ronnie Donaldson, a Stellenbosch graduate and recently appointed lecturer at the University of the North:

> Calvyn stood up and said in Afrikaans, he said, and I'll never forget his words, it was so nicely phrased, he said "dit is 'n oppervlakkige gekrap na niks" in Afrikaans, meaning a superficial scratching of the surface for something of use, or something like that. And my mouth just fell open, here was this guy saying this to the professor. And I thought, ok, this is a nice guy ...[15]

Conference spaces offered opportunities for the performance of expertise (Craggs and Mahony 2014) and the intellectual challenging of apartheid geographies.

## *Teaching Spaces*

Studying or working at a university in South Africa could provide the opportunity to learn about and contest apartheid, beyond the everyday exclusions and inequalities that were ever-present. Keyan Tomaselli (2001, p. 140) argued that 'my time as a [geography] student at the University of Witwatersrand between 1968 and 1973 gave me, for the first time, the conceptual tools to understand what I long remember feeling about apartheid – but was until then unable to articulate.' At some institutions, in South Africa as in Tanzania, it was influences outside geography – from history, politics and sociology for example, that inspired students. For example at the University of Natal, Durban in the 1970s, radical lecturers like Rick Turner (Politics) and Fatima Meer (Sociology) inspired geographers such as Jeff McCarthy and Dianne Scott who took their courses as part of their studies.[16] Turner was assassinated by South African government agents in 1978, bringing the violent realities of apartheid close to campus and influencing his colleagues and students further.

However, many we interviewed were critical of their geographical education which was seen as conservative, outdated, dull, and ignoring the realities of contemporary South Africa.[17] This was especially true of those studying at HBUs in the 1970s and 1980s, where as Priscilla Cunnan put it, 'we often felt like we were being taught by senior (white) staff who didn't make it at other universities'.[18] The HBUs were dominated by Afrikaner staff, many of whom were trained at Stellenbosch, thus the conservative curriculum taught at that institution, close to the apartheid regime, was reproduced in institutions such as The University of the Western Cape, Durban-Westville, and the University of the North. A continued focus on the University of the Western Cape provides a window onto student and staff experiences in these institutions and the ways in which these were increasingly challenged by anti-apartheid activist staff and students in the 1980s.

Geography at UWC was founded in 1961 by Jacobus van Zyl, the first Head of Department, and Tielman J. Roos, both Stellenbosch graduates. In the 1960s and 1970s, geography remained conservative and traditional, teaching descriptive and later quantitative geography. Discussing his experience as an undergraduate geography student in the early 1970s, Vernon Domingo was scathing:

> this was a geography curriculum that was firmly based in the apartheid ideology. We focused mostly on physical geography with very little focus on social settings. In urban geography we were introduced to American models of urban structure (concentric, sector, and multiple nuclei models) with never a mention of the urban structure right outside the campus, with never any attempt to address what was right in front of us and the massive and glaring inequality between the non-white townships and the white residential areas. Same was true when we had a short section on political geography, having to do with issues such as heartland theory and location of national capitals.[19]

By the early 1980s, UWC, and other similar institutions had started to appoint black lecturers and technicians to more junior roles, and this began slowly to shift the tenor of the education offered. UWC also saw a wider transformation in this period into a radical space. In his 1987 inaugural lecture, the Rector Jakes Gerwel declared that the university should become 'an intellectual home of the left' (quoted in Wolpe 1995, p. 286). Within the UWC geography department things began to change too, albeit unevenly. Protest spilled into the classroom: Roos's teaching was disrupted by student activist Cecil Esau who was later imprisoned on Robben Island (Gilfellan et al. 2016, p. 286). Van Zyl, whose 1962 book on the South African City had virtually ignored the violence of apartheid, furnishing 'only a somewhat sterile picture relative to the unfolding events and struggles in the region' (Beavon and Rogerson 1981, p. 172) continued teaching political geography throughout the 1980s. However, as Gilfellan et al. (2016, p. 283) report, his material was increasingly contested:

> One afternoon Van Zyl found his third-year political geography students with an alternative set of course readers. The students had discarded Van Zyl's course reader and declared his study materials obsolete and out of tune with reality. This incident left an indelible mark on Van Zyl's psyche and he lamented this incident for many years.

When Harold Wesso took over the teaching of political geography in the late 1980s the course had a much more radical edge – he was, according to Gilfellan, 'openly decolonised, Africanised'.[20] Wesso's colleague Michael Dyssel noted that van Zyl was relieved: 'he realised he wasn't the one to teach about Black politics'.[21] Wesso was, in his teaching

> more explicit in addressing issues around the ANC that was unbanned at that stage, changes within white politics, changes within Congress politics … and also Pan-Africanist stuff … around daily political things, along with protest on this campus along with brutality of the police, along with sell-outs [informers] at the university.[22]

One of Wesso's former students Manfred Spocter remembers his teaching as transformational:

> Dr Wesso, brilliant man .... So at honours level that was where things used to really intrigue me. So Dr Wesso would come in and light up his pipe and you would talk. Wow this is great! So sitting at the feet of the master, and more importantly he was a coloured man. And a Doctor! My goodness! And you were engaging with Wesso, and Wesso wrote his PhD on the decolonisation of the geographical mind, something like that ... which was extremely progressive at the time. So we would learn Marxist theory, we would start way back when. We would start with Van Humboldt and Ritter, and Darwinism and see how these things fit together, and how the South African government worked, not just geography but other subjects as well, to serve their own Christian nationalist interests.[23]

Despite these changes though, the influence of the senior Afrikaner staff remained until the 1990s, often felt in seemingly small decisions which illustrated a much wider chasm between these staff and the students as well as the new generation of lecturers. For example, with protests raging on campus, and police on the streets outside, attitudes to attendance seemed out of step: 'staying in a three-hour lab session until five, against the backdrop of the political situation, I think people looked at that and thought this is nonsense', reflected Michael Dyssel.[24] Dyssel also noted that similar inflexibility also affected fieldtrips (Figure 6.2):

> We would go on field trips to Namibia just a small group of students in a minibus ... and when Sam Nujoma [founder of SWAPO, the anti-colonial and anti-apartheid movement] was released, we were in Walvis Bay which is the major port of Namibia, some 300 km from Windhoek the capital. And some of the more politically conscious students .... Asked me "Mike look is there no way, this is a historical moment, Namibia is awaiting independence. Sam Nujoma will step onto the tarmac, ... is there no way we can take this drive to Windhoek? We can still do some of the stuff we were going to do in other locations." And needless to say old Professor van Zyl said "No".[25]

We can see classrooms and field trips, then, like conferences, and tearooms, as spaces of geography that provided opportunities for the inequality of the apartheid system to be manifested but also challenged. Throughout the 1980s, the subjects taught, and the people doing the teaching, began to change, though this was uneven across the sector. Most differences were seen in the HBUs where, from the late 1970s, non-white junior lecturers began to be appointed. In some places though, there was striking continuity. Stellenbosch saw very little change to its staffing and curriculum throughout the 1980s and 1990s (Donaldson 2020). And even in places where teaching and topics changed substantially, such as at UWC, senior Afrikaner colleagues continued to have a substantial influence in decision-making and on the atmosphere of the department. The next section explores the more overt challenging of apartheid on university campuses.

**Figure 6.2** University of the Western Cape students on a field trip, 1984, Gilfellan second from left. Courtesy of Calvyn Gilfellan.

## Campus Politics and Activism

### *Campus Protests*

The experience of studying on campus in South Africa could be radicalising. Priscilla Cunnan, a geography student at Durban-Westville, remembered that

> The reality of the country we were living in became more real at Durban-Westville … I learnt about geography in a different context [beyond the classroom] … We'd have people who would video things happening in the townships … there would be these lecture venues that people knew weren't locked. We'd have to wait until 9pm, and then you know slowly two-by-two go in there, knowing that if we ever got caught we would be in serious trouble.[26]

Geography students were involved in producing some of the films like those watched by Priscilla Cunnan at UDW. Keyan Tomaselli, a geography student at Witwatersrand in the late 1960s made

> *Wits Protest* (from 1970–1974) [which] covered the almost daily clashes between police, assorted right-wing thugs, and my student colleagues at the University of Witwatersrand .... This movie footage of police attacks on student protestors was screened for months on end between 1972 and 1974, in the evenings on campus, at churches, at other venues ... With the help of my close friend Alan Mabin [another geography student, who was jailed briefly in 1972 as a result of his activism], our films made on these protests were shown, edited, and immediately in unedited form, at campus film screenings and other events. (Tomaselli 2001, p. 140–141)

Protests on and off campus were a common part of life under apartheid from the 1940s onwards, though the pattern of protest and violence was not evenly distributed across institutions, and some individuals remained outside of these currents. Hannes van der Merwe, newly graduated from Stellenbosch and working at Fort Hare in 1975–6 reflected that as

> a young lecturer embroiled in rolling out course materials and course frameworks apparently I was sufficiently isolated from student politics not to anticipate the total closure of Fort Hare in 1976 – part of the general national student uprising in that year.[27]

The all-white Afrikaans-medium Stellenbosch, to which van der Merwe returned the following year and where he remained for the duration of his career, was largely isolated from disruption. However, many of the HBUs as well as some of the white 'open' universities were regularly affected. Durban Westville was often the site of protests and violent crackdowns (Figures 6.3 and 6.4). At Witwatersrand, when protests kicked off, the atmosphere on campus was, according to Gordon Pirie, then a lecturer in the geography department, 'horrible. Screaming, yelling, police dogs' but also 'exciting' because it brought leaders of the struggle to the university.[28]

The University of the Western Cape was shut for much of 1973 on account of student protests, and in the 1980s protests were also a regular occurrence (Gilfellan et al. 2016). Geography lecturer Michael Dyssel described the atmosphere on campus:

> Robert Sobukwe Road [then Modderdam Road, where the university entrance was] was obviously a war zone, we would have a mass meeting here and feelings, emotions, boil over, some of the students would have been jailed again ... many a time students would run to the gate, ... picking up a stone and attacking the guys that would have postures on the high buildings opposite the university, who would be armed with rifles, albeit ... with rubber bullets, and a whole contingent in their armoured vehicles on the other side ...[29]

These events significantly disrupted university life and study, often leading to the suspension of studies for months on end. Cunnan at UDW in Durban remembers 'in my first year the university was closed from July to October, and, then

**Figure 6.3** Student protest at the Indian University College (the precursor to the University of Durban Westville) c. 1960s. Source: SP Olivier Collection, No: 1152/630. University of KwaZulu-Natal, Special Collections, Gandhi-Luthuli Documentation Centre.

we wrote our final year exams in November, and that was it, and miraculously I passed!'[30] Ngaka Mosiane had a similar experience studying at the University of Bophutswana (set aside for Tswana speakers in the Bantustan of the same name): 'in the first couple of days after classes started for my very first semester the university was closed and we were told to go home because there was an attempted coup.' Mosiane notes that these closures could also have further consequences, especially for poorer students and those forced to live away from campus: 'I lived off-campus, with all the attendant challenges of lacking transport to go "home" at night, thus taking risks by walking alone; not knowing what will happen the next day [with the strike situation].'[31] Studying in such conditions was hugely trying. Flip Schoeman's nearly complete doctoral work was destroyed by fire after a protest on campus at the University of the North burned the Soil Science department

**Figure 6.4** Protests at the University of Durban-Westville Campus, May 1972. Source: Van Niekerk collection, No: 1152/518. University of KwaZulu-Natal, Special Collections, Gandhi- Luthuli Documentation Centre.

to the ground. He never completed his studies and left academia (Tait et al. 2016). However, these experiences could also be important for understanding South Africa, as Cunnan argued: 'that was the benefit in a strange way of being at that university. My friends who went to the white university [the University of Natal, Pietermaritzburg, which was not the centre of sustained action] never had that experience, because they never had any protest marches, or boycotts.'

Another common experience for geography staff and students was fear of spies and government informers. Four of my interviewees had clear evidence that they were being watched.[32] This cast suspicion far wider. Colleagues Calvyn Gilfellan and Michael Dyssel at the University of the Western Cape both recounted rumours that their senior white Afrikaner colleagues were working for the government:

> you would hear all these rumours. That Myburg [urban geographer at UWC] was a spy for … special branch …. the security apparatus everywhere. And the students knew that Myburg was part of that. Of course I was more, "no he can't be, the pictures he is taking are for history." Because they would see him taking pictures.[33]

> There were rumours that some of the older folk they fed info to the security police. … the political consciousness … at that stage made us very aware of even the slightest thing that might have a hint of 'listen we can't trust this at all.'[34]

## *Geographers as Activists Beyond Campus*

Calvyn Gilfellan was heavily involved with anti-apartheid activism, described by his colleague Michael Dyssel as 'an activist literally with hands and feet and everything on the ground, without any reservations around his beliefs.'[35] Gilfellan used geographical skills acquired in his role as a technician in the department in his activism. These included photo-development and aerial photo analysis. Gilfellan elaborated:

> if there was a consumer boycott for instance, and there needed to be areas identified on a map of Paarl where people who would undermine that boycott would normally go, those were the kinds of things that they would expect, not even ask, but expect you to do. Or more underground would be where are the installations... one of our student colleagues was supposed to blow up a bridge where the security police would take you when they arrested you.[36]

And whilst generally aerial photographs of sensitive areas like military installations were restricted, blurred, or removed from circulation, the Geography Department of the University of the Western Cape had a full unredacted set.[37] Gilfellan's geography training therefore included lots of 'things that I could use in academia but also in the struggle'.[38]

Another strand of Gilfellan's activism was his work, alongside his colleague Michael Dyssel in community education. The two worked together in community projects, especially providing school-level education during the widespread disruption to teaching in the mid-1980s as students boycotted schools in response to the increasing hostilities and violence of the apartheid regime in the townships. Gilfellan and Dyssel worked in the evenings and weekends providing education for young people in the sprawling settlement of Bellville which neighboured the campus. The University of the Western Cape had by the mid-1980s become a base for radical scholars and community work was also seen as an important part of the university's wider mission, which refuted the 'political and ideological grounds' on which the university was established – apartheid – and aimed to serve the 'Third World communities', including those bordering the campus (Wolpe 1995, quoting the UWC mission statement, 1982).

Working with school children could be both a pragmatic solution to the challenges they faced in gaining an education under apartheid, and part of a broader mission to bring pride and dignity to the community as one element of the wider struggle. The Black Consciousness Movement leader Steve Biko saw community education as crucial in 'promoting a spirit of self-reliance and black consciousness among all black people in South Africa' (1978 [1969–1972], p. 38; Wilson 2011). As an inward-looking process which included the revalorisation of Black identities, cultures, and experiences, Black Consciousness strove to 'make the black man come to himself; to pump back life into his empty shell: to infuse him with pride and dignity …' (Biko 1978 [1969–1972], p. 29). And education was important in this process, as Biko argued: 'the interrelationship between the consciousness of the self and the emancipatory programme is of paramount importance' (1978 [1969–1972], p. 49). Dyssel reflected on what they thought they were doing in their community work and with their undergraduate students: 'decolonisation wasn't really a buzz word at that stage … even if it wasn't a programme, even if it was haphazard, that bit and piece can maybe plant a seed and conscientise three or four students, and then we have achieved our goal'.[39] Gilfellan made explicit his political and intellectual influences when reflecting: 'it was this whole notion of Paolo Freire's (2017 [1970]) *Pedagogy of the oppressed* where the idea of the organic intellectual you could live it out here'.[40] Drawing on Marxist thought, Freire (2017 [1970]) argued that knowledge could be co-created between teacher and student and that education was key to revolutionary liberation from colonialism:

> pedagogy … must be forged *with*, not *for*, the oppressed … in the incessant struggle to regain their humanity. This pedagogy makes oppression and its causes objects of reflection by the oppressed, and from that reflection will come their necessary engagement in the struggle for their liberation. (Freire 2017 [1970], p. 22)

Gilfellan's activism got him arrested and detained on more than one occasion, where his reputation as a geographer was well known by other inmates who would draw on his expertise to pass the time, asking 'about the mountains, you know you look out at the mountains and they ask you, you know, is it sedimentary rocks; "chief, the cloud that is coming is it cumulonimbus or nimbostratus?!"'. While in prison in July 1988, Gilfellan received a message from his Geography II class, signed by all 44 students and smuggled into his cell:

> We just want to let you know that we are thinking a lot about you and hope that you are doing well … We hope to surprise you with the Statistics test results so you can see that the time and work that has gone into it has not been in vain. We hope to see you again soon because we miss that friendly smile on campus. Love and Blessings.[41]

The university and department also released separate statements demanding his release, and Gilfellan described the latter as 'very heart-warming given the political tension in the department'.[42] Gilfellan wasn't the only geography staff member detained at UWC. The radical geographer Harold Wesso was also imprisoned. One of his students remembered that he missed

> one or two lectures because he was locked up. And then he came to the third lecture and it seemed like he had come straight from the charge office because he was all dishevelled and said "sorry I have not been here, you probably know what has happened" …[43]

When Gilfellan returned to work after some months in prison, some of his white colleagues didn't mention his detention at all, though some surprised him. Tielman Roos 'was probably the most sympathetic one, though he was the most conservative one. He was openly Broederbond ["a secret and exclusively male ... organization dedicated to the advancement of Afrikaner interest" (Donaldson 2020, p. 4)], but he was also very religious.'[44] Gilfellan found out later that Roos had applied for permission to visit him in prison, though he was released before the visit could take place.[45] In a clear indication of the complexity of apartheid politics as played out within geography departments across the country, Roos took early retirement shortly after this incident, in 1989, because, as a colleague recalled

> he felt threatened ... he said that he could not teach at a university where he had to look over his shoulder all the time, where he was an identified member of the Broederbond. Somebody had written on his door, you know, that he was to be killed or something.[46]

The University of the Western Cape wasn't the only department where staff were actively involved in the wider struggle, often working through community organisations and on urban issues such as housing, planning, segregation of group areas, and access to infrastructure. Alan Mabin at the University of Witwatersrand co-founded Planact, which, in collaboration with community-based organisations lobbied local government for services and infrastructure for poor and informal settlements from the mid-1980s onwards (and the organisation is still going strong).[47] Keyan Tomaselli, who had made the student protest films in the early 1970s alongside Mabin, was by the late 1980s collaborating with the Westville Residents Support Group in Durban, alongside his wife and fellow academic Ruth Tomaselli. Brij Maharaj was Secretary of the Merebank Community Centre, also in Durban. At the University of Natal, several geographers including Robert Haswell, Michael Sutcliffe, Paul Wellings, and Jeff McCarthy were variously engaging with the progressive trade unions, Black civic society organisations, the anti-apartheid coalition the United Democratic Front (UDF), and the African National Congress (ANC).

When Michael Sutcliffe returned to the University of Natal Durban as lecturer in 1982, following his PhD at Ohio State University in the US, he recalled that 'no sooner had I got back in my office, than a comrade came into my office then and said "Hi, I know you can analyse things with computers, data"' and requested his support for anti-apartheid projects.[48] Doing this analysis brought Sutcliffe into contact with the UDF and ANC, and he went on to be heavily involved in the movement, running a community research unit, establishing NGOs within and beyond the university which provided a screen for uMkhonto we Sizwe (MK, the armed wing of the ANC), as well as working to improve living standards, and contributing to studies aiming to support South Africa's democratic transition:

> there were probably a hundred studies we ran on workers, of living conditions of people, and I would analyse them, write the stuff up, we would always have publicity stuff, but we would have organising behind it .... if the police were putting a road-block in, tear gassing people, we would go in behind the scenes, do a survey of people, what do you think of these things etc., and we'd then tell the newspapers, and they would publish it and they could say, such and such survey of so and so grand academic shows 99.9% of people don't want to be teargassed! Not quite as simple as that but you know what I mean.[49]

In collaboration with others, Sutcliffe founded the Built Environment Support Group, which, similar to Planact in Johannesburg, worked to fight forced removals, as well as 'recording and analysing the apartheid violence and its impact on communities and undertaking community and trades unions surveys' (Sutcliffe 2020, p. 15; e.g., Sutcliffe 1983a, 1983b, 1983c, 1984, 1986). By the late 1980s the political situation in South Africa was incredibly tense, and often violent. Some of those – often young people – involved in monitoring state violence as part of Sutcliffe's activist research were killed.[50] Many more were imprisoned. At the University of Witwatersrand, the anthropologist David Webster well known to colleagues in geography and planning, was killed by state assassins in May 1989.[51] This was the context in which those opposing apartheid were working.

## Geographical Research and Apartheid

### *Research for (Separate) Development in South Africa*

This type of activist geography co-existed alongside the majority of geographical research into the 1980s that either actively supported or ignored apartheid. The discipline in South Africa had a long history of engaging – problematically – with Black South African communities and with the materialisation of apartheid on the ground. Whilst 'Demographical-anthropological' work focusing on black South Africans (often known in apartheid parlance as 'Bantus') had all but disappeared after the 1930s, Fuggle (1967, p. 348) reported that there was another 'spate of articles' in that area between 1952–1956. This period coincided with the appointment of the Tomlinson Commission, a 'governmental commission to investigate socio-economic aspects of the South African Bantu' (Fuggle 1967, p. 348–349) and reflected public debate about the 'distribution and settlement' of black South Africans in the wake of the formation of the bantustans – the 'notionally self-governing "tribal" states' formally known as 'homelands' (Dubow 2014, p. 64) – in 1951. Work continued on this theme into the 1960s, but attention shifted towards the applied urban and regional planning required to establish and develop these bantustans (Fuggle 1967). In this work, some South African geographers can be understood to have been directly supporting one of the key

– and most hated – tenets of the apartheid policy of separate development (Ramutsindela 2002).

Central planning and scientific management were central to the apartheid state, especially after Verwoerd took over at the Department of Native Affairs in 1950, where his huge 'feat of social, demographical and spatial engineering' relied on the work of cartographers, planners, and urban and regional development experts (Dubow 2014, p. 113). Indeed, David Smith (1982) has characterised apartheid planning as one of the most ambitious exercises in applied geography ever attempted. Much of this expertise was drawn from Afrikaner universities which were close to government, especially Stellenbosch, and from their students for whom 'new positions opened in the government's Natural Resources Development Council and its successor, the Directorate of Planning' (Barnard 2016, p. 83). Jan Moolman, formerly a member of staff in the geography department at Stellenbosch was a member of the Tomlinson Commission, which in 1954 recommended substantial government spending to ensure a better standard of living for black people, but in the context of the large-scale implementation of grand apartheid, the complete separation of 'racial' communities through the bantustans (Barnard 2016). The Geography Department at Stellenbosch included units dedicated to applied research in support of national development, for example the Institute for Cartography and Institute for Planning Research. Focussing on mapping and the production of atlases, as well as consultancy, geographers working for these units provided 'a number of inputs to national S[outh] A[frican] and Namibian planning initiatives for government and government-related agencies', including development plans for the 'homelands' for black Africans in both South Africa and Namibia (then under South African control), and reports into the needs of different 'racial groups'.[52]

Research, reports and evaluations of policy related to 'national' development of the new bantustans paralleled in many ways the regional planning, rural development and new settlement policies of developmental governments in other parts of Africa. They drew on similar theories of urban and rural development, economic growth and modernisation. 'Failed top-down projects proliferated' and the programme called for the design and construction of not one but 10 new capitals, one for each 'homeland' (Beinart 2012, p. 15). The linked 'Betterment' policy resettled, sometimes forcibly, rural populations into larger villages, and in the process led not only to social dislocation, but also damage to agriculture as people were moved away from fertile fields (Beinart 2012).

Hannes van der Merwe, who was involved with some of these development plans in South Africa, recalled, 'We were always quite proud of our involvement with national analytical research and reporting'. Reflecting on his career, van der Merwe noted that apartheid absolutely shaped everything about it: 'I studied at, performed research and teaching at an all-white Afrikaans-language university reputed to be architecturally responsible for the development of the apartheid policy.' Nevertheless, of the work he completed alongside colleagues, van der Merwe was more circumspect:

> in my own experience in my discipline, I never encountered overt political influence on any of what we did or professed. I am satisfied that the conduct of my teachers before me and that of me and my colleagues once we took over was driven by relatively liberal values that could answer most challenges.[53]

It was possible for many white geographers developing their careers against the backdrop of apartheid to see their own work, even when feeding into government policy, as separate from politics.

## *Apartheid: An Absent Subject*

This applied planning research aside, much of geography in South Africa ignored apartheid and the experiences of the majority of South Africans into the 1980s. Rogerson and Parnell (1989, p. 15) recorded that geographers often seemed to be 'absorbed with problems and issues exclusive to white society' – for example residential housing patterns in white areas of the city, leisure, consumption, and the retail geographies of shopping malls. Drawing heavily on American urban models, researchers drew comparisons between North American and South African cities. Jeff McCarthy remembers research about Johannesburg conducted at the University of Witwatersrand in the 1970s finding that it displayed similar urban structures to Chicago, a conclusion reached by examining the 'whites only' data sets and thereby ignoring the experiences of Black South Africans.[54] 'The bleached character of much human geographical enquiry' produced a geography that at best ignored, and at worse diverted attention from, the violence wrought by apartheid on the landscape of the country (Rogerson and Parnell 1989, p. 15). An increased interest in urban spatial organisation with the turn to quantitative geography led to a focus on patterns and processes, locations, and distances, rather than people, which was, as Beavon and Rogerson (1981, p. 164) noted with understatement, 'unfortunate … against the sombre shadows thrown by Sharpeville [the 1960 massacre of people protesting against the pass laws] and the rigorous enforcement of the policies and programmes which fall under the rubric of apartheid'. There were exceptions to this pattern, such as the early book by the geographer Ron Davies with sociology colleagues at the University of Natal, Durban, exploring the 'racial ecology' of that city (Kuper et al. 1958). However further research did not build on this, perhaps because several of Davies' co-authors left South Africa shortly afterwards as a result of their opposition to apartheid (Waters 2015). In his presidential address to the South African Geographical Society in 1981, Keith Beavon (1982, p. 3) noted only two references to townships in South Africa in the keyword index of the *South African Geographical Journal*, spanning the 63 years it had been published (both to Soweto). He argued that these spaces, where millions of black South Africans lived, were 'terra incognita' for urban geographers (Beavon 1982). One might add that whilst this was certainly true of many, these spaces were known intimately by the black South African geography staff and students who were forced by apartheid policies to live and study in them.

Many South African geographers – as well as those writing about South Africa from the UK, including the Oxford geographer Anthony Lemon – cast their work as apolitical. Lemon's *Apartheid: The Geography of Separation* (1976) described the spatial patterns formed as a result of apartheid policies but was criticised for utilising apartheid terms of reference. For example, Lemon recalled, he had a section titled 'Coloureds and Indians: An unresolved problem' and was attacked by a reviewer for 'identifying with apartheid by labelling the section in that way' – reflecting a 'problem' that was in the minds of apartheid planners only.[55] Beavon and Rogerson (1981, p. 164) argue, 'in the heyday of positivist human geography … during the 1960s and early 1970s, the legitimacy of apartheid ideology was rarely challenge a priori by geographers.' Ramutsindela (2002, p. 6) goes further, arguing that 'the state and the discipline shared common ground in legitimizing racial differences, social relations, and the spatial organization of society.'

Why was there so little critical attention played to apartheid by geographers until the 1980s, despite a generalised disapproval for the system professed by many white liberal geographers, especially at the English-speaking universities? An obvious answer is that many powerful geographers were agnostic or actively supportive of the regime. In addition, there were barriers to researching the apartheid system and the impacts it had on the non-white majority of South Africa. This research was not welcomed by local and national government and was therefore challenging (Beavon and Rogerson 1981; Rogerson 1985; Pirie and Mather 2016). Such research could also negatively impact your career.

It was also difficult to access areas of the cities designated as black Group Areas, especially if you were white. Dianne Scott remembered an incident at a Durban conference fieldtrip in the 1980s which underlined this challenge. Led by urban geographer Ron Davies, two buses of students and their delegates were making their way to the informal settlement of Malukazi, adjacent to the black township of Umlazi on the outskirts of Durban, only for the convoy to be stopped by the police, Davies arrested, and the buses turned back.[56] It was forbidden by the Group Areas Act to visit areas designated for other 'racial groups' except under particular circumstances. Stephen Rule, working at Vista University in Soweto, another black township, this time in Johannesburg, was able to publish a paper on the area, but only with the help of his black students, who were collecting the data as part of a research methods course (Rule 1993). He remembered: 'I would go and drop them off in areas to do the research, and sit in the car, once or twice I would go and knock on doors but it would not be the safest for a white guy to do that. I could be mistaken for a policeman or a spy, so it wasn't that easy.'[57]

Dianne Scott's PhD (completed in 1995 after part-time study) at the University of Natal, Durban is a rare example of research focused on Clairwood, an Indian area of the city by a white scholar. The research was fun, and she was made welcome in the area, though she reflected she could not have conducted the same research in a black township. But Priscilla Cunnan, categorised as 'Indian', did manage to conduct her Masters research at Durban-Westville in an informal settlement housing poor black migrants on the edge of the city in the 1980s. Cunnan remembered:

> it turned out to be one of the nicest experiences, because the people embraced me ... I had this battered old Mini that really needed, you know, a push to start, I would just park it, nobody dared steal it ... people would say 'we'll look after your car, don't worry', and I would just walk into the settlement. Residents were also very kind and patient with me while I conducted my research.[58]

Despite these challenges, then, research *was* possible in some places, by some scholars, reflecting the complexities of the apartheid system, as well as the possibilities open to those willing to persevere.

The challenges of contemporary research, as well as the fear of assisting the apartheid state with work on current issues often meant that researchers turned to more historical subjects (Pirie and Mather 2016).[59] But here too there were barriers. Often access was easier for those classified as 'white'. Brij Maharj, a young lecturer at the University of Durban-Westville had to hide his true research intentions when applying for archival access, which was sometimes denied, leaving him working with only partial records.[60] The Witwatersrand lecturer Gordon Pirie, categorised as 'white', was provided full access to archival materials by the South African Department of Co-operation and Development for his research, but, on reading the draft paper critical of zoning in black townships, they refused permission for the materials' inclusion in the article (and tried to block publication altogether) (Rogerson 1985; See Pirie 1984, with its footnote noting 'Certain archival material has had to be excluded from the paper following refusal by the South African Department of Co-operation and Development to allow publication.').

Another issue was the wider censorship of literature by the South African government's publications control department (Rogerson 1985). Publications likely to be withheld included those by 'banned' people, meaning that it was impossible to quote authors and critics in exile within texts published in South Africa, as well as those seen as promoting Marxism (Rogerson 1985). Richard Peet's *Radical Geography: Alternative Viewpoints on Social Issues* (1977a) was banned because of its support for communism. A 'temporary embargo' was also placed on David Smith's *Living under Apartheid* (Smith 1982). And self-censorship was perhaps an even bigger problem. Promotions and government research funding was often suspected of being related to towing the government line, leading many academics to pursue research unlikely to concern the authorities (Rogerson 1985).

Finally, along with isolation caused by censorship, South Africa was seen, particularly after the violent repression of protests in Soweto in 1976, as a 'no-go' area for international scholars (Beavon and Rogerson 1981, p. 160; Crush et al.

1982). This meant that there were few academics from the rest of the world conducting research in South Africa. Whilst they might have been in a freer position to write critically about apartheid, many internationally felt that to research in South Africa was to legitimise the South African government regime.

## *Growing Critique*

This parlous situation in which apartheid was often ignored and sometimes actively supported by South African geographers began to change, in some places, and for some academics, from the late 1970s. This was fuelled by the importation of radical scholarship from the United States, as well as by the increasingly violence and activism on the streets in the country following Soweto (Ramutsindela 2002).

Following the turn from British and towards American geography in the late 1960s, young South African researchers often looked to the US for their PhD study. Alongside a desire to study with the new stars of the discipline, North America was a tempting destination for South Africans, as it had been for many others across the continent, as there were far more scholarships available than in the UK, and they paid more generously.[61] Amongst those to travel were D. S. Rajah (in the late 1960s) and Vernon Domingo (in 1975, following his banning from the University of South Africa for leading student protests at UWC). Both went to Clark University, where the radical geographers Richard Peet, Ben Wisner, and Jim Blaut were based at various points in the late 1960s and 1970s, and where the radical journal *Antipode* was founded in 1969 (see Peet 1977b). The mid to late 1970s saw a whole raft of students follow: Chris Rogerson (a Brit who had been working in South Africa, to Queen's, Canada), Alan Mabin (Simon Fraser, Canada), Dave Lincoln (Wilfred Laurier, Canada), and Jeff McCarthy, Michael Sutcliffe and Felicity Sutcliffe to work with Kevin Cox at Ohio State in the US (Crush 1993).[62] In the early 1980s, Dhiru Soni travelled to Johns Hopkins University in Baltimore, US, on a scholarship between his masters and PhD at Durban-Westville, to work with David Harvey.[63]

Jonathan Crush, who had researched and worked in South Africa and did his PhD in Canada recalled:

> Many went back to South Africa in the 1980s and began rewriting the nature and purpose of South African Geography, what it was about, what it wanted to do. In Canada and the US we were exposed to ways of thinking that we wouldn't have been in South Africa. In particular: (a) neo-marxist critiques of development theory and development geography; (b) radical geography critique associated with scholars such as Bill Bunge, David Harvey, Dick Peet, and Ed Soja; and (c) the greater scope for inter-disciplinary engagement and coursework in North American graduate programmes.[64]

Whilst D. S. Rajah, a trailblazer in completing his PhD at the University of Natal, Durban in the mid-1960s before visiting Clark on a scholarship, did not publish based on the more radical currents he encountered in the US, his visit catalysed the critical ideas and career trajectories of a whole new generation of scholars at Durban-Westville where he lectured. When he returned, he introduced new readings, including David Harvey's *Social Justice and the City* (1973) which chimed with the conditions in South African cities. Dhiru Soni, Cecil Seethal, and Brij Maharaj all trace their own travels to the US, which they cast as crucial to their career development, as building from Rajah's visit and subsequent teaching.[65] Soni recalled 'I remember him saying to the group [he was teaching] as I was on my way out to go to the States [in the early 1980s], he says "I can guarantee one thing, when Soni comes back he's gonna be someone else." And it was the truth'.[66] For Soni, his time at Johns Hopkins

> influenced a lot of what happened thereafter, because I had the opportunity to interact not only with David Harvey, but also other, what I would call radical geographers. And quite obviously they were quite keen to know what was happening in South Africa ... When I came back I realised that our education was actually one that was created in the image of apartheid. That we were not allowed to be critical, whether it was colonialism or apartheid itself.

In addition to these external influences, the situation internally in South Africa spurred on new research into the violence and inequality of apartheid. When Jeff McCarthy returned from Ohio State to take up a lectureship at Rhodes University he recalled:

> it hit me over the head politically, I mean ... the poverty was just unbelievable to me, ... Grahamstown was too small to escape that ... you saw everything, including ongoing warfare actually. I got there in 1979 and you know my intellect and my existential thing came together, and I had been reading Castells and all this stuff on urban social movements, and I found out about the black civic organisation PEBCO (Port Elizabeth Black People's Civic Organisation) which I went on to work with.[67]

## *People's Geography*

By the early 1980s, then, there was a growing critique of South African geography from younger scholars keen to produce more critical, emancipatory research. This work emanated mainly from the 'white' English speaking universities, and from some of the HBUs, such as the University of Durban-Westville. The new research coalesced around the need for a 'People's Geography' which, drawing together historical, urban, development and environmental geography especially, would be 'vitally relevant to a future non-racial, democratic South Africa' (Rogerson and Parnell 1989, p. 13). New areas of research concentrated on two main approaches, both drawing on radical traditions. The first focused on recovering and highlighting the Black experience, drawing on Marxist but also more liberal traditions. Developing primarily out of a vibrant department at Witwatersrand, this strand drew on the writings of social history, or people's history, developing at the same university, and also on scholarship by those like the South African Shula Marks and E.P. Thompson in the UK. This strand of research was interdisciplinary in nature, also taking inspiration from sociology, social anthropology, economics, and political science (Beavon 1982, p. 17). The second strand, emerging primarily from the University of Natal, Durban, focused more on contemporary issues and urged engagement with political and community groups (Wellings and McCarthy 1983). Whilst these new approaches were to change South African geography substantially over the next decade, it is worth remembering here that many geographers continued just as before. As Beavon and Rogerson (1981, p.174) put it, they retreated 'into a comfortable academic laager, cocooned by a web of concepts, categories and methods which serve not only to rationalize the existing *status quo* but to insulate the mind against the staggering injustices of daily life in southern Africa.'

The first trend in 'Peoples Geography' saw an explosion of critical historical work, much of it urban, but also focusing on labour and migration (e.g. Rogerson and Beavon 1982; Crush 1986a, 1988; Maharaj 1989; Soni 1993). This body of work sought to provide 'a critical reinterpretation of colonialism and imperialism' that emphasised the 'importance of locating southern Africa within the broader historical context of the expansion of capitalism in order to explain meaningfully internal social, political and spatial dynamics' (Beavon and Rogerson 1981, p. 168). Seen as an antidote to the 'antiseptic landscapes of much existing historical geographic writing', this historical work sought to 'understand how South Africa's brutal landscapes were forged in subordination and struggle, and how the geography of privilege and deprivation has saturated the experience of both overlords and underlings in that society' (Crush 1986b, p. 3).

In addition to work with a historical focus, some research sought to uncover the experiences of the working class, mainly black, urban communities (e.g. Rogerson and Beavon 1980; Pirie 1982; Hart and Pirie 1984). For Ramutsindela (2002, p. 7), 'the apartheid city as a concept stands out as one of the most notable achievements' of South African geography in this period. Alongside published research, Witwatersrand lecturers Alan Mabin and Gordon Pirie sought, drawing on Bill Bunge's ground-breaking work in Detroit, to introduce their (mainly white middle class) students to areas of the city which were unknown to them (McCarthy 1982; Mabin 1984). According to Pirie, these 'East' and 'West' Johannesburg Geographical Expeditions 'made a big difference to shifting mindsets amongst students, and produced lots of interesting dissertations. And it spilled internally into new kinds of knowledge in the department and a new interest in critical enquiry.' Of course, as Pirie reflected, 'we sat inside the minibus looking through the windows and were able to go home to our nice comfortable beds.'[68]

The second trend, emerging from a group of scholars centred on the University of Natal, Durban, focused more on the present and grew in part from political commitments outside of the university. It drew heavily on the radical tradition particularly in US Geography, and trod the line between activism and research. A leading proponent of this direction, Jeff McCarthy, remembers working with the banned Durban Housing Action Committee (DHAC) and visiting its Secretary Pravin Gordhan in the evenings when he was under house arrest to discuss the work of David Harvey and Manuel Castells.[69] This branch of research provided studies of communities, political action, housing struggles, and strikes, as well as state violence, and often developed through work with trade unions, community groups, and housing activists. Rather than the humanities focused historical urban work emerging from Witwatersrand, which tended to draw on archives, oral histories, and literary analysis, this work more often drew on the social sciences approaches of questionnaires and continued to draw upon quantitative analytical skills developed as part of the turn to spatial science after the late 1960s.

Whilst the two branches of research could be gathered together under the umbrella of a People's Geography, those advocating the latter approach argued that 'current moves towards a geography *of* working class southern Africa needs to be situated within a broader geography *for* working class southern Africa.' (Wellings and McCarthy 1983, p. 337). Without more explicit links to 'radical political economy and political practice, contemporary trends in human geography could well degenerate into an "emotional tourism"', Jeff McCarthy argued in 1982 (p. 59). Working to 'assuage the collective liberal conscience', research about the Black experience, even drawing on radical history, risked blunting the demand for action and failing to engage with the actually existing Marxist revolutionary practice of the political opposition in South Africa (Wellings and McCarthy 1983, p. 340; McCarthy 1982). Moreover, it could be seen as irrelevant: as McCarthy rather cuttingly argued, 'which black working-class activist needs a geographer of white middle class background to explain to him or her about the "symbolic realm" of conditions in the *apartheid* city or townships?' (McCarthy 1982, p. 62). From this perspective, *Living under Apartheid* (Smith 1982), held up by government censors, could be critiqued for focusing on 'white consciousness areas' like 'riots, domestic servants, black housing' rather than the 'need for political organisation and mobilisation' (Sutcliffe 1982, p. 178). However, the question of 'who speaks' was just as live in the survey questions and answers of the Durban geographers as it was in the archives and analyses of those working the liberal-radical tradition at Witwatersrand (Crush 1993). Despite these criticisms, taken together these two research directions played a useful role in 'de-bunking and de-mythologising' the *apartheid* state (Rogerson and Parnell 1989, p. 16).

## *Decolonising South African Geography*

Alongside research providing a geography for (and of) the people, there was a third important aspect of academic debate emerging in South Africa at this time, which focused on the need to transform the discipline of geography, particularly in relation to its dependence on theory produced in the UK and North America. As early as 1973, Alan Mabin and Keyan Tomaselli, then undergraduate students, decried 'the ever-present tendency to employ foreign and more particularly American experience as a norm in South African geography … the result of concentration on exogenous directions of study has been scant attention paid to the particular nature and special problems of South African urban development' (1973, p. 5). Whilst drawing on their experience and study in the US, several of those South Africans returning to lecturing roles at home also recognised this need (e.g. McCarthy 1982).

A key voice in these debates was Jonathan Crush, born in England but raised in Southern Africa. In 1982 he developed this argument alongside colleagues, now explicitly utilising the language of decolonisation. 'There emerged in southern Africa' they argued,

> a 'colonized' geography, one which accepted that the only real theory was that created in the Euro-American heartland and which applied the received theories to mythologize and rationalize the economic and social inequalities that constitute the stark panorama for human geographical research in the region… the process of political decolonization must be matched by the decolonization of the existing colonial geo graphies and attitudes concerning South Africa… [this] requires recognition of the fact that southern African social formation may be an actual source for theorizing and not a mere sink for Euro-American thinking. (Crush et al. 1982, pp. 197–198)

This work anticipated twenty-first century debates about southern theory and comparative urbanism (many driven by South African scholars such as Jenny Robinson, who studied for her undergraduate and master's degrees at the University of Natal, Durban, under Jeff McCarthy).

Alongside these debates about decolonising theory in South African geography, ran perhaps even more important arguments about decolonising the structures of geography, diversifying the disciplinary community in South Africa, and fundamentally rethinking the curriculum and the organisation of teaching and learning. Much of the impetus for these debates emerged from HBUs, primarily the 'Indian' University of Durban-Westville and the 'Coloured' University of the Western Cape.

This found its fullest intellectual expression in the PhD and publications of Harold Wesso, completed at the University of the Western Cape (where he also lectured) in 1992 (Wesso, 1992, 1994; Wesso and Parnell 1992). Primarily a study of the history of geography in South Africa, Wesso's thesis concluded with a discussion of the possibilities for the decolonisation of geography, which he argued would involve,

> a deliberate effort to Africanise the discipline and to empower people geographically … the institutional structures as well as the practice of geography is completely

> dominated and controlled by whites. The concept of "Africanisation" recognises the imperative of transforming, not only the institutional structures of the discipline of geography in South Africa, but also the nature and scope of the discipline. (Wesso 1992, p. 342–343)

Wesso highlighted the need to change the institutions and structures of geography, but also 'the curriculum and the whole way in which learning and teaching is organised' (1992, p. 343). His PhD foreshadowed twenty-first century debates and protests, emerging in South Africa and globally, about the need to 'decolonise the university' as an institution (see Elliott Cooper 2017; Bhambra et al. 2018; Hendricks 2018).

As Wesso's PhD noted, a key aspect of decolonising geography was examining who could work, publish and develop a successful career in academic geography. But this issue was not reflected on in any depth in the new people's geography emerging from the white liberal universities in the 1980s. Despite the flurry of radical work emerging from these institutions in this period, in these papers there is barely a mention of the whiteness of the discipline and challenges of inclusion. In their introduction to a special issue of *Tijdschrift voor economische en sociale geografie* on 'Decolonizing the Human Geography of Southern Africa' (1982) the editors Jonathan Crush, Hendrik Reitsma and Christian Rogerson did not discuss questions of who was writing and reading South African geography. A review of the special issue heralded the papers as 'the foundation upon which a revolutionary, homespun, "decolonised" human geography of southern Africa may take shape' (Wellings 1983, p.360). Yet all the contributors were white men (Jonathan Crush, John Taylor, Hendrik Reitsma, Gordon Pirie, David Lincoln, David Simon, Christian Rogerson, Keith Beavon). Similarly David Smith's *Living under Apartheid* (1982) only featured chapters by white contributors. Reflecting on this in retrospect, Smith noted wryly it was '*Living under Apartheid* by people who didn't suffer the blunt end of it …'.[70] When they met at the Association of American Geographers Conference in Baltimore in 1989, Dhiru Soni 'confronted' Smith about this issue.[71] Both agreed the book would have been stronger with more diverse voices, and the follow up *The Apartheid City and Beyond*, published in 1992 included six chapters by Black scholars. Smith remembered that: 'this was deliberate, and I got praise from some people for the extent to which, although they were quite thin on the ground, I got people who weren't classified as white in that book.'[72]

## Working for Transformation

### *Challenging Geography's White Institutions*

It took until the end of the 1980s before articles began to discuss the issue of what Rogerson and Parnell called South Africa's 'lilywhite' geographies (1989, p. 16, see also Mabin 1989; Rogerson 1990; Simon 1994); and even then, the issue was usually mentioned rather than engaged with in any depth. Crush argued in 1993 (p. 60) that decolonising the discipline was 'taken extremely seriously' by South African geographers from the early 1980s. However, this was a partial vision of decolonising, which understood the dangers of imported theories and topics, raged against the colonialism and racism of the South African government, and attempted to shatter 'the boundaries between campus and community' (Crush 1993, p. 60), but which did not often reflect in any detail on the experiences of geographers of colour and the institutional structures which reinforced the whiteness of the discipline. The divisions between 'racial' groups imposed on the university sector along with the rest of the country produced a divided geographical community, in which it was possible to write critically about apartheid without engaging meaningfully with non-white colleagues in other universities, often just a few miles across town. Ramutsindela is clear: 'even the self-proclaimed progressive geographers paid little or no attention to strategies for black intellectual empowerment in the 1980s' (2002, p. 7). And as Alan Mabin argued at the time, this blind spot – 'the comparative failure of geography to produce publishing black geographers' – had consequences for the ability of geographical scholarship to make a theoretical and practical impact on 'pressing scholarly, pedagogic, and political concerns in contemporary South Africa' (1989, p. 124).

By the late 1980s, a new generation of Black geographers were increasingly challenging the conditions in which they worked and the barriers they faced. Dhiru Soni, by 1992 a senior lecturer in the geography department at the University of Durban-Westville was key in making many of these arguments. As he remembers:

> at that stage it was important for us to begin to research, to publish etc., and of course it was a very stifling environment, whilst many of our colleagues in the department who were white and Afrikaner were publishing on a regular basis, we found it hard to publish even one article. At that point I wanted to know what would it mean for me to publish an article in a South African journal of geography … [Both the Afrikaans and English speaking societies] in a sense denied us membership. The Afrikaner society quite obviously didn't even consider any kind of application,[73] whereas the English-speaking body required some other white geographer to recommend membership … At that stage I became very annoyed about what was happening to us. The fact that they were not only stifling us but making sure that we did not progress.[74]

Gordon Pirie, who worked at Witwatersrand for many years reflected that the South African Geography Society and its journal (the *South African Geographical Journal* – SAGJ) were controlled 'tightly by an elite of top scientist-scholars. Its base at Wits [Witwatersrand] afforded huge privileges to the Department there … Outsiders would surely have felt this. The SAJG Editorship & Board was the gatekeeper that all are.'[75] The editors and board were overwhelmingly white. Jeff McCarthy went further: 'Bear in mind who it was that controlled the journals, it was all of this, nudge nudge wink wink kind of stuff, subliminal racism.'[76]

Barriers to publication and therefore progression also included the huge differences in resources between departments, including both access to materials such as books, and time, with heavy teaching loads in the HBUs, and an expectation or desire to undertake community work and activism. Calvyn Gilfellan reflected that he and his Black colleagues at UWC didn't publish a lot, reflecting that 'you compromise on your theoretical work because you try to strike a balance between your academic work, research and community work. And somehow, our senior white guys expected us to do that difficult community work.'[77] Another barrier was a lack of enabling environments within departments. Soni remembered at Durban-Westville 'there was lots of engagement within the department with Afrikaner and other colleagues and it wasn't pleasant.'[78] In comparison, the geography department at Witwatersrand – restricted mainly to white academics and students by apartheid regulations – is remembered as a hugely collegial, fun place to work, with lots of socialising within and outside the university (Pirie and Mather 2016).[79] Reflecting on the dressing up, pranks and other social events in the department, Pirie and Mather (2016, p. 78) noted recently that '[b]eing housed in a well-resourced university was a privilege … Lightness had shades, of course, served clubby purposes, and was of its time.'

Some of those working in the purportedly open (but still almost entirely white) universities, like Witwatersrand and University of Natal Durban were reluctant to collaborate with colleagues working at apartheid institutions designed for different racial groups for fear of supporting the government regime. Michael Sutcliffe, a geographer at the ('white') University of Natal, Durban initially had little to do with his colleagues at the University of Durban-Westville on the other side of the city. A vocal critic of apartheid, Sutcliffe saw UDW as 'run by the Broederbond basically. So … people … were viewed with a bit of suspicion, because the question was who was selling out who.'[80] Reflecting recently, he commented 'you really in hindsight felt very sorry for academics in those institutions.'[81] But that there was so little cooperation between scholars based at universities in the same cities (the University of Cape Town and the University of the Western Cape, the University of Natal Durban, and the University of Durban-Westville, the University of Witwatersrand and Vista University for example – see Figure 6.1 for locations) seems extraordinary.

Whilst the South African government worked hard to uphold the separation of apartheid, through the invention of a segregated university system and policies that attempted to maintain these boundaries, geographers and other academics were also implicated in upholding these separations. It is striking that, whilst increasingly critical research was emerging from geography departments in the 1980s which contested apartheid, apart from a few individual connections, no sustained effort was made to build networks of solidarity between geographers working at different types of institutions. As Dhiru Soni notes, after 1993 there were more opportunities to be one community, but before then, 'it was them and us'.[82] And these divisions in some respects outlasted the apartheid regime: discussing the later merger between the University of Durban-Westville and the University of Natal to become the University of KwaZulu Natal (UKZN) in the late 1990s, Brij Maharaj reflected that lots of the discussions still reflected 'a subminimal fear of contamination, now that's a coded way of saying race.'[83]

Soni, lecturing at UDW in the 1980s worked hard to engage colleagues at the University of Natal, reaching out to the radical Jeff McCarthy, then working at the Pietermaritzburg campus, requesting PhD supervision for his younger colleague Brij Maharaj. McCarthy agreed; as he remembers,

> It started with me and Dhiru [Soni] in my case ... Dhiru was a bold person and you had to have bold people breaking out. So he was important in that way ... and I must say Dhiru helped me in a lot of ways and I'm grateful for that.

In this example, it took courage and initiative from a Black geographer to initiate connections with colleagues at a well-resourced white university.

Soni, along with colleagues including Cecil Seethal, Brij Maharaj, Harold Wesso, and Lindisizwe Magi also demanded change within the institutions of South African Geography – its two geography societies and their journals. Discussions between Soni and radical geographers – lead by the American Jim Blaut – abroad (at the Baltimore Association of American Geographers Meeting in 1989) provided important international support for their demands (Soni and Maharaj 2005). Soni recalls:

> at that supper meeting I explained what was happening in South Africa and they were quite annoyed, the fact that some of the so called progressive white academics at that stage from the English-speaking sector were attending conferences and speaking about apartheid in a very negative way, yet they couldn't understand that they were not helping us .... it was suggested that I confront both of these organisations [the two geographical societies] and threaten them with an international boycott that they were prepared to mount for me.[84]

Around the same time, key representatives from the Black geography community were invited to discuss 'whether the South African Geographical Society might be able to help black geographers overcome the legacy of discrimination' in South Africa.[85] Whilst this can be seen as the first real move by the white dominated South African Geography Society to engage and support black geographers, this invitation was viewed as 'patronising', linked to the 'impending boycott', and drew a 'crisp and brazen' response (Soni and Maharaj 2005, p. 1029). There followed a series of meetings, where initially the responses from the Society representatives were felt to be inadequate 'tit bits', 'piecemeal change' rather than the 'radical' break that was required.[86] Jeff McCarthy, also involved in these discussions recalls: 'I think the view of my white colleagues would have been an assimilationist view ... let's bring them in to our organisation. Dhiru being a brave guy wasn't going to buy that.'[87]

Alongside these meetings ran increasingly critical presentations at national conferences and inaugural lectures from several of those involved, including Wesso (1989), Soni (1990; 1991), Magi (1990) and Maharaj (1992). In 1991 in Potchefstroom, 'almost like home for the Afrikaner society', Soni 'challenged' the audience, questioning why, although Black scholars were now used to 'meeting, talking and intermingling with other geographers' they were 'never allowed to publish'.[88] In 1990, analysis had shown that the journals of the South African geography societies had published 'only a negligible 0.08 percent of black geographers' (Magi 1990, p. 1029). Whilst Soni and Maharaj got support from a new generation of white scholars, Soni remembers that there was in general an explosive response to his paper:

> it became a very acrimonious meeting, I don't think I got the kind of flak that I received from the progressive geographers from the Afrikaners, I think they realised they had done wrong, but it was the progressives who became extremely defensive, accusing me of telling lies.[89]

Maharaj, pitching in in support of his UDW colleague recalled that:

> a very senior Professor who I'm not going to name came up to me afterwards and said you've just destroyed your career, because I made some strong points. So I said I don't know what era you're living in, because this is the Mandela era.[90]

A comment from one British geographer, Anthony Lemon, provides evidence of how these demands were viewed by some in the South African and global geographical community: 'There were a couple of sort of, angry young men ... they were quite belligerent in the South African Geographical Society I think.'[91]

This type of response may account for the length of time it took from the initial demands to the dissolution of the two existing Societies and the foundation of a new inclusive one in 1994, just as the country itself was ushering in a new democratic era.[92] Whilst he has published little about this, or indeed about critical geography more broadly, Lindisizwe Magi played a key role as a crucial unifying figure in the new Society of South African Geographers, a 'figure of some gravitas, who was able to take a position in the society ... [to] pull it together.'[93]

The early 1990s saw a gradual shift in the conferences and publishing landscape towards a more diverse and unified community of scholars. A conference review from 1993 entitled 'Changing of the Old Guard' reflected that

> It was ... pleasing to see an increase in presentations from the HBUs as well as from practitioners ... another important feature of the Conference was the presence of a new, younger breed of committed geographers, engaged in very relevant research, and who were very keen to straddle previous divisions amongst academics (especially those of colour, language and region). (Maharaj 1993, p. 4)

Soni felt that this new diversity was

> the greatest joy. It afforded people [the opportunity] to intermingle, to begin to realise their true worth. And I think that publications, the level of discussions, critical thought etc. really emerged, and that was for me a vindication.[94]

## *Consultancy and Critique*

In the early 1990s, after the unbanning of the ANC and the release of Nelson Mandela, the attitude of many South African geographers towards contemporary research and working with the government (or the government in waiting) began to shift. So did the opportunities and funding landscape, as South Africa opened up to international donors and researchers. This section moves on from the transformations within geography's institutions, to explore other impacts of the broader shift towards democracy on the discipline. It focuses first on topics and approaches to research, before moving on to discuss some of the roles that geographers played in the transformation of the country after 1994.

Whilst the transition to democracy opened up many more opportunities for Black scholars in South Africa's universities as students and staff, one key impact of this political change was many Black academics leaving the sector entirely, 'threatening to erode the availability of the critical intellectual mass in the discipline' (Ramutsindela 2002, p. 7). Both Harold Wesso and Calvyn Gilfellan, colleagues at the University of the Western Cape and, in their different ways, key contributors to the decolonisation of geography, took up roles for the new ANC government in the years after 1994: Gilfellan, whose research focused on heritage and tourism, left to work in the heritage sector for the local government, eventually becoming Chief Executive Officer of the Castle of Good Hope in Cape Town, whilst Harold Wesso became Director of the Western Cape Provincial Government's Reconstruction and Development Programme. They were not unusual. That so many young Black researchers left the universities after the change to democracy produced an increasing challenge to the sector as it aimed to transform.

Many of those who stayed in academia shifted their research direction, often away from historical research, and towards applied contemporary work, much as their colleagues had done in the 1960s and 1970s in other parts of Sub-Saharan Africa. Chris Rogerson, Jonathan Crush and Sue Parnell, key names in the 1980s 'people's historical geography' all mentioned this explicitly in relation to their own trajectories. As Jonathan Crush noted:

> The possibility of fundamental political and socio-economic engagement in transformation in the late 1980s and early 1990s, made historical work and historical geography seem irrelevant at best. Geographers began to cast around for ways to become relevant to the political and social changes taking place ... Other academics seemed to be wrestling with the same issue of policy relevance in their respective

> fields. And most, as far as I can see, abandoned the past. I myself have got boxes and boxes full of archival materials I've never touched and never will.[95]

Parnell recalled 'I remember explicitly somebody saying to me, "you know about why Apartheid was what it was, help us fix it, work on local government and the contemporary stuff." That's how I got into it ... and that wasn't unusual.'[96] But she also highlighted other factors beyond a desire and opportunity to work for the new democratic state. The early 1990s saw 'the end of the wave of social history intellectually, it was on a decline, it was the end of Marxism intellectually' and so a big impetus behind the new 'people's geography' of the 1980s begun to fade away. Moreover, consulting and working directly for government was well-rewarded: 'what nobody will tell is that there was a lot of money for policy work, both for direct research and some people personally made a lot of money out of it.'[97] A large proportion of geographers in South Africa now do applied research, often in a formal consulting capacity (Parnell 2007).

In another echo of the experiences of universities elsewhere in Africa after independence, funding came not only from the South African government but also poured in from international donors desperate to support the fledgling 'Rainbow Nation'. 'With the end of the boycott', Jonathan Crush recalled, 'geographers and other academics also began to arrive in South Africa in search of partners for their research projects. The feeling of isolation of the 1980s quickly dissipated.'[98] Rogerson remembers that in the early 1990s 'suddenly we had all the world aid agencies here .... Everybody wanted us ... It was bonanza time in terms of research possibilities ... it was just amazing.'[99] This wasn't an even process; suspicion of scholars at formerly 'white' Afrikaans-language universities like Stellenbosch meant that consultancy opportunities did not present themselves for these scholars.[100] At Stellenbosch, rather than a transformed agenda in the 1990s and early 2000s, research continued much as before, with Ronnie Donaldson's PhD the exception in focusing on urban post-apartheid transformation (Donaldson and Ferreira 2020).

As in other parts of Africa, this dominance of applied research lead by donor and government interests had consequences for both the originality of the research emerging from South African universities and for its ability to be critical (Ramutsindela 2002; Parnell 2007; Mamdani 2016a; 2016b; Maharaj 2019). It also led to concerns about the transparency and accountability of democratic decision-making. Indeed, Parnell noted that in many cases after 1994, young inexperienced Black policy makers were shadowed by older, often white consultants (like her!) and goes as far as to argue that 'in certain domains, including urban reconstruction, the reigns of technical power have been held in the hands of a few local and international academics, consultants and advisors' (2007, p. 118). Nevertheless, what Maharaj (2019) calls the 'policy-consultancy nexus' afforded many opportunities for South African geographers to seek to influence the work of government and private companies in exciting new ways. The enthusiasm and the elation of the early 1990s moment is captured by Jeff McCarthy:

> Mandela [was out] … it was fabulous. And then because they didn't trust the old civil servants, in the ANC they used people like us to do all the policy work, and this was again amazing. And so I combined this with university work and the university was very happy about that … all they wanted from me was to be a good social scientist, nothing more, nothing less.[101]

Many geographers were involved in the remodelling of South African cities in various ways. Between 1997 and 2007, Sue Parnell acted as a consultant for the UK, Swedish, and German Development agencies, the Cape Town and Johannesburg city councils, The South African Presidency, The South African Treasury, and the United Nations Development Programme, amongst other contracts, focusing broadly on urban policy in South Africa (Parnell 2007). In 1994 Jeff McCarthy became head of the Institute for Social and Economic Research (ISER) at the University of Durban-Westville, where with colleagues he influenced urban reconstruction policy, including on housing and upgrading informal settlements. McCarthy worked with old friends such as Pravin Gordhan (with whom he had discussed David Harvey when Gordhan was under house arrest in the late 1980s) who was now in government, on global urban development strategies.[102]

Like Gilfellan and Wesso, Michael Sutcliffe left the university sector entirely in the 1990s, becoming a member of the Provincial Legislature from 1994 to 1999, focusing particularly on local government reform. Between 1999 and 2009, he chaired the Municipal Demarcation Board which, in the context of a racially segregated South Africa, sought to redraw the boundaries of the nation in a more equitable and inclusive way by uniting formerly 'white', 'black', 'indian', and 'coloured' areas into new more functional and integrated municipalities (Khosa et al. 1997).[103] As his former colleague Dianne Scott noted, 'it was really the spatial reorganisation of the whole country.'[104] In 2002, Sutcliffe became city manager in eThekwini (Greater Durban). His tenure included the development of infrastructure for the South African FIFA World Cup 2010, a new integrated public transport system, and the delivery of over 150,000 homes. It was not without controversy though, with allegations of corruption and the attempted redevelopment of a central market strongly criticised by some community activists and academics (including Sutcliffe's former colleague at Durban-Westville, Brij Maharaj, and Patrick Bond; see Bond 2011; Maharaj 2020).

## Conclusion

This chapter has explored the role of geography and geographers in different forms of anti-apartheid and decolonisation work in the last two decades of the twentieth century. It showed how through research, teaching, activism, and community work scholars worked to transform geography, and perhaps to decolonise it, whilst also demonstrating the sometimes limited and partial decolonisation that was imagined and practised. Reflecting on the period, Dianne Scott articulated the need to understand

> the power and vicious hold the apartheid government had on the universities, academics and of course the society at large. The barriers this pernicious system created were huge and activist academics ... had a monumental task in trying to change or reconfigure the system, risking their freedom and their lives.[105]

This was difficult and sometimes dangerous work, against the backdrop of a racist, violent, and powerful regime.

Harold Wesso argued in 1992 (p. 345) that the Twentieth Century saw 'The Anglicisation, the Afrikanerisation, and the Americanisation of geography'. Over this time, the first generation of mainly British founding geographers were followed by a generation of Afrikaner academics in the 1950s and 1960s, and the discipline embraced quantification and American theories which 'resulted in the total neglect of African needs and African culture.' To what extent did this change in the 1980s and 1990s as the discipline began to change, intellectually and institutionally? In 1989, Mabin had argued that 'the legacy of discrimination which has ensured that almost every publishing geographer is white must surely be addressed internally if the discipline is to escape the charge that it reproduces *apartheid*' (1989, p. 124). However, in 2000 huge gaps in publication records remained, with these particularly stark between black African and white authors (Fairhurst et al. 2003). Whilst there were increasing opportunities for Black academics, many geography departments in the formerly white institutions remained overwhelmingly white. By 2000, of eleven departments in historically advantaged universities, only one, Witwatersrand, employed a black African academic; and Cape Town, Free State, Natal Durban, Potchefstroom, Rand Afrikaans, Rhodes, and Stellenbosch had no academics of colour on their staff at all (Fairhurst et al. 2003, p. 48). In contrast, most of the departments in the HBUs employed a diverse group of academic staff (Fairhurst et al. 2003). Whilst universities across the sector looked to position themselves in relation to the post-1994 government's transformation plans (Wolpe 1995), geography departments tended to be responding to external demands rather than driving transformation (Ramutsindela 2002).

Against this backdrop, Maano Ramutsindela argued that 'the tagging of individuals as "progressive" needs to be handled with care as unfolding events and practices at universities, and geography departments in particular, throw up a series of questions about transformation' (2002, p. 9). Moreover despite two decades of discussion about the decolonisation of geography, which was first discussed explicitly in the early 1980s (Crush et al. 1982), South African geography had yet to engage in any detail with wider African scholarship and theorising about decolonisation 'e.g. Biko, Diop, Eze, Fanon, Mamdani, Mazrui, and Nkrumah', nor to link South Africa again with the continent of Africa and the South more broadly (Ramutsindela 2002,

p. 9). The large-scale protests under the banner of #FeesMustFall and #RhodesMustFall which erupted in 2015 highlighted the extent to which these decolonisation processes remain contested and incomplete. We return to these contemporary movements in the conclusion. Before that, the final substantive chapter turns to the ongoing connections between British and African geography in the years following decolonisation. Chapter 7 builds on some of the themes of this and earlier chapters by highlighting the impacts on professional lives of repressive regimes, including notably the apartheid state, and the (sometimes forced) mobility *and* immobility produced by these dynamics. It also demonstrates the ongoing entanglement of African and British geographies, long after constitutional decolonisation.

## Notes

1 Calvyn Gilfellan, Interview with Calvyn Gilfellan and Ellsworth McPherson, Cape Town, South Africa, 18 October 2019.
2 The South African Geographical Society never had any formal bar, whilst this was lifted at the South African Society for Geography in 1976.
3 Ellsworth McPherson, Interview with Calvyn Gilfellan and Ellsworth McPherson.
4 Interview with Brij Maharaj, Durban, South Africa, 29 October 2019.
5 Interview with Sue Parnell, Bristol, UK, 19 September 2019.
6 Interview with Ronnie Donaldson, Stellenbosch, South Africa, 21 October 2019.
7 Interview with Brij Maharaj.
8 Ellsworth McPherson, Interview with Calvyn Gilfellan and Ellsworth McPherson.
9 Calvyn Gilfellan, Interview with Calvyn Gilfellan and Ellsworth McPherson.
10 Calvyn Gilfellan, Interview with Calvyn Gilfellan and Ellsworth McPherson.
11 Personal correspondence with Keyan Tomaselli, November 2019.
12 Calvyn Gilfellan, Interview with Calvyn Gilfellan and Ellsworth McPherson.
13 Interview with Michael Dyssel, Cape Town, South Africa, 17 October 2019.
14 Calvyn Gilfellan, Interview with Calvyn Gilfellan and Ellsworth McPherson.
15 Interview with Ronnie Donaldson, Stellenbosch, South Africa, 21 October 2019.
16 Interview with Dianne Scott, Cape Town, South Africa, 23 October 2019; Interview with Jeff McCarthy, via Zoom, 18 November 2019.
17 E.g. Interview with Calvyn Gilfellan and Ellsworth McPherson; Interview with Priscilla Cunnan, London, UK, 4 July 2019; Interview with Cecil Seethal, Pietermaritzburg, South Africa, 30 October 2019; Interview with Alan Mabin, via Zoom, 11 November 2019; Personal correspondence with Vernon Domingo, August 2019.
18 Interview with Priscilla Cunnan.
19 Personal correspondence with Vernon Domingo.
20 Calvyn Gilfellan, Interview with Calvyn Gilfellan and Ellsworth McPherson.
21 Interview with Michael Dyssel.
22 Interview with Michael Dyssel.
23 Wesso was not actually Dr until his PhD was awarded in 1992, after Spocter studied under him. Interview with Manfred Spocter, Stellenbosch, South Africa, 21 October 2019.

24 Interview with Michael Dyssel.
25 Interview with Michael Dyssel.
26 Interview with Priscilla Cunnan.
27 Personal correspondence with Hannes van der Merwe, November 2019.
28 Interview with Gordon Pirie, London, UK, 15 November 2019.
29 Interview with Michael Dyssel.
30 Interview with Priscilla Cunnan.
31 Personal correspondence with Ngaka Mosiane, December 2019.
32 Calvyn Gilfellan, Brij Maharaj, Jeff McCarthy and Michael Sutcliffe. Interview with Calvyn Gillfellan and Ellsworth McPherson; Interview with Brij Maharaj, Durban, South Africa, 29 October 2019; Interview with Jeff McCarthy, via Zoom, 18 November 2019; Interview with Michael Sutcliffe, Durban, South Africa, 27 October 2019.
33 Calvyn Gilfellan, Interview with Calvyn Gilfellan and Ellsworth McPherson.
34 Interview with Michael Dyssel.
35 Interview with Michael Dyssel.
36 Calvyn Gilfellan, Interview with Calvyn Gilfellan and Ellsworth McPherson.
37 Ellsworth McPherson, Interview with Calvyn Gilfellan and Ellsworth McPherson.
38 Calvyn Gilfellan, Interview with Calvyn Gilfellan and Ellsworth McPherson.
39 Interview with Michael Dyssel.
40 Calvyn Gilfellan, Interview with Calvyn Gilfellan and Ellsworth McPherson.
41 Thank you to Caitlin Graaf for help with this translation. Letter courtesy of Calvyn Gilfellan.
42 Personal correspondence with Calvyn Gilfellan, November 2019.
43 Interview with Manfred Spocter.
44 Calvyn Gilfellan, Interview with Calvyn Gilfellan and Ellsworth McPherson.
45 Calvyn Gilfellan, Interview with Calvyn Gilfellan and Ellsworth McPherson.
46 Ellsworth McPherson, Interview with Calvyn Gilfellan and Ellsworth McPherson.
47 https://planact.org.za/.
48 Interview with Michael Sutcliffe.
49 Interview with Michael Sutcliffe.
50 Interview with Michael Sutcliffe.
51 Personal communication with Alan Mabin, January 2021.
52 Personal correspondence with Hannes van der Merwe, November 2019. For example, Page D. (Van der Merwe J.H.: *Cartographer*) 1976: *Kaokoland: Evaluasie van hulpbronne en Ontwikkelingsvoorstelle.* [Kaokoland: Resource Evaluation and Development Proposals] Institute for Planning Research, University of Stellenbosch. Page D. (Van der Merwe J.H.: *Cartographer*) 1979: *'n Raamwerk vir ontwikkeling van Kavango, Vol 2, Atlas.* [A Framework for the Development of the Kavango, Vol 2. Atlas] Institute for Planning Research, University of Stellenbosch. Page D (Van der Merwe J.H.: *Cartographer*) 1981: *Development plan for the Ciskei, Vol 3, Atlas.* Institute for Planning Research, University of Stellenbosch. South Africa 1983. *Report of the Constitutional Committee of the President's Council regarding the needs of the Griquas.* P.R. 2/83. Government Printer, Cape Town.
53 Personal correspondence with Hannes van der Merwe.
54 Interview with Jeff McCarthy.

55 Interview with Anthony Lemon, Oxford, UK, 15 June 2018.
56 Interview with Dianne Scott.
57 Interview with Stephen Rule, Cape Town, South Africa, 18 October 2019.
58 Interview with Priscilla Cunnan.
59 Interview with Christian Rogerson, Hermanus, South Africa, 16 October 2019.
60 Interview with Brij Maharaj.
61 Interview with Jeff McCarthy.
62 Interview with Dhiru Soni, Durban, South Africa, 31 October 2019; Personal correspondence with Jonathan Crush, January 2020; Personal correspondence with Vernon Domingo.
63 Interview with Dhiru Soni.
64 Personal correspondence with Jonathan Crush.
65 Interview with Dhiru Soni, Interview with Brij Maharaj; Interview with Cecil Seethal.
66 Interview with Dhiru Soni.
67 Interview with Jeff McCarthy.
68 Interview with Gordon Pirie.
69 Interview with Jeff McCarthy.
70 Interview with David Smith, London, UK, 7 June 2019.
71 Interview with Dhiru Soni.
72 Interview with David Smith.
73 A booklet produced at the time of the merger of the societies in 1994 notes that at the AGM of the Afrikaans dominated South African Society for Geography in 1976, 'the "white" qualification was scrapped', however this clearly made little difference to how the Society was viewed or operated in practice. See Barnard, W.S. 'On old and new tasks, the society for geography' 15 August 1994 in *Society of South African Geographers, 29 September 1994 Merger Celebration Programme*. Society for South African Geographers Archive, Stellenbosch University Archives.
74 Interview with Dhiru Soni.
75 Interview with Gordon Pirie.
76 Interview with Jeff McCarthy.
77 Calvyn Gilfellan, Interview with Calvyn Gilfellan and Ellsworth McPherson.
78 Interview with Dhiru Soni.
79 Interview with Gordon Pirie; Interview with Chris Rogerson.
80 Interview with Michael Sutcliffe.
81 Interview with Michael Sutcliffe.
82 Interview with Dhiru Soni.
83 Interview with Brij Maharaj.
84 Interview with Dhiru Soni.
85 Letter to C. Seethal, 29 May 1989 from representatives of the South African Geographical Society, quoted in Soni and Maharaj (2005).
86 Interview with Dhiru Soni.
87 Interview with Jeff McCarthy.
88 Interview with Dhiru Soni. For the paper, see Soni (1991).
89 Interview with Dhiru Soni.
90 Interview with Brij Maharaj.

91 Interview with Anthony Lemon.
92 South African Geographical Society Newsletter, June 1994. South African Geographical Society Archive, Stellenbosch University Archives.
93 Interview with Sue Parnell.
94 Interview with Dhiru Soni.
95 Personal correspondence with Jonathan Crush.
96 Interview with Sue Parnell.
97 Interview with Sue Parnell.
98 Personal correspondence with Jonathan Crush.
99 Interview with Christian Rogerson.
100 Interview with Christian Rogerson.
101 Interview with Jeff McCarthy.
102 Interview with Jeff McCarthy.
103 www.demarcation.org.za.
104 Interview with Dianne Scott.
105 Interview with Dianne Scott.

## References

Ajayi, J.F.A., Lameck, G., and Johnson, G.A. (1996). *The African Experience with Higher Education*. Athens: Ohio University Press.

Barnard, W.S. (2016). *Encountering Adamastor: South Africa's Founding Geographers in Time and Place*, Vol. 1. Stellenbosch: African Sun Media.

Beavon, K.S. (1982). Black townships in South Africa: terra incognita for urban geographers. *South African Geographical Journal* 64 (1): 3–20. https://doi.org/10.1080/03736245.1982.10559651.

Beavon, K.S. and Rogerson, C.M. (1981). Trekking on: recent trends in the human geography of southern Africa. *Progress in Human Geography* 5 (2): 159–189. https://doi.org/10.1177/030913258100500201.

Beinart, W. (2012). Beyond 'homelands': some ideas about the history of African rural areas in South Africa. *South African Historical Journal* 64 (1): 5–21. https://doi.org/10.1080/02582473.2012.642225.

Bhambra, G., Gebrial, D., and Nişancıoğlu, K. (2018). *Decolonising the University*. London: Pluto Press.

Biko, S. (1978 [1969–1972]). *I Write What I Like: A Selection of His Writings Edited by Aelred Stubbs C.R.* London: Heinemann.

Bond, P. (2011). Limits of Interurban entrepreneurial competition in Durban: structure and agency under Neoliberal-Nationalist municipal management. *Association of American Geographers Conference*, Seattle.

Craggs, R. and Mahony, M. (2014). The geographies of the conference: knowledge, performance and protest. *Geography Compass* 8 (6): 414–430. https://doi.org/10.1111/gec3.12137.

Crush, J. (1986a). Swazi migrant workers and the Witwatersrand gold mines 1886–1920. *Journal of Historical Geography* 12 (1): 27–40. https://doi.org/10.1016/S0305-7488(86)80006-8.

Crush, J. (1986b). Towards a people's historical geography for South Africa. *Journal of Historical Geography* 12 (1): 2–3. https://doi.org/10.1016/S0305-7488(86)80003-2.

Crush, J. (1988). Tin, time and space in the valley of heaven. *Transactions of the Institute of British Geographers* 13 (2): 211–221. https://doi.org/10.2307/622507.

Crush, J. (1993). The discomforts of distance: post-colonialism and South African geography. *South African Geographical Journal* 75 (2): 60–68. https://doi.org/10.1080/03736245.1993.10586406.

Crush, J., Reitsma, H., and Rogerson, C. (1982). Decolonizing the human geography of Southern Africa. *Tijdschrift voor Economische En Sociale Geografie* 73 (4): 197–198. https://doi.org/10.1111/j.1467-9663.1982.tb00955.x.

Donaldson, R. (2020). Introduction to reflections on a century of geography at Stellenbosch University. *South African Geographical Journal* 102 (3): 273–281. https://doi.org/10.1080/03736245.2020.1795917.

Donaldson, R. and Ferreira, S.L.A. (2020). A century of human geography at Stellenbosch University: reflections on urban and tourism geography. *South African Geographical Journal* 102 (3): 327–343. https://doi.org/10.1080/03736245.2020.1795916.

Dubow, S. (2014). *Apartheid, 1948-1994*. Oxford: Oxford University Press.

Elliott-Cooper, A. (2017). 'Free, decolonised education': a lesson from the South African student struggle. *Area* 49: 332–334. https://doi.org/10.1111/area.12375.

Fairhurst, U.J., Davies, R.J., Fox, R.C. et al. (2003). *Geography: The State of the Discipline in South Africa (2000–2001)*. Pretoria: Society of South African Geographers.

Freire, P. (2017 [1970]). *Pedagogy of the Oppressed*. London: Penguin.

Fuggle, R.F. (1967). South African geography: an analysis. *The Professional Geographer* 19 (6): 345–351. https://doi.org/10.1111/j.0033-0124.1967.00345.x.

Gilfellan, C., Dyssel, M., and McPherson, E. (2016). Worlds apart: a transformative history of geography at the University of the Western Cape. In: *The Origin and Growth of Geography as a Discipline at South Africa Universities* (ed. G. Visser, R. Donaldson, and C. Seethal), 277–298. Stellenbosch: Sun Press.

Hart, D.M. and Pirie, G. (1984). The sight and soul of Sofiatown. *Geographical Review* 74 (1): 38–47. https://doi.org/10.2307/214759.

Hendricks, C. (2018). Decolonising universities in South Africa: rigged spaces? *International Journal of African Renaissance Studies - Multi-, Inter- and Transdisciplinarity* 13 (1): 16–38. https://doi.org/10.1080/18186874.2018.1474990.

Khosa, M.M. and Muthien, Y.G. (1997). 'The expert, the public and the politician': drawing South Africa's new provincial boundaries. *South African Geographical Journal* 79 (1): 1–12. https://doi.org/10.1080/03736245.1997.9713616.

Kuper, L., Watts, H., and Davies, R. (1958). *Durban. A Study in Racial Ecology*. New York: Columbia University Press.

Lemon, A. (1976). *Apartheid: The Geography of Separation*. Farnborough: Saxon House.

Mabin, A. (1984). WEJGE: the genesis of an exploration in geographical urban learning. *South African Geographer* 12 (1): 69–79.

Mabin, A. (1989). Does geography matter? a review of the past decade's books by geographers of contemporary South Africa. *South African Geographical Journal* 71 (2): 121–126. https://doi.org/10.1080/03736245.1989.9713521.

Mabin, A. and Tomaselli, K. (1973). Perspectives of South African towns and cities. *Fifth South African Students Geographical Conference*, Stellenbosch.

Magi, L.M. (1990). *Geography in Society: The Site of Struggle*. Inaugural Lecture University of Zululand.

Maharaj, B. (1989). Residential immobility — the case of tenants in 'outhouses'. *South African Journal of Sociology* 20 (4): 241–248. https://doi.org/10.1080/02580144.1989.10430707.

Maharaj, B. (1992). *Towards the reconstruction of South African geography*. Annual meeting, Chatsworth Geography Teachers Association.

Maharaj, B. (1993). Changing the old guard. *South African Geographical Society Newsletter* 4–5.

Maharaj, B. (2019). The rise and decline of radical geography in South Africa. In: *Spatial Histories of Radical Geography: North America and Beyond* (ed. T. Barnes and E. Sheppard), 315–328. Chichester: John Wiley and Sons.

Maharaj, B. (2020). Contesting violent displacement: the case of Warwick market in Durban, South Africa. *International Development Planning Review* 42 (1): 13–33. https://doi.org/10.3828/idpr.2019.25.

Mamdani, M. (2016a). Undoing the effects of neoliberal reform: the experience of Uganda's Makerere institute of social research. In: *In North-South Knowledge Networks: Towards Equitable Collaboration between Academics, Donors and Universities* (ed. T. Halvorsen and J. Nossum), 109–134. African Minds: Cape Town.

Mamdani, M. (2016b). Between the public intellectual and the scholar: decolonization and some post-independence initiatives in African higher education. *Inter-Asia Cultural Studies* 17 (1): 68–83. https://doi.org/10.1080/14649373.2016.1140260.

McCarthy, J. (1982). Radical geography, mainstream geography and Southern Africa. *Social Dynamics* 8 (1): 53–70. https://doi.org/10.1080/02533958208458315.

Parnell, S. (2007). The academic—policy interface in post-apartheid urban research: personal reflections. *South African Geographical Journal* 89 (2): 111–120. https://doi.org/10.1080/03736245.2007.9713880.

Peet, R. (1977a). *Radical Geography: Alternative Viewpoints on Social Issues*. Chicago: Maaroufa Press.

Peet, R. (1977b). The development of radical geography in the United States. *Progress in Human Geography* 1 (2): 240–263. https://doi.org/10.1177/030913257700100203.

Pirie, G. (1982). Mostly Jubek: urbanism in some South African English literature. *South African Geographical Journal* 64 (1): 63–71. https://doi.org/10.1080/03736245.1982.10559655.

Pirie, G. (1984). Ethno-linguistic zoning in South African black townships. *Area* 16 (4): 291–298.

Pirie, G. and Mather, C. (2016). Geography at the University of the Witwatersrand. In: *The Origin and Growth of Geography as a Discipline at South Africa Universities* (ed. G. Visser, R. Donaldson, and C. Seethal), 71–93. Stellenbosch: Sun Press.

Ramutsindela, M.F. (2002). The philosophy of progress and South African geography. *South African Geographical Journal* 84 (1): 4–11. https://doi.org/10.1080/03736245.2002.9713750.

Rogerson, C.M. (1985). Censorship and apartheid geography. *Environment and Planning A* 17 (6): 723–728.

Rogerson, C.M. (1990). Editorial: a New South Africa – a New South African geography? *GeoJournal* 22 (3): 227–229.

Rogerson, C.M. and Beavon, K.S.O. (1980). The awakening of 'informal sector' studies in southern Africa. *South African Geographical Journal* 62 (2): 175–190. https://doi.org/10.1080/03736245.1980.10559633.

Rogerson, C.M. and Beavon, K.S.O. (1982). Getting by in the 'informal sector' of Soweto. *Tijdschrift voor Economische En Sociale Geografie* 73 (4): 250–265.

Rogerson, C.M. and Parnell, S.M. (1989). Fostered by the laager: apartheid human geography in the 1980s. *Area* 21 (1): 13–26.

Rule, S.P. (1993). Propinquitous social diversity in Diepkloof, Soweto. *South African Journal of Sociology* 24 (1): 9–13. https://doi/org/10.1080/02580144.1993.10429872.

Simon, D. (1994). Putting South Africa (n geography) back into Africa. *Area* 26 (3): 296–300.

Smith, D.M. (ed.) (1982). *Living under Apartheid: Aspects of Urbanization and Social Change in South Africa*. London: Allan and Unwin.

Smith, D.M. (ed.) (1992). *The Apartheid City and Beyond: Urbanization and Social Change in South Africa*. London: Routledge.

Soni, D.V. (1990) The need for a people's geography. *Stanger and Northern Districts Geographical Conference*, South Africa.

Soni, D.V. (1991) Knocking at the door: the future of geography and geographers in South Africa. *South African Geographical Society Conference*, Potchefstroom.

Soni, D.V. (1993) Life, hope and despair: a socio-political anatomy of a squatter struggle in Durban, South Africa. *Association of American Geographers Annual Conference*, Atlanta.

Soni, D.V. and Maharaj, B. (2005). Voices from the periphery: solidarity and political practice (in memory of a good friend and comrade). *Antipode* 37 (5): 1027–1032. https://doi.org/10.1111/j.0066-4812.2005.00551.x.

Sutcliffe, M. (1982). Review of D. M. Smith (ed.) (1982) 'Living under apartheid'. *South African Geographical Journal* 65: 178–179.

Sutcliffe, M. (1983a). Household survey for cato manor resident's association. *Built Environment Support Group* 83 (4).

Sutcliffe, M. (1983b). Household surveys in Lamontville, Klaarwater, Hambanati, Chesterville, Shakaville and the hostels. *Built Environment Support Group* 84 (15).

Sutcliffe, M. (1983c). "Housing for all in cato manor": an empty slogan or people's response. *Community Research Unit, Built Environment Support Group* 83: 4.

Sutcliffe, M. (1984). Survey of housing conditions for Durban central resident's association. *Built Environment Support Group* 84 (16).

Sutcliffe, M. (1986). *The Lamontville Township Operation: 'Safety' or 'Security'?* University of Natal.

Sutcliffe, M. (2020). Do they make a difference: (ex) politicians with a background in planning? In: *Planners in Politics: Do They Make a Difference?* (ed. L. Albrechts), 147–173. Chelternham: Edward Elgar.

Tait, N., Donaldson, R., Hattingh, P., and Ramudzuli, M. (2016). Geography at the iconic "Bush University" of Turfloop: from University of the North to Limpopo University. In: *The Origin and Growth of Geography as a Discipline at South Africa Universities* (ed. G. Visser, R. Donaldson, and C. Seethal), 255–276. Stellenbosch: Sun Press.

Tomaselli, K.G. (2001). Contradictory subjectivity: movies, apartheid, and postmodernism. *Cultural Studies? Critical Methodologies* 1 (2): 139–156. https://doi.org/10.1177/153270860100100201.

Waters, G. (2015). Liberalism interruptus: Leo Kuper and the Durban School of oppositional empirical sociology of the 1950s and 1960s. *Transformation: Critical Perspectives on Southern Africa* 88 (1): 43–61. https://doi.org/10.1353/trn.2015.0020.

Wellings, P. (1983). Decolonising the Human Geography of Southern Africa edited by Jonathan Crush, Henk Reitsma, and Chris Rogerson - Iijdschrift voor economische en social geografie, Vol. lxxiii, No. 4, 1982. Amsterdam, Royal Dutch Geographical Society. Pp. 197–268. Dfl. 12.50. *The Journal of Modern African Studies* 21 (2): 360–364. https://doi.org/10.1017/S0022278X00023454.

Wellings, P.A. and McCarthy, J.J. (1983). Whither southern African human geography? *Area* 15 (4): 337–345.

Wesso, H. (1989). People's education: towards a people's geography in South Africa. *South African Geography Society Conference*; Pretoria.

Wesso, H. (1992). *The colonisation of the geographical mind: a critical contextual analysis of the institutionalisation and establishment of geography as an academic discipline in South Africa.* Unpublished PhD Thesis. University of the Western Cape.

Wesso, H. (1994). The colonization of geographic thought: the South African experience. In: *Geography and Empire* (ed. A. Godlewska and N. Smith), 316–332. Oxford: Blackwell.

Wesso, H. and Parnell, S. (1992). Geography education in South Africa: colonial roots and prospects for change. In: *Geography in a Changing South Africa: Progress and Prospects* (ed. C.M. Rogerson and J. McCarthy), 186–200. Cape Town: Oxford University Press.

Wilson, L. (2011). *Steve Biko.* Athens: Ohio University Press.

Wolpe, H. (1995). The debate on university transformation in South Africa: the case of the University of the Western Cape. *Comparative Education* 31 (2): 275–292.

# Chapter Seven
# Legacies of Decolonisation in African and British Geography

## Introduction

John Pugh, a former colonial surveyor and one of the founding geographers at what was then University College Ibadan returned from Africa in 1956 and moved straight into a position at King's College, University of London. He saw this as a definite promotion – 'translation from the "pioneer fringe" to a major university' – and remained at King's for the rest of his career.[1] He brought back with him a PhD on Nigerian geomorphology and continued to publish on the physical geography of Nigeria, as well as Britain, in his role at King's, and to examine the PhDs of African geographers studying in London.[2] But Pugh also brought back memories and attitudes. Bill Morgan remembered that his colleague at King's College 'used to surprise students with his account of the disease problems one could encounter in tropical conditions'.[3] On the annual surveying fieldtrip one former student reminisced:

> After taking measurements for most of the day, the evenings were spent reducing the observations to seemingly endless braced quadrilaterals, interspersed with stories of what John called 'real surveying' or 'getting your legs brown'. It was the stuff of heroes, of the lone surveyor, his faithful booker and a gang of assistants setting off into the African bush for months on end, and returning only when the task was complete. His students were enthralled as this ex-colonial surveyor told tales of great daring and endurance.[4]

*Decolonising Geography? Disciplinary Histories and the End of the British Empire in Africa, 1948–1998*, First Edition.
Ruth Craggs and Hannah Neate.

Pugh left before Nigeria became independent but decolonisation increasingly made the positions of British academics in African universities less certain throughout the 1960s. Some left willingly, building their careers back in the UK or overseas, whilst others were forced out. Deteriorating conditions in many African institutions in the 1970s and 1980s made them less attractive propositions for foreign geographers, and increasingly challenging workplaces for African scholars. This period saw a loosening but not a severing of connections between the British and African academic world. Legacies of the era of decolonisation continued to shape African and British geography into the 1990s.

The chapter begins by exploring the motivations and experiences of British geographers leaving African geography departments for positions in the UK and elsewhere. It demonstrates that decolonisation reduced opportunities for British and other foreign lecturers in African universities, alongside, and as a consequence of, the increasing opportunities for African geographers discussed in Chapter 3. However, this wasn't a straightforward or uniform process, and British geographers found roles in universities in some African geography departments for many years after constitutional decolonisation.

The second part of the chapter explores the experiences of African geographers working in Ghana, Nigeria, Tanzania, and Uganda in the 1970s and 1980s, exploring some of the challenges they faced and the ways that they utilised ongoing connections with the West creatively to allow them to get by against a backdrop of funding crises caused by the oil shocks and later structural adjustment. This section demonstrates the challenges facing African geographers seeking to develop their careers in the later twentieth century, and the ingenuity of these scholars in making opportunities for research and teaching against the odds. It also highlights the long legacies of international friendships formed against the backdrop of constitutional decolonisation for solidarity and survival decades later.

The final section explores the impact of African experiences on the development of the discipline of geography in the UK, as British geographers returned from the continent. They carried with them longstanding academic interests in the places where they had worked, affective and intellectual relationships, and material links with people and places, as well as memories of colonial experience, that shaped their subsequent careers and teaching. These connections underline the importance of the biographical approach followed in this book for grasping the colonial legacies still active in the discipline of geography today. While research exists on academic mobility (e.g. Byram and Dervin 2009; Pietsch 2013), this 'return' has not been examined in any detail, reflecting broader trends within literatures to focus on the outward movement of experts from the global north and their impact in the south rather than those travelling in the opposite direction (Hodge 2007; Healey and Upton 2010). The historiography of empire and decolonisation has traditionally discussed the 'metropole' and colony apart (Stoler and Cooper 1997), if indeed the former is examined at all, but as this chapter shows, the legacies of decolonisation in both African and British geography stretched far beyond constitutional independence and continued to connect scholarly communities across the globe into the 1990s and beyond.

## Leaving Africa

### *Pastures New*

Very few British academics remained employed at African universities for their whole careers. Just as the reasons for taking roles at the Universities of Ghana, Ibadan, Makerere, and Dar es Salaam reflected everything from the vagaries of the job market to various forms of political commitment, so too did the decision to return to the UK or continue careers elsewhere. Most had seen a role at an African university as a stepping-stone rather than a permanent position, and most returned to roles in the UK, though positions elsewhere in the Commonwealth, including in Australia, New Zealand, and Canada, as well as the United States were also common, demonstrating the continuing pertinence of the colonial British academic world into the 1960s (Pietsch 2013).[5]

John Pugh's move from Ibadan to King's College London was facilitated by the small network of geographers that was so central to staffing African universities in the 1950s. Dudley Stamp and R. O. Buchanan of the LSE (where at this time, geography was taught jointly with King's) put forward Pugh's name.[6] These connections continued to matter: Steel recruited Mansell Prothero, formerly of Ibadan for his geography department at Liverpool; and when Pugh was head of department at King's, he brought to his department Bill Morgan, who he had also known at Ibadan. Prior to his move to King's, Morgan and Roland Moss – another former Ibadan geographer – worked and published together at the University of Birmingham (see Morgan and Moss 1965; Moss and Morgan 1967). As Tamson Pietsch (2013, p. 83) notes, 'Personal relationships developed in … common rooms and laboratories were carried with academics when they migrated', and academic geographers thus pursued their careers, and secured positions, in the UK based on contacts and experience forged in Africa. Africa remained part of British geography's academic world through the 1950s and 1960s. Whilst Pugh was pleased to secure this position at King's College, there were also negatives of a move to the UK. In a letter to a future colleague he noted, 'I have quoted my present salary scale, but realize that any transfer to Britain must involve some decrease'.[7] Returning provided better career opportunities but was often also associated with a lowering of living standards, something commonly reported by colonial officials following their return to Britain (Buettner 2016).

Whilst many geographers wanted to return to the UK after a few years working in Africa, by the end of the 1960s, these positions abroad were anyway diminishing in number. Decolonisation and Africanisation reduced the demand for British academics in African universities. Whilst short term positions often funded by development budgets continued to bring substantial numbers of British and other global north academics to African universities into the 1970s, as highlighted in Chapters 4 and 5, open ended contracts reduced significantly. A twenty-year period of opportunities in the British colonial and post-colonial world was coming to an end. At the same time, the prestige of these overseas appointments was also reducing. Hugh Clout, then a young member of the geography department at University College, University of London, remembers being warned against taking a role in one of the new universities in the former colonies, which could damage a promising career.[8] It wasn't necessarily a level playing field when it came to returning and finding an academic job. As one British academic wrote to his colleague at the Inter-University Council for Higher Education Overseas (IUC), 'In some fields such as development studies, third world politics, tropical medicine and so forth, an academic from an African university may be able to demonstrate a comparatively rare expertise and be accepted quickly into a senior position in higher education in Britain', however others found it more difficult.[9] A British academic at the University of Dar es Salaam noted in 1968 that there were 'fears' amongst 'expatriate staff over the supposed difficulty of returning to academic life in Britain'.[10] By the early 1970s, there were certainly inklings of doubt about the long-term impacts of spending prolonged periods of an academic career overseas demonstrated in IUC correspondence:

> I imagine you will have heard of [the geographer] Paul Temple's success with the second chair at Birmingham, and the grapevine suggests that another former overseas Professor is on the shortlist for a UK chair. It is good to be able to demonstrate that overseas service is no bar to furtherment of a UK career,[11]

noted one IUC employee in correspondence with Bryan Langlands, the British Head of Geography at Makerere, in 1973.

Though the IUC was still recruiting for overseas universities in the 1970s, they were also increasingly being called upon to provide funding to assist academics hoping to return to work in Britain through their extensive network of contacts and through funded resettlement fellowships.[12] As well as providing a mechanism for re-employment in emergency situations (see also Kolinsky 1983), Resettlement Fellowships were designed to 'give overseas staff a chance to bring themselves up to date again in their subject if they had been somewhat isolated overseas'.[13] Implicit in this statement was a hierarchy of networks and knowledge making, with higher education institutions in Africa – including those that had been established as late-colonial projects – positioned as remote and cut off from centres of power and knowledge. For geography, a discipline deeply embedded in cultures of fieldwork, for British researchers with particular interests in Africa there was a tension between the perceived need to get 'out there' in order to stake a claim on expertise, and the need to return to the UK to exert influence and shape the academic field.

## *Reluctance and Compulsion*

In light of political instability in Ghana, Nigeria, and Uganda, the Chairman of the IUC reported in 1972 that 'Increasing use is being made in the current circumstances, of the IUC provisions for helping staff forced to leave their overseas posts for political reasons outside their own control'.[14] The IUC was actively involved in assisting staff at Makerere University after Idi Amin announced plans to remove British influences from Uganda in December 1972. The threats were part of a wider campaign of violence that hit the university directly, with the removal and killing in 1973 of the then Vice-Chancellor, Frank Kalimuzu the most high profile and horrific act.[15] Changes forced on the university included the imposition of new contracts on local (rather than more favourable 'expatriate') terms for foreign staff, alongside direct threats to academics who spoke out against the regime or its impacts at the university. For British academics, it was reported to the IUC in 1974 that 'as long as the expatriates kept a low profile things were generally reasonable' with the campus largely untouched by the military.[16]

As a senior member of staff and through his involvement with the Expatriate Staff Association, the British Head of Geography Bryan Langlands was in active correspondence with the IUC and the British High Commission throughout the period 1972–1974, writing in 1972:

> There is a wide spectrum of opinion, varying from those who have clearly made up their minds that they would now prefer to leave Uganda, to those who feel that given some guarantee for their personal security, they would prefer to stay as long as the Ugandan government feels some need for their services. Even for the individual there may be a personal conflict between a desire to ensure the proper working of the university as an institution which has commanded his loyalty for a long time, and a possibility that his continued presence could be less advantageous financially and otherwise as compared with the short term employee who decides to leave now.[17]

The Geography department at Makerere, which still employed a substantial number of British lecturers in the early 1970s, was impacted by Amin's regime, though as was often the case, the foreign staff were more protected than their Ugandan colleagues.[18] Amin's expulsion of Ugandan Asians from 1971 had already forced Gulzar Jamal, a young Ugandan Research Assistant and MA student in the geography department to leave for Canada.[19] A 'substantial core' of staff members left in 1972–1973, leaving the department decimated (Baker 1976, p. 188). The Amin regime responded to personnel shortages in academia, health, and other public roles in the form of recruitment drives that targeted countries such as Egypt, Ghana, and those in the Eastern bloc.[20] Such politically driven recruitment drives were publicised in the press, though appear to have had no impact on the functioning of the Geography Department, where instead the response was to appoint 'recent graduates as full-time special assistants' to support the staff that remained (Baker 1976, p.188). These graduates of the department had to postpone research opportunities in light of the 'heavy student admissions', negatively impacting their own research careers. Kenneth Baker (1976, p. 188), former Chair of Geography at Makerere, reported in that in 1976 'the staff-student ratio was far less favourable than the 1:15 which is indicated as a Ugandan norm'. The Kenyan Professor Joseph Ouma, a later Head of Geography at Makerere, was more direct in decrying the 'scholastic damage that Geography, as a discipline [in Uganda]... incurred in the 1970s'.[21] The rapid loss of academic staff – both foreign and Ugandan – had deep and long-lasting impacts for Makerere University. According to Sicherman (2005, p. xv), '[t]he Amin coup ushered in two decades of decline so disastrous that Makerere, like the nation, lay in virtual ruins when the forces led by Yoweri K. Museveni took power early in 1986'.

Of the 119 British staff at Makerere in July 1972 less than 50 remained by June 1973.[22] Not all of those who left were happy to do so, with many very attached to Uganda (Graham 1989). The large-scale departure of staff from Makerere constituted 'the first major emergency of this kind' which the IUC had faced in its 27 year history to 1973.[23] A call for assistance was sent out to all UK-based Vice-Chancellors early that year asking for help offering jobs to academics who left Uganda suddenly in this period. Responses were sympathetic though varied, with most sending a polite acknowledgement, and a smaller number offering to interview for posts or to provide positions in departments albeit often short-term and sometimes unsalaried.[24] The IUC also corresponded with Ahmadu Bello University, Haile Sellassie I University and the University of Benin in an attempt to try to assist British staff whose preference was to stay in Africa.

By June 1973 eleven British staff at Makerere had been awarded resettlement fellowships. In addition, seventeen found new posts in the UK. Elsewhere we see how established patterns and networks of Commonwealth and post-colonial institutions continued to hold importance with five Makerere academics getting jobs in Australia, two in Canada, and others securing roles at the University of Dar es Salaam, the University of Hong Kong, the University of Khartoum, University of Nairobi, University of Papua New Guinea, and the University of Zambia.[25] British academic staff were not the only group receiving assistance in finding new scholarly homes. A letter from the Deputy Secretary of the Committee of Vice-Chancellors and Principles of the Universities of the United Kingdom from December 1972 indicated that by that time 'fifty-four Asian students expelled from Uganda' had 'been admitted this autumn to universities in the United Kingdom either through the Central Council machinery or by universities directly'.[26]

For some, including Bryan Langlands, the practicalities and logistics of shifting an academic life from Uganda back to Britain was a particular concern. Writing to Ian Maxwell, Deputy Director of the Inter-University Council in 1973, Langlands estimated (very precisely!) that:

> If I were to leave now I would have to shift 1577 bound books, 2243 soft backed books, 1101 journals and 1592 papers and pamphlets. I have also a bibliography on Uganda of 16000 file cards, and probably 4000 other cards. I have a manuscript of a book and 80 original maps for that. I have also a valuable collection of antique maps. A lot of my books are also valuable African first editions. Possibly it is the thought of the formidable task of moving these which adds to my stability now![27]

Due to his length of service at Makerere (joining the geography department in 1953) Langland's personal geographical library can probably be considered an extreme example – a reflection of 'a distinctive personality' – though covering costs for flights and shipping personal items was a very real concern for those who left in haste (O'Connor 1989, p. 108). In the end Langlands remained at Makerere University until 1976 – much longer than most (including many Ugandans) (O'Connor 1989).

In his obituary, David MacMaster (1989, p. 366), a former Makerere colleague recorded how Langlands 'was evicted from Uganda virtually at gun point' in 1976. This followed his 'appointment as Chairman of a Commission of Inquiry into Makerere University Affairs' which examined the deaths of students on campus, primarily Kenyans (Hoyle 1989, p. 187). The inquiry was disrupted by the murder of a key witness; and Langlands 'was abruptly expelled before his report was completed' (Hoyle 1989, p. 187). Providing evidence of the complicated legacy of British academics in colonial universities, Langlands, who was criticised as colonial in his attitudes towards and marking of East African students in Chapter 3, was later hailed as 'a man of courage and determination' as he tried to protect African students and 'stood up for what he believed to be right in the affairs of the university' against the violence of the Amin regime (O'Connor 1989, p. 108).

Langlands applied for roles in other Commonwealth countries, and at one point indicated a preference for working in Fiji rather than returning to the UK.[28] On leaving Uganda Langlands was awarded an OBE in 1978, for services to higher education overseas. However, a resettlement year at the University of London 'had not been happy … too much had changed and, marked by recent trauma, he viewed contemporary England much as an outsider' (MacMaster 1989, p. 366). Langlands was eventually appointed Director of Studies and Head of Department in the newly created School of Environmental Sciences, Ulster Polytechnic. This was a move to another recently founded institution 'anxious to sustain its own identity and standards', and against a backdrop of political violence and tension, though this time in Northern Ireland (MacMaster 1989, p. 366). Langlands' Ugandan experience was seen as valuable in this context. He died aged 60 in a plane crash in 1989 whilst travelling home from the annual meeting of the Institute of British Geographers.

Similar sentiments of not quite fitting in are echoed in the obituary for Paul Temple, who lectured at Makerere for seven years (1960–1967) before becoming the Head of Department at the University of Dar es Salaam (1967–1973). Whilst

Temple was held up by the Inter-University Council as a success story, taking up a Chair at the University of Birmingham at 37,[29] his obituary following his death in 1986 aged only 49 recorded that 'he did not find the move from East Africa back to England at all easy. It was as if his heart had not travelled with him' (Doornkamp 1986, p. 244). For some, particularly those who had spent significant periods of their careers overseas, returning was a complex and not always welcome prospect.

The geographer Swanzie Agnew was also forcibly removed from lecturing roles in both South Africa in 1959, and later Malawi, in 1975. Agnew, one of very few female geographers in South Africa in the 1950s, was born in South Africa but married to a Scottish Baronet who she met studying botany at Edinburgh University. In 1953 the couple moved to University College Fort Hare, formerly the South African Native College, founded in 1916 by British missionaries to educate black Africans from across the continent. In the 1950s, Fort Hare was the only higher education institution in the country to teach more than a handful of black students (Maddrell 2009). Sir Fulque Agnew became Registrar, and Swanzie Agnew a member of the geography department.[30] Though never permanently appointed due to her position as a woman and wife of the Registrar, instead working on a series of temporary contracts to fill departmental vacancies, Swanzie Agnew was nevertheless influential. She was in post for six years, many of them as Head of Department. The Agnews were popular figures with students, identified as amongst the most open and liberal of the white staff, committed to integrated education and in their opposition to apartheid (Williams 2001). The couple were vocal opponents of the government's attempts to bring Fort Hare under state control within the apartheid university system in the late 1950s.[31] As reported in *The Times* on 28 September 1959, the Agnews were two of seven white members of the academic staff dismissed following the government takeover. The article noted that 'in the absence of official comment it can only be assumed that the dismissals have been ordered because the Department of Bantu Education considers that the staff affected are not in sympathy with the ideal of *apartheid*.'[32] This was confirmed a week later, when the Minister of Bantu Education declared that he had 'disposed of their services' because he would 'not permit a penny of any funds of which I have control to be paid to any persons in these institutions who are known to be sabotaging the Government's policy of apartheid' (Horrell 1961, p. 277, quoted in Williams 2001, p. 526). While British nationals were ejected, some South African staff were put on trial, and it is important to recognise in this case, as in the case of Uganda, the greater freedoms available to foreign rather than local staff when the situation became unbearable or they were removed. Nevertheless, the Agnews' ejection lead to both 'financial and emotional strain', with many tears shed (Williams 2001, p. 530).

Following a five-year period, 1960–1965, when she became academic head at the Royal Ballet School in Richmond, UK, Agnew returned to Africa – taking up a post as Professor of Geography at the University of Malawi.[33] Agnew was at the

University of Malawi for a decade, and returned to Scotland having resigned her post in response to President Banda's regime, 'after being made to live and work in increasingly intolerable conditions'.[34] As Avril Maddrell has noted 'to count Swanzie Agnew's influence on geographical knowledge purely in terms of publications is at best limited. She played an important role through her principled life and teaching of university geography contra regimes which were repressive and racist' (2009, p. 252). We can put Swanzie Agnew, Kenneth Baker, and Bryan Langlands (both Makerere), and Michael Barbour (Ibadan) in this category of British geographers who spent the large part of their careers overseas and ended their careers without illustrious publishing records. However, they made palpable impacts in other ways.

Whilst not all British geographers wanted to leave Africa, the unfolding political situation across many African countries undoubtedly took the shine off such positions. For example, one British academic working at Dar es Salaam reflected that by the mid-1970s:

> the authoritarian nature of the state became very apparent ... In a sense that I went out there hoping for something new and the whole "new" had been sort of tinged with authoritarianism and the whole ethos had become a bit rusty. So the appeal of the country as a place where experimentation was going on in a way which was different from other newly developed countries disappeared. So for expats anyway I think the appeal probably diminished.[35]

## Isolation and Creativity

### *Growing Isolation*

Whilst British academics were free to leave African institutions when conditions became difficult, most Africans did not have the same luxury and those that remained in universities in Ghana, Nigeria, Uganda, and Tanzania struggled against the austerity brought about by oil shocks, structural adjustment, and political instability from the 1970s to the 1990s (Ajayi et al. 1996; Mkandawire 1995). With this austerity came increasing isolation from international networks. One way to track this is to explore subscriptions to the International Geographical Union (IGU). The IGU provided welcome mobility and visibility for African academics, and an international network which did not centre on the UK or North America. But their archive tells a story of repeated failures to pay subscriptions. As noted in Chapter 3, in 1976, around the time he was deported from Uganda, Brian Langlands paid $700 of the country's IGU arrears.[36] By the middle of the 1980s, Nigeria, too, had begun to default on its dues. A combination of donations from an oil company in addition to personal payments from Nigerian geographers were needed to keep Nigeria in the Union throughout the decade, and again came to rescue in the mid-1990s, as 'Bose Iyun, Deputy President of the IGU was almost prevented from standing again due to Nigeria's parlous position and lack of payments. One letter read:

> Enclosed is a money order for $375.00 to enable Prof Fola ['Bose] Iyun to retain her chance at re-election for vice-presidency. I do not want to think of it as a contribution towards Nigeria's delinquent annual dues. Though I guess that's what it is.[37]

These failures to pay reflected wider problems in many African universities which were mounting from the late 1970s. Alongside domestic instability (in Uganda, Nigeria, Ghana, and South Africa), World Bank instigated Structural Adjustment Programmes cut social spending in sub-Saharan Africa by more than a quarter between 1980 and 1986, and this included swingeing cuts to higher education subsidies (Zeilig 2009). The World Bank report *Accelerated Development in sub-Saharan Africa*, known as the 'Berg Report' (1981) suggested prioritising funding for primary over secondary, further and higher education, and some of the biggest donors in the earlier twentieth century – the American foundations and European development agencies – also shifted their funding priorities away from higher education (Ajayi et al. 1996). By the early 1990s, universities across the continent were starved of resources (Alidou et al. 2000; Amutabi 2002; Kerr and Mapanje 2002; Musisi and Nansozi 2003). Just as Cold War and decolonising geopolitics had aligned to pour funding into African higher education in the 1950s and 1960s, from the 1970s onwards, geopolitical and geoeconomic contexts that reached far beyond the university and the continent of Africa were to fundamentally shape the African university landscape in the last decades of the twentieth century.

In Nigeria, Akin Mabogunje retired from the University of Ibadan at 50, in 1981; by then, he remembers, 'things had deteriorated … the universities were being starved of funds by the military, and so the quality of research had started nose-diving'.[38] By this point, in the context of spiralling inflation 'the remuneration of university teachers became hardly adequate to keep body and soul together' (Mabogunje 1998, p. 16). Reflecting on the materiality of this deterioration highlights the infrastructural issues and inequalities that produced hugely challenging working conditions. Comparing the 1970s to the conditions experienced from the 1980s onwards, Albert Aweto, awarded his PhD in 1978 from Ibadan and still working in the department in 2017, explained:

> Of course then the facilities were better, at least the infrastructure, the power outage was rare if it occurred at all … water was always available. So many of the problems students now contend with, like power, electricity outage, water, were non-existent then … But all that began to change with the inception of the structural adjustment programme which resulted in the massive devaluation of our national currency. It

> became difficult for our library to import books and journals from abroad so there was no continuity in terms of the journals. You may be looking for a particular issue of the journal and it is not on the shelf, it was not even ordered because of the problem of obtaining foreign exchange to purchase journals from abroad.[39]

These 'morale-shattering downturns' and 'the deplorable state of teaching aids and facilities' were 'very demoralizing to both staff and students', recorded Areola and Faniran in 1998 (pp. 33; 41). When the University of Iowa geographer Gerard Rushton visited Ibadan to work on a collaborative research project in 1982 with 'Bola Ayeni, Ayeni remembers 'things were so bad. There was no water, there was no electricity. Every morning I would get buckets of water – I lived in town – to carry them into campus, so they would wash their faces.' According to Ayeni, the trip improved Rushton's view of the Nigerian's work: 'he said if someone can stay in the environment I know, and write [that is impressive]'.[40] But many didn't. Speaking in 2004, Udo reflected on the damaging impact on African scholarship of the 'ongoing brain drain by Africans who are fed up with poor pay, decaying infrastructure, police brutality and general social insecurity' (p. 98). Hountondji (1990) argues that this brain drain is one important aspect of a wider intellectual extraversion affecting African universities.

This impact was not limited to Ibadan, or indeed Nigeria. Structural adjustment hollowed out universities across Africa, restricting funding, bringing in student fees, and shifting understandings of their value from that of public good to individual career investment (Mamdani 2007; Luhanga et al. 2000). At Makerere, in Geography, staff shortages were compounded by other factors such as 'the inflationary costs of hiring vehicles [which] made their impact upon the provision and, in particular, upon the mounting of residential fieldwork' (Baker 1976, p188). At the University of Dar es Salaam, the former Head of Geography Len Berry was shocked when he visited in 1987, to find that his former colleague Adolfo Mascarenhas was compelled to run a small bus company with his wife, Ophelia, also a prominent academic at the university, alongside their university work, to make ends meet.[41]

Since the late 1970s (and earlier in Uganda, where the regime of Idi Amin hugely damaged higher education), many academics had been forced to take up a range of supplementary jobs in order to get by (Kwesiga 2002; Mamdani 2007). As conditions worsened in the 1980s and 1990s, consultancy opportunities, which had existed for decades, became an increasingly important source of income for African scholars based on the continent (Wight et al. 2014). As Serunkuma (2018, p. 125) reflects, for academics having to 'eke an existence in several side-kick engagements … consultancy [offered and] offers an easy respite to the material challenges.'

Twenty-first century critiques of the 'galloping consultancy culture' (Mamdani 2011, n.p.) in African universities highlight the negative consequences of the dominance of these modes of research:

> Consultants presume that research is all about finding answers to problems defined by a client. They think of research as finding answers, not as formulating a problem. The consultancy culture is institutionalized through short courses in research methodology, courses that teach students a set of tools to gather and process quantitative information, from which to cull answers. (Mamdani 2011, n.p.)

Moreover, academics working in these types of roles (either explicitly as consultants for international organisations or implicitly as unequal partners gathering data in research projects lead from the global north) often lack agency in defining important areas of research and are not involved in the creation of new knowledge (Wight et al. 2014). Most searingly, Mamdani (2011, n.p.) has described this role as similar to that of 'native informers'. This derogatory description fails to highlight the political and economic contexts in which African researchers (and indigenous communities working with colonial authorities in an earlier period) made their decisions about research and career directions (Serunkuma, 2018). Nevertheless, longstanding critiques of applied research, of which consultancy is the most extreme version, suggest that the austerity of the late 1970s onwards has reproduced and reinforced colonial patterns, entrenching inequalities in and between institutions in the North and Africa, and undermining efforts to decolonise universities economically or epistemologically (Hountondji 1990).

### *Solidarity and Creativity*

Despite these extreme challenges of austerity and repression, individual campaigns, connections and solidarity made a difference in individual cases, allowing African academics to continue to travel, research, teach, and publish. One important example was a 'twinning relationship, which started in 1988 between the University of Ibadan and the University of Iowa' which allowed for staff and student exchanges, joint research and library donations (Udo 2004, p. 98). This built on long-standing connections and friendships developed through the visiting scholar scheme nearly two decades earlier: Mike McNulty of Iowa had visited Ibadan for a semester in 1970–1971 and made longstanding friendships with a number of Ibadan colleagues (Udo 2004).

The valuable twinning relationship was brought to an abrupt end when Sani Abacha became head of state in Nigeria in 1993 (in a return to full military rule) and 'the US took a tough stance against Nigeria, including stopping that funding'.[42] But connections continued and hard work in Nigeria and the US resulted in the establishment of a GIS lab. Ayeni went to the US to physically carry the computers back to Nigeria with Iowa geographer Rex Honey.[43] The resulting Masters programme in GIS was, according to Udo, 'perhaps the most visible and memorable achievement' of the twinning (2004, p. 98). Reflecting at the end of the century, the Nigerian geographers Areola and Faniran (1998, p. 41) described the fortitude and innovation of their colleagues: 'academic staff have continued to push along in the true age-old tradition by bringing to bear on their teaching and research activities all their resources of ingenuity and ability to improvise to make the best of a bad situation.'

There are similar examples across African geography departments as individuals used creativity, ingenuity, and their connections, to leverage funding, expertise, and other support. In the late 1980s, George Benneh of the University of Ghana worked with his Danish colleague Henrik Jeppeson to secure funding of 1.6 million krone (about 70 million cedis) for a remote sensing unit from the Danish development agency DANIDA.[44] The two had met when Jeppeson visited Legon in 1966–1967, as part of 'The Danish Geographical Society's Ghana expedition, where the former Danish buildings on the Gold Coast were mapped and described.'[45] As Benneh recalled, Jeppeson 'met me in the senior common room of Commonwealth Hall and asked me whether I visited Denmark when I was in Britain, and I said no. I said the first opportunity you invite me to come. And I did. And lecture, based on my fieldwork'.[46] This encounter led to repeated visiting positions for Benneh in Denmark (in 1973, 1983, and 1996),[47] and, in the late 1980s, to an invitation from Jeppeson, now Vice-Dean of Science at the University of Copenhagen, to a conference organised by DANIDA at which the funding was secured.[48] Remembering the event in his autobiography, Benneh recalled:

> When I arrived in Copenhagen in the evening, Jeppeson told me that if I wanted the Remote Sensing Centre in my department, then the next day would be my big opportunity because all the big guns in DANIDA would be there. That night I had to sit down and write my speech. Lo and behold, the next day we went to the conference and not only were DANIDA officials there, but their minister for overseas development as well! I read my paper and argued very strongly for a remote sensing applications unit in Legon … during the break the minister came to see me, and after congratulating me, told me how much he had enjoyed my presentation. He said that I had made a very strong case and they would support the application. (Benneh 2016, pp. 47–48)

## Legacies in the UK

Connections with African colleagues and departments also shaped British geography. In 1972 a disciplinary review of UK geography noted that there were 'about 50 staff members with recent experience of teaching in tropical African universities' in British geography departments (Steel and Watson 1972, p. 142). Africa dominated as a result of broader colonial dynamics: 'there may well be more Africanist geographers' noted Ben Farmer a year later, 'because of the teaching opportunities till recently available in ex-British territories' in Africa, which were 'by and large denied or absent' in most of South East Asia (Farmer 1973, p. 11). Although not everyone continued to conduct research in Africa once they returned to Britain, the opportunities afforded by the development of colonial higher education shaped the regional interests of geographers in post-war Britain and resulted in the production of significant research outputs. An obituary of Dick Hodder records that 'his years in Nigeria provided experience which would source the analysis of a number of books on African economic and political development'; and Africa remained central in the work of Hodder's former colleagues at Ibadan: Prothero, Moss, Morgan, and Barbour amongst others.[49] Tony O'Connor, who worked at Makerere and Ibadan amongst other African universities before taking up a position at University College London recorded that 'the impact of my African experience was huge.'[50] For many returnees their time in African universities resulted in connections that were sustained through a range of academic practices: external examination (for example John Pugh in Ghana and at Ibadan); research visits and visiting professorships (for Bill Morgan in Nigeria; Rowland Moss in West, East, and South Africa); consultancy for international agencies (such as Mansell Prothero for the WHO and Len Berry for the World Bank and UNDP); and supervision of African PhD students.[51]

## *Textbook Africa*

Textbooks were the most obvious way in which experiences in the (former) colonies impacted upon the teaching of geography in Britain, as well as in Africa. Carr-Saunders (1948, p. 6), speaking to the Geographical Association in 1947, had cautioned that academic research might have to come after 'the preparation of material for urgent teaching needs'. Many turned these teaching materials, alongside their academic research, into textbooks. For example, Pugh co-authored *Land and People in Nigeria* (1955) with his first colleague at Ibadan, Keith Buchanan, and *West Africa* (1969) with Morgan.

Buchanan and Pugh's 1955 *Land and People in Nigeria* provides a good example of the typical output of staff at African universities in the 1950s and 1960s. This textbook was used on university courses, including at Ibadan, but also in schools in Nigeria and the UK. It includes chapters on physical geography and climate, resources and land use, transport, communications and services, and, intriguingly, 'The Future'. Though the volume was published just five years before independence, and soon after the implementation of a new constitution (providing more but still limited power to Nigerian politicians), this final chapter, just two pages long, makes no explicit mention of the challenges and opportunities of political change. Rather, it focuses on the need for more modern intensive farming methods and population redistribution to feed and manage a rapidly growing population, and the requirement for increased education and health provision to guarantee agricultural efficiency. The chapter advocates a middle way in colonial development, with a continuing role for white communities and the colonial power. This was required, according to the authors, because of the 'attitudes and abilities of Nigerians', who, due to a lack of education under colonialism, were not entrepreneurial or willing to invest in modern developments. Indeed, the index only records five pages on politics ('Political organisation') which describe the system of colonial governance, just one more page than is accorded to 'fruit', which is also supplemented by separate entries for 'Bananas, coconuts, fruit canning, fruit squash manufacture, melons, pineapples.' The politics of race, development and decolonisation are almost entirely absent from the book, even if occasional sentences – for example, mention that the increasing trend of social segregation of the white population from the black is 'fraught with danger' – demonstrate a concern for these issues (Buchanan and Pugh 1955, pp. 249, 251, 99). *Land and People in Nigeria*, 'a classic in Third World regional Geography', continued to be used in Nigeria for much longer than the UK, remaining in 2004, according to Reuben (Ken) Udo, 'the best regional text on Nigeria' (2004, p. 90). In representing the geography of African countries to more than one generation of Nigerian school and university students (as well as students in the UK), these textbooks, very often authored by British academics, made a substantial contribution to the disciplinary imagination of Africa in this period (Myers 2001).

British lecturers at other colonial university colleges also wrote standard texts, national atlases, and later more systematic accounts (Steel and Prothero 1964; e.g. Prothero 1969; Agnew and Stubbs 1972; Morgan 1973). A good example is *The Lands and Peoples of East Africa* (1960) by Gladys Hickman and William Dickens. Hickman, who was seconded to Makerere College (1960) and the University College of Ghana (1958–1961) from Bristol University, drew on her experiences on secondment and of travelling extensively – 'including from Ghana to Timbuktu by VW Beetle' – to produce this and other texts (Geographical Association 2006 cited in Maddrell 2009, p. 143). Following Dickens' death Kenneth Baker and Robert Steel advised on the final text. With a background in geographical education, Hickman's contributions to academic geography are not often recognised. Hickman's *The Lands and Peoples* was a textbook primarily aimed at African students who were, at the time, still sitting the Cambridge Overseas School Certificate, though Hoyle noted in his 1975 review of the new edition that 'the book deserves to be much more widely used by those in other parts of the world making a special study of East Africa' (p. 246). By 1970 it was in its seventh impression (Maddrell 2009). It was also reprinted in 1986, reflecting how geographical teaching and learning about East Africa in the late twentieth century continued to be influenced by the secondments and academic exchanges that took place in the late colonial period.

Another book by Hickman, *The New Africa* (1973, with later editions in 1976, 1980 and 1990) was aimed at the British market. A criticism of *The Lands and Peoples* related to how, in discussing development, 'the political dimension [was] largely ignored' (Hoyle 1975, p. 246). *The New Africa*, was again based on Hickman's travels, and, whilst it included a discussion of apartheid under a heading 'Separation and discrimination' this book was also strongly criticised:

> it includes some information on the workings of apartheid, but continually modifies any adverse impressions. For instance, after describing the journey black Sowetans have to make into 'white' Johannesburg each day she adds: 'However, their journey is perhaps no worse than that which thousands of Londoners voluntarily undergo daily commuting to work.' (Naidoo 1985, p. 23)

'Perhaps Ms. Hickman has seen Londoners travelling in between railway carriages and hanging out of open doors and windows while their train is in motion', the article's author – the acclaimed South African children's writer Beverley Naidoo at that point in exile from the apartheid regime in London – adds caustically (1985, p. 23). Explicit discussions of apartheid were not standard fayre of regional overviews, national atlases and country profiles, whether in geography textbooks or academic texts. Indeed, when Anthony Lemon studied African geography as an undergraduate Oxford,

> the classic geography textbook on South Africa at that time was simply called *South Africa* (1961) by Monica Cole ... Chapter forty-four if I remember is called 'the Bantu Territories' or something like that and that's it, you know, it's about as apolitical as you could get, ... virtually ignores apartheid and the huge impact that it has on social, economic and political geography, it's quite extraordinary.[52]

## *Area Studies, Development Studies, and Development Geography*

Alongside teaching resources, returnees were central in the foundation of several new research centres at British universities. Following the 1961 publication of the government commissioned Hayter Report, selected universities were chosen to act as centres for 'area studies' (Sidaway 2013). Bill Morgan, who moved to the geography department at the University of Birmingham from Ibadan in 1959 'played a significant role in the founding [in 1963] and early development of the Centre of West African Studies' there, alongside another Ibadan colleague, Roland Moss (Moss 1992, p. xii; Fage 1964). Morgan was also one of the founder members of the African Studies Association of the United Kingdom (in 1963), and an early chair of the Developing Areas Study Group, founded in 1973 (Coppock et al. 1973; Gilbert 1974; Dixon 1981). In these roles, Morgan was not unusual. The geographer David McMaster, formerly of Makerere University College, Uganda was also central to the other Centre for Africa Studies founded in response to Hayter, in Edinburgh, and in the early 1970s geographers filled three principal officer roles at the African Studies Association and were 'the second largest group' of the organisation's overall membership (Steel and Watson 1972, p. 142).[53] Indeed, the 'Africa-wide experience' also made it possible for Tony O'Connor 'to act as Secretary/Vice-Chair of the UK Standing Committee on University Studies of Africa, and later the European Council on African Studies'.[54]

In the 1960s, area studies, and the study of 'developing areas', became more formalised in the UK. This included at the new 'plate glass' universities founded in the 1960s such as the Universities of East Anglia and Sussex which included the first Development Studies programmes in the country. In its first decade, the University of East Anglia's development programme was staffed by an array of academics, including several geographers, who had experience working at African universities, adding another layer of complexity to the story of how colonial officials and scientific experts became development professionals (Kothari 2005).[55]

In part the shift to area studies and development reflected broader geopolitical and educational shifts, a desire on behalf of the UK government not to lose its competitive advantage in expertise about developing areas as it withdrew from the colonies, and a concomitant rise in interest in area studies in the US reflecting Cold War concerns (Child and Barnes 2019). But the founding of such centres and associations, as well as the building up of African or South East Asian expertise in particular departments also reflected the interests of a large group of academics, geographers amongst them, who had worked in the colonial universities. They in turn taught a whole new generation of Africanists and development geographers who studied in these departments in the 1960s and 1970s, and who sometimes then travelled to and worked at African universities for short periods themselves in the 1970s and 1980s.

Working in Africa sometimes catalysed a shift from pure to applied research and, at the same time, from scientist to development 'expert'. Len Berry, at the Universities of Khartoum and then Dar es Salaam, shifted his focus considerably as a result of these appointments: '[Africa] had a major impact on my trajectory, I could have ended up doing hydrological and geomorphological kinds of things.' Instead a growing interest in water supply and desertification meant 'I really got very focused on the human side of stuff … really moving from being a geoscientist nerd, to being more of an activist person.'[56]

Within geography departments, regional expertise in (parts of) Africa were institutionalised through the experience of those who had worked in African universities, enriching the academic lives of both the teaching staff and many students who took their courses.[57] John Pugh and Bill Morgan formerly of Ibadan contributed to the regional geography of Africa courses at King's College London. These covered:

> the physical environment, its evolution and distinctive features, peoples and colonisation, natural resources and the distribution of economic activities, the interaction of physical and human elements in the evolution of the cultural landscape.[58]

A student who took the course in the early 1970s remembers that 'it was like the old regional geography, in that each topic was covered thematically. So you'd have agricultural change, mining …, the role of transport, urbanisation, and so on …', though by the 1970s the focus on post-colonial development was more pronounced, and the course covered 'the scope for development now that they [African countries] are independent'.[59]

Regional courses continued to be taught into the late 1970s at the University of London and beyond, despite the fact that by 1972, twenty-four percent of departments claimed to have already 'given up the teaching of regional geography' altogether, in favour of 'Systematic geography based on quantitative methods' (Steel and Watson 1972, pp. 150–151). Staff with African experience did not fall on one side of the disciplinary debates between regional geography and new quantitative methodologies that dominated the 1960s. For example, of those who had worked at Ibadan, Pugh and Barbour were often seen as representing traditional regional description, but Hodder – who returned from Nigeria in 1963 and spent most of the rest of his career at School of Oriental and African Studies (SOAS) – was associated with the new interdisciplinary area studies and a focus on economic and political change. Michael Chisholm, Terry Coppock, and Ed Soja, at Ibadan in the 1960s, along with their Nigerian colleague Akin Mabogunje, were associated with the new quantitative and theoretical geography.[60]

Peggie M. Hobson taught in South Africa at the University of Cape Town (1953–1955), followed by the University of Natal, Durban (1955–1963) (Bartholomew 1989). After returning to the UK in the mid-1960s Hobson was given a job overseeing a female hall of residence at University College London. To the surprise of the geography department the job came with a part-time senior lectureship, which did not please the head of department – particularly as Hobson had a limited publishing profile and a demeanour and approach to teaching that was seen to be more akin to a school teacher, than an upcoming geographer.[61] It was challenging for some returnees who were older and brought with them approaches to teaching and research that were seen as outdated. Hobson's course on South Africa was criticised by younger colleagues for whom regional geographical approaches were no longer adequate. Amongst this it's also important to consider attitudes towards age and gender: it would not have been easy for Hobson to join a department as a female member of staff 10 or more years older than some of her predominantly male colleagues. Here, gender and time spent overseas, contributed to ambivalent and negative responses from some colleagues.[62]

In addition, Peggy Hobson's course said virtually nothing about the politics of apartheid, and as such her interpretations were also unpopular amongst the more politically active students, who found the course out of sync with the situation they knew was unfolding in South Africa.[63] But in this Hobson wasn't unusual: despite the pace of political change and the increasingly political nature of debates about Africa in this period, King's College London students of the time 'don't remember [lecturers] … talking about politics [or] … the contrast between rich and poor'. Studying geography at Oxford in the mid-1960s, Tony Lemon took a course on Southern Africa and doesn't remember 'writing any essay on South Africa that was really about apartheid'. The course drew on Monica Cole's textbook *South Africa* which as we've already heard virtually ignored apartheid; Lemon reflected that 'I guess the course largely did as well.'[64]

## *African Connections, Colonial Nostalgia*

The fact that many lecturers had experience of Africa was a draw for many students when choosing their courses: 'the reason why I took the … course in the second year, was because I knew … that they had some sort of connection with Africa' remembered one King's College London student. Moreover, this fact affected the way that the material was represented in class: 'I think because they'd worked there, they knew Nigeria, for example, was different from Ghana, and they were interested in the difference between Nigeria and Ghana, in why they were different'.[65]

Randall Baker who studied undergraduate geography at the University of Wales in Swansea (1962–1965) articulated the powerful influence of being taught by geographers with experience in Africa:

> During my studies at the University of Wales I had "specialized in Africa" and had caught the spirit of enthusiasm for that continent from some of my teachers. At that time there were many professors who somehow had been involved in the colonial experience and who continued to conduct their research in Africa, bringing first-hand experience and excitement to the classroom. (Baker and Polonskyi 2015, p. 15)

It wasn't just staff with experience and enthusiasm for the continent that sparked his interest. Baker also recalls how his student peers also provided connections and insights into the emerging postcolonial world:

> Our student body at the University of Wales had a sprinkling of Nigerians, Ghanaians and Gambians, and we shared with them the ardour that accompanied the coming to independence of their particular homelands, which in some cases was actually as we studied. At one point, the international student society seemed to be having a "coming out" party for a new country almost every month. How was it possible *not* to be swept up by this enthusiasm? (Baker and Polonskyi 2015, p. 16)

Following the award of a Rockefeller Scholarship, Baker went to participate in the African Studies Programme at Makerere in 1965 and he attributes his time at Makerere as the starting point for a continent-hopping career.[66] Gina Porter, who was taught by Mansell Prothero and Robert Steel as an undergraduate at the University of Liverpool, provides another example of this phenomenon. She remembers:

> Liverpool was very Africa focused. So my undergraduate teaching … had been Africa focused. There was a whole group of people who'd worked in sub-Saharan Africa, and postgrads too …. So I suppose that I was pretty much steeped in it … and there were African research students too. So I had a friend from Cameroon, from Sudan …[67]

As a result of this teaching, Porter developed an interest in Africa, and specifically northern Nigeria which has spanned her whole academic career, leading directly to the decision to work in Nigerian universities, including at Ibadan, for more than a decade from 1974, before she returned to the UK and the University of Durham.[68] In this, she, and many others, benefitted from '[Robert] Steel's active support and assistance'.[69]

A wide range of different threads fed into the new geography of 'developing areas' emerging in the 1970s. These quite different approaches to teaching about 'development' or 'Africa' often co-existed within one department or even course (albeit not always entirely happily!).[70] At Liverpool in the late 1960s and early 1970s these debates were vigorous and exciting:

> There was a whole younger cohort in the department who were very critical of Robert Steel and people like that, of course, and that was also what made it very interesting. All those debates about what was development about? What is it about? And what should people's position be if they went out to work in those locations? Do they have a place at all?[71]

As some of the British radical scholars who had worked at Dar es Salaam in the 1970s on questions of underdevelopment and dependency returned to positions in UK Universities, they joined a new generation of geographers increasingly critical of the development project and drawing on Marxist and later feminist and post-colonial critiques. But many of this new generation gained their first insight into and interest in African geographies through the academics returning from positions in late colonial and newly independent countries. Debby Potts provides one such example: as an undergraduate she learnt about African geography from Tony O'Connor who had completed his PhD research at Makerere in the 1960s, and she was then appointed as a lecturer at SOAS by Dick Hodder who had been at Ibadan in the 1950s (teaching Akin Mabogunje's wife Titi to drive whilst he was there, Mabogunje 2011). At SOAS Potts inherited Hodder's African development course, though she revised it substantially. Potts was one of a younger generation of increasingly radical scholars exploring urbanisation, migration and development in southern Africa. She was heavily involved in the anti-apartheid campaign in the UK, including writing resources for school geography students on the subject (Potts 1985; Figure 7.1; Norcup 2015) and being a member of the non-stop against apartheid picket outside the South African Embassy on Trafalgar Square (Brown and Yaffe 2014).

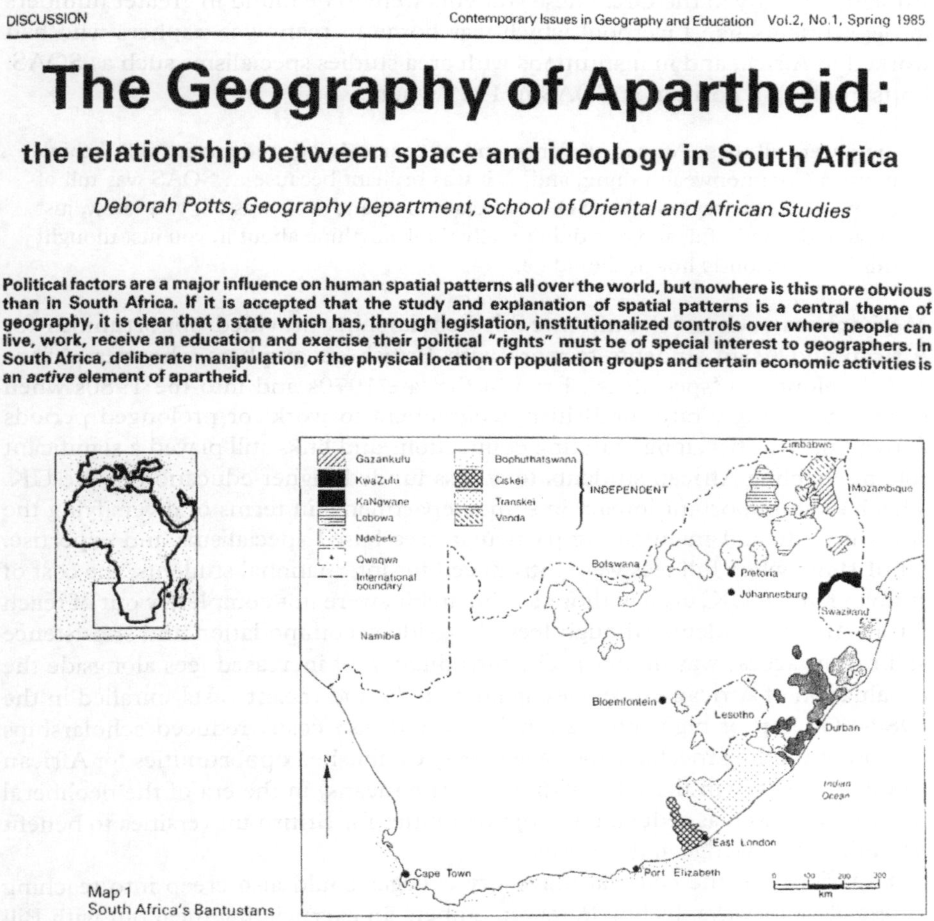
DISCUSSION Contemporary Issues in Geography and Education Vol.2, No.1, Spring 1985

# The Geography of Apartheid:

## the relationship between space and ideology in South Africa

*Deborah Potts, Geography Department, School of Oriental and African Studies*

**Political factors are a major influence on human spatial patterns all over the world, but nowhere is this more obvious than in South Africa. If it is accepted that the study and explanation of spatial patterns is a central theme of geography, it is clear that a state which has, through legislation, institutionalized controls over where people can live, work, receive an education and exercise their political "rights" must be of special interest to geographers. In South Africa, deliberate manipulation of the physical location of population groups and certain economic activities is an *active* element of apartheid.**



Map 1
South Africa's Bantustans

**Figure 7.1** Part of an article for school children on apartheid. Potts, D. (1985) 'The Geography of Apartheid' *Contemporary Issues in Geography and Education*, 2: 2–7. Courtesy of Geography Workshop Productions and CIGE/Association for Curriculum Development in Geography www.geographyworkshop.com.

But Potts was heavily influenced in her career by both Hodder and O'Connor, part of the first generation of geographers writing about 'developing areas' through more traditional regional, economic and modernisation lenses. She took from O'Connor especially a deep knowledge of the continent, an appreciation of the value of extended fieldwork in Africa, and a commitment to the research and careers of African scholars.[72]

Returnee British geographers brought with them not just knowledge and expertise but also extended networks that encouraged and supported overseas students to study in the UK. These students were to be found in greater numbers in departments like Liverpool which was home to many geographers who had worked in Africa, and in institutions with area studies specialisms such as SOAS. Potts' took up a position at SOAS in 1979 where:

> there were all these African students, and of course Asian students as well, I think it was a Commonwealth thing, and … it was brilliant because … SOAS was full of people from the regions that SOAS specialised in and so it was all, you know, just great and wonderful, and one didn't really think anything about it, you just thought this is … obviously how it should be.[73]

Mechanisms such as Commonwealth Scholarships provided opportunities for African and Asian students to work with British geographers who were Africa and development 'specialists'. Even in the late 1970s and into the 1980s when it was increasingly rare for British geographers to work for prolonged periods African universities, long-standing connections and links still played a significant role in enabling African students to access funded higher education in the UK. This had an important impact in some departments in terms of diversifying the student body and maintaining particular area-based specialisms and expertise. Until 1980, when full fees were introduced for international students, the cost of studying in the UK, even without a scholarship, were not completely out of reach for all African students (though fees alongside accommodation and subsistence still meant access was limited). The introduction of increased fees alongside the devaluation of African currencies against the Pound meant costs spiralled in the 1980s. On top of high international fees and visa costs, reduced scholarships and other funding mechanisms have hugely diminished opportunities for African students to study in the UK (without private means) in the era of the neoliberal university, as well as reducing the opportunities for British universities to benefit from their knowledge and expertise.

Experience in the colonial university colleges could also creep into teaching in less obvious ways. Debby Potts remembers an excruciating meal out with Bill Morgan (King's College, University of London) and George Kay (University of Staffordshire, and formerly the University of Rhodesia) in the 1980s, where the two older men 'talked over' her and reminisced about when they were surveying in Nyasaland (now Malawi) with the help of 'native porters'. The porters were local men who had been imprisoned for failing to implement a colonial era agricultural modernisation scheme (which turned out to be very problematic – see Beinart 1984) and were used as free labour. It was very 'boys own stuff' and depressing for a young critical development geographer.[74]

Returnees sometimes brought with them not only knowledge and interest in Africa, but also memories, recounted and passed on to students. Particularly for those who worked in the colonial colleges in the earlier years before independence, these memories could be of colonial service, and could reproduce notions of geography as a masculine, heroic, expeditionary science, and

Africa as exotic and sometimes threatening (Blunt 1994; Driver 2001). These reflected longstanding understandings of the spaces of empire, and the role of the geographer within them. A consideration of the careers of academics returning from positions in the (post)colonies demonstrates how colonial disciplinary practices were actively performed to a new generation of students far beyond the timelines of formal decolonisation.

The association of some returning geographers with the British empire was passed on through lectures and informal conversations, though the 'colonial' – understood positively and negatively – could also be conveyed (or assumed) by younger colleagues or students by the very fact of a career path that had passed through Africa.[75] Some of Barbour's Nigerian colleagues felt him to be committed to the project of decolonisation.[76] Yet when he returned to the UK, initially to University College, before taking up a permanent position at the University of Ulster, Barbour was viewed by some as a 'colonial'. It was his years at Ibadan, and Khartoum before that, combined with a particular manner and an old-fashioned approach to scholarship, that produced this indictment.[77]

Monica Cole's time spent in South Africa at the University of Cape Town (1947–1948) and the University of Witwatersrand (1948–1951) led to her becoming a 'leader in the field of geo-botany' specialising in savannahs, with her work underpinned by extensive first-hand data collection and fieldwork.[78] Cole's work had the potential to help exploit untapped resources: following her move to Witwatersrand in 1948, 'She became interested in vegetation anomalies of the Highveld and Lowveld, and in the possibility of using plants as indicators of mineralization.' (Jackson 1994, p. 244). Later, from '1967 to 1971 Monica Cole was involved in a major copper prospecting venture in a wide area of Namibia [then under South African control] and Botswana. She recognized the herb Helichrypsum Leptolepis, as a copper indicator and so assisted in tracing the copper-bearing beds' (Jackson 1994, p. 245).

Cole was exceptionally successful, and a female pioneer in a male dominated field. The Ghanaian geographer Elizabeth Ardayfio-Schandorf had long looked up to Cole, and got the chance to meet her whilst studying in the UK:

> One of my astounding experiences then was meeting Professor Monica Cole, the only female geographer whose textbook on Geomorphology I read in the 1960s. What was even more astounding was coming face to face with such an eminent female geography scholar long after my undergraduate class-mates had nicknamed me after her. Meeting her personally in England did not only make such incredible impression on me, but also inspired such a great admiration and honour which eventually turned out as an academic booster.[79]

Yet whilst she inspired a generation of female geographers – including in Africa – Cole's South African connections were often controversial once she returned to the UK (Maddrell 2009). As Head of Department at Bedford College (which became Royal Holloway University of London) she was a forceful and divisive leader (Maddrell 2009). Reflecting on the controversies from the present day it is sometimes difficult to disentangle the personality clashes from the political concerns, which included Marxist critiques of the role of capital in the university, as well as anti-apartheid feeling.[80] What is clear however, is that Cole was a strong advocate of forging links with government agencies and industry, including the exploitative mining industry in apartheid South Africa. She also made 'public justification of apartheid in lectures' (Maddrell 2009, p. 273) and owned shares in some of the companies mining in South Africa.[81] One of Cole's obituaries, otherwise silent on these controversies, noted in somewhat elliptical fashion, 'The mining industry gave support and encouragement' for her research (Jackson 1994, p. 244). Indeed, noted another, 'Monica was a well-known visitor to the mining-houses of Johannesburg and Melbourne.' (Potter and Catt 1994, p. 374).

In 1971 Cole's conduct was scrutinised in the Bedford College Students' Union magazine, *Inner Circular*, in an article calling for her removal from her post, due to her competence, her support for connections between industry and universities, and her position on South Africa:

> Her political views as regards the situation in South Africa are also of interest. The companies that are involved with the geography department at present are all mining companies with workings or interests in South Africa. They all actively support (by continuing business) the S.A. regime showing their disregard for any facet of human life beyond the exploitation potential of their employees. That they turn a blind eye to the nature of the atrocities committed in South Africa is not surprising when one looks at their profits. Prof Cole is so completely in sympathy with the S.A. government that she has worked at Witwatersrand University, which operates apartheid as do all South African institutions, and she employs and is employed by companies that aid S.A. in its pursuance of fascism.[82]

While the call in the student magazine didn't result in Cole being removed from post, subsequent complaints by members of her own department resulted in her being moved into the role of director of research in geobotany, a position that she held until retirement in 1987.

Despite this long-standing controversy, Cole's support for the overtly racist regime in South Africa is not something that has tainted her career significantly. Indeed, in 1987 'she received the RGS-IBG Murchison Medal for major contributions to the geography of South Africa and to the understanding of savannas'.[83] It is still possible today for physical geography undergraduate or postgraduate students to receive a 'Monica Cole Research Grant' from the Royal Geographical Society with the IBG. The grants provide assistance in 'undertaking original fieldwork overseas', capturing the essence of Cole's driving force behind her research.[84] However, we see here in Cole an example (and she is by no means the only one) of a geographer whose academic and scholarly credentials are more commonly known than the attitudes towards the apartheid regime and its big business supporters. Moreover, Cole's successful research in South Africa relied upon these connections with the South African government and mining industries which, directly and indirectly, funded and facilitated the research for which she is so well known. Through mechanisms such as ongoing grant awards, Cole continues to have an impact today, and that impact was embedded in her connections with, and attitudes towards, South Africa. We are provided here with a clear example of the ongoing need to consider under whose name institutions like the Royal Geographical Society with the Institute of British Geographers choose to further geographical knowledge (see also Griffiths and Baker 2019).

## Conclusion

As decolonisation continued apace in Africa in the 1960s, increasingly, the opportunities for British academics to undertake lengthy careers at African universities reduced, though two-to-three-year positions funded by overseas development aid continued into the 1980s. In this way, decolonisation had a material impact on the careers of a number of British geographers working in Africa. Many returned from Africa willingly, keen to build their careers in the UK or elsewhere in the world. Others left regretfully, sometimes sad to leave but accepting of the changes decolonisation was bringing to the university sector, a few forced out against their will by the violent regimes of Idi Amin in Uganda and the Nationalist Party in South Africa. Where university conditions worsened, British academics were lucky to have the opportunity to leave the country, unlike many of their local colleagues. Nevertheless, returning was not always easy; though some thrived and utilised their African development knowledge to carve out successful careers, others struggled to fit in, and were sometimes viewed as colonial relics by their younger colleagues in the UK.

The situation for African geographers working in Africa in the 1980s and 1990s was very difficult, and those involved have documented these challenges in detail (e.g. Areola and Faniran 1998; Areola and Okafor 1998; Mabogunje 1998). Here we have highlighted some of the ways that African geographers made creative use of international connections forged through the period of decolonisation, sometimes decades later, to manage these challenges and continue to develop their careers and departments.

Decolonisation also had long lasting legacies on geography in the UK. Returning to positions in Great Britain and Northern Ireland, geographers brought back longstanding academic interests in the places where they had worked, which contributed to the shape and regional focus of what was to become development geography, as well as to the area studies project in the UK. They produced textbooks and courses that reflected these interests and passed them on to another generation of geographers. They applied quantitative approaches to Africa, as well as regional tropical accounts that continued to shape representations of the continent throughout the second half of the twentieth century. These were complemented by the more radical approaches to development highlighted in Chapter 4. They also brought back affective experiences and academic networks which shaped UK geography in important but sometimes hard to grasp ways. Their presence supported clusters of African students studying in the UK and their ongoing relationships with their former departments could be mutually beneficial. But these emotional ties could also be problematic, importing a masculine, colonial, and exclusionary vision of geographical work (especially fieldwork) which the discipline is still grappling with today.

Whilst much research about empire and decolonisation focuses on the period up until the 1960s, as this chapter highlights, professional careers encompassing late colonial development and decolonisation often lasted late into the twentieth century. Thus the account here contributes to recent calls to explore empire and its legacies beyond the 1960s (Drayton 2016), and to wider debates – for example in relation to the development industry (Kothari 2005; Hodge 2007) – about second careers and expertise across the periods of colonialism and independence. Whilst the generation of British geographers who worked in African departments around the time of decolonisation are no longer active in the discipline, their influence remains, both in African geography departments and in the UK. This legacy is still alive today, and the careful and nuanced exploration of this ambiguous history can inform contemporary debates about 'decolonising geography'. It is to these connections that we turn in the final concluding chapter.

## Notes

1 John Pugh to R.O. Buchanan, 2 January 1956, Personal papers of John Pugh, private collection, courtesy of Mike Pugh, hereafter JP papers.
2 John Pugh, Curriculum Vitae, JP papers.
3 Speech at Pugh's memorial service by Bill Morgan, courtesy of Keith Atkinson, a King's College London undergraduate (1957–1960), 1999.
4 Speech at Pugh's memorial service by Keith Atkinson, 1999, courtesy of Keith Atkinson.
5 For example at least six British Ibadan faculty took up positions elsewhere in the Commonwealth beyond the UK.
6 Pugh to R.O. Buchanan, 2 January 1956.
7 Pugh to Wooldridge, 14th December 1955, JP papers.
8 Interview with Hugh Clout, London, UK, 30 November 2016.
9 Letter from D.M. Thompson, Imperial College of Science and Technology to I.C.M Maxwell, Inter-University Council, 19 August 1972, National Archives (hereafter NA), BW 90 2188.
10 Letter from Osborne, Prof of Physics University College Dar es Salaam to Office for Overseas Development, 8 October 1968, NA BW 90 2188.
11 Letter from J.M. Theakstone, IUC Head of Division 'B' to Langlands, 24 January 1973, NA BW 90 1385.

12 Letter from J.M. Theakstone, IUC Head of Division 'B' to Langlands.
13 Inter-University Council for Higher Education Overseas Document F: I.U.C./C.35/63, 9 October 1963, NA BW 90 2188.
14 IUC Chairman's Report to the Council for the Academic Year 1971–1972, Robert Steel Papers, University of Liverpool, hereafter RS D91/4/8/3.
15 Letter from E. J. Richards Vice Chancellor of Loughborough University of Technology to E.C Griffiths, Director of the IUC, 28 February 1973, NA BW 90 1385.
16 Report of Visits to IUC by Professor T R Milburn Professor of Botany Makerere University, 7 January 1974, NA BW 90 1387. For Ugandan Asians and other Ugandans not supportive of Amin, this was certainly not the case.
17 'The position of British subjects at Makerere' Letter from Bryan Langlands to Acting High Commissioner, British High Commission, Kampala, 14 December 1972, NA BW 90 1385.
18 Inter-University Council for Higher Education Overseas Visit by Mr. Donald Mann to the IUC 11 December 1972 at 2.30pm: Professor of Mathematics – Makerere University, NA BW 90 1385.
19 World Bank Profile on Amin and Gulzar Shivji (Shivji was Gulzar's married name): https://ismailimail.blog/2010/01/11/world-bank-profile-on-amin-and-gulzar-shivji [last accessed 31 January 2022]; For Gulzar Jamal's early work in the geography department see: Jamal, G. (1972) *Uganda squared; a systematic analysis of geographic data on Uganda arranged by grid-squares* Occasional Papers Makerere Geography Department no. 24. Makerere University College.
20 'Amin woos S'poreans' *Straits Times* 14 December 1972.
21 Letter from Prof Joseph Ouma, Head of Geography, Makerere to 'Leader of Delegation' National Geographical Committee to 24th Congress IGU Aug-Sept 1980. Exact date not provided. International Geographical Union Archive, Archiv für Geographie, Leibniz-Institut für Länderkunde, hereafter IGU: II-18–8 Uganda.
22 Letter from J B Butterworth, Chairman of the Inter-University Council to UK Vice-Chancellors, 6 March 1973, NA BW 90 1385.
23 Letter from J B Butterworth.
24 Makerere University and the present Uganda situation: resettlement of expatriate staff, NA BW 90 1386.
25 Makerere University and the present Uganda situation: negotiations with ODA over continuation of IUC help to Makerere; resettlement of ex-patriates and financial assistance; BESS and pension contributions, NA BW 90 1387.
26 Note to Vice-Chancellors and Principles from B.H. Taylor Deputy Secretary The Committee of Vice Chancellors and Principles of the Universities of the United Kingdom, 15 December 1972, NA BW 90 1387.
27 Letter from Langlands to Ian Maxwell, Deputy Director Inter-University Council, 23 January 1973, NA BW 90 1385.
28 Makerere University and the present Uganda situation: resettlement of expatriate staff; BESS policy, NA BW 90 1386.
29 Letter from J.M. Theakstone, IUC Head of Division 'B' to Langlands.
30 Swanzie Lady Agnew of Lochnaw, Obituary, *The Telegraph*, 12 October 2000.
31 E.g. *Daily Despatch*, East London, 16 March, 1957; *Manchester Guardian*, 27 Match 1957. They comprised two of the 14 signatories of a public letter of protest to the Secretary of Education, Arts and Sciences published in South African and British newspapers in March 1957, seconded a motion in the College Senate rejecting the

proposed 'Separate University Education Act, 1957' and joined protest marches along with students (Williams 2001).

32 'Bantu university dismissals', *The Times* 28 September 1959 p. 9.

33 Letter from Swanzie Agnew to Chancy D Harris (IGU Secretary-General) 19 July 1975, IGU: II-8–6 Malawi.

34 Swanzie Lady Agnew of Lochnaw, Obituary, *The Telegraph*, 12 October 2000.

35 Interview with a British geographer who worked at Dar es Salaam in the 1970s, University of Glasgow, UK, 17 August 2012. Interview conducted by Professor Joanne Sharp.

36 Letter from Langlands to IGU Treasurer, 19 August 1976, IGU: II-11–7 Uganda.

37 Letter from Lawrence C. Nkemdirim to Eckhar Ehlers, Secretary-General and Treasurer, IGU, 5 March 1996. IGU: II 22–20 Nigeria.

38 Interview with Akin Mabogunje, London, UK, 7 July 2016.

39 Interview with Albert Aweto, Ibadan, Nigeria, 19 June 2017.

40 Interview with 'Bola Ayeni, Ibadan, Nigeria, 8 July 2017.

41 Interview with Len Berry, via Zoom, 5 March 2021.

42 Interview with 'Bola Ayeni.

43 Interview with 'Bola Ayeni.

44 Letter, George Benneh to Vice-Chancellor, University of Ghana, 28 December 1989. University of Ghana Archives, hereafter UG: UGI/3/2/3/56/138.

45 Mindeord: Tidligere NAT-dekan, Henrik Jeppeson, er død (In Memorium: former NAT Dean Henrik Eppeson has died) 29 July 2014. https://www.science.ku.dk/presse/nyhedsarkiv/2014/henrik_jeppesen [Last accessed 31 January 2022].

46 Interview with George Benneh, Legon, Ghana, 16 January 2017.

47 Academy of Europe, Member CV: George Benneh. https://www.ae-info.org/ae/Member/Benneh_George/CV [last accessed 31 January 2022].

48 Interview with George Benneh.

49 Obituary, Dick Hodder, *The Times* 24 November 2006.

50 Personal communication with Tony O'Connor, March 2018.

51 Personal communication with Tony O'Connor.

52 Interview with Anthony Lemon, Oxford, UK, 15 June 2018.

53 See CAS@50 brochure, http://www.sps.ed.ac.uk/__data/assets/pdf_file/0015/83022/CAS@50_brochure.pdf last accessed 15 March 2017.

54 Personal communication with Tony O'Connor.

55 Randall Baker (Makerere) and Ian Thomas (Dar es Salaam) were both geographers in the Development programme at UEA, whilst they were joined by David Feldman an agricultural economist who had also been at Dar es Salaam.

56 Interview with Len Berry, via Zoom, 26 February 2021.

57 Personal correspondence with Tony O'Connor.

58 KCL Course outline (1973), University of London Archives, AC/8/24/7/2.

59 Interview with DN, graduated from King's College London in 1976, London, UK, 24 June 2016.

60 Interview with Akin Mabogunje.

61 Interview with Hugh Clout, London, UK, 30 November 2016.

62 Interview with Hugh Clout.

63 Interview with Debbie Potts London, UK, October 2018; Interview with Hugh Clout.

64 Interview with Anthony Lemon.
65 Interview with DN.
66 Interview with Randall Baker, Llandrindod, UK, 5 September 2019.
67 Interview with Gina Porter, via Zoom, 27 April 2017.
68 Interview with Gina Porter.
69 Gina Porter, personal communication, May 2017.
70 Interview with Hugh Clout.
71 Interview with Gina Porter.
72 Interview with Debby Potts.
73 Interview with Debby Potts.
74 Interview with Debby Potts.
75 Interview with Hugh Clout.
76 Interview with Akin Mabogunje.
77 Interview with Hugh Clout.
78 Monica Cole Research Grant: Royal Geographical Society www.rgs.org.
79 Elizabeth Ardayfio-Schandorf, personal correspondence, 22 June 2022.
80 'A question of competence' *Inner Circular*, November 1971 Royal Holloway University of London Archives (Hereafter RHUL) Bedford College Papers BC AS/203/1/1–2; Interview with Hugh Clout. Thank you to Felix Driver for alerting us to this material.
81 'Open letter to staff re: Bedford support for Apartheid' *Inner Circular* November 1971 RHUL Bedford College Papers BC AS/203/1/1–2.
82 'Open letter to staff re: Bedford support for Apartheid'.
83 Monica Cole Research Grant: Royal Geographical Society www.rgs.org.
84 Monica Cole Research Grant: Royal Geographical Society www.rgs.org.

## References

Agnew, S. and Stubbs, G.M. (eds.) (1972). *Malawi in Maps*. London: University of London Press.

Ajayi, J.F.A., Lameck, G., and Johnson, G.A. (1996). *The African Experience with Higher Education*. Athens: Ohio University Press.

Alidou, O., Caffentzis, G., and Federici, S. (2000). *A Thousand Flowers. Social Struggles against Structural Adjustment in African Universities*. Trenton: Africa World Press.

Amutabi, M. (2002). Crisis and student protest in universities in Kenya: examining the role of students in national leadership and the democratization process. *African Studies Review* 45 (2): 157–178. https://doi.org/10.1017/S0002020600031474.

Areola, O. and Faniran, A. (1998). Undergraduate training at Ibadan. In: *50 Years of Geography in Nigeria: The Ibadan Story* (ed. O. Areola and S.I. Okafor), 19–46. Ibadan: Ibadan University Press.

Areola, O. and Okafor, S.I. (eds.) (1998). *50 Years of Geography in Nigeria: The Ibadan Story*. Ibadan: Ibadan University Press.

Baker, R. and Polonskyi, M. (2015). *Cold War in a Hot Place*. Sofia: East-West Publishing.

Baker, S.J.K. (1976). Geography in Uganda. *The Geographical Journal* 142 (1): 188.

Bartholomew, J.C. (1989). Obituary Peggie M. Hobson. *Scottish Geographical Magazine* 105 (1): 64. https://doi.org/10.1080/00369228918736754.

Beinart, W. (1984). Soil erosion, conservationism and ideas about development: a southern African exploration, 1900–1960. *Journal of Southern African Studies* 11 (1): 52–83. https://doi.org/10.1080/03057078408708088.

Benneh, G. (2016). *My Time, My Nation*. Accra: Winmat Publishers Limited.

Berg, E. (1981). *Accelerated Development in Sub-Saharan Africa: An Agenda for Action*. Washington: International Bank for Reconstruction and Development/World Bank.

Blunt, A. (1994). *Travel, Gender and Imperialism: Mary Kingsley and West Africa*. New York: Guildford Press.

Brown, G. and Yaffe, H. (2014). Practices of solidarity: opposing apartheid in the centre of London. *Antipode* 46 (1): 34–52. https://doi.org/10.1111/anti.12037.

Buchanan, K.M. and Pugh, J.P. (1955). *Land and People in Nigeria*. London: University of London Press.

Buettner, L. (2016). *Europe after Empire: Decolonization, Society and Culture*. Cambridge: Cambridge University Press.

Byram, M. and Dervin, F. (eds.) (2009). *Students, Staff and Academic Mobility in Higher Education*. Cambridge: Cambridge Scholars Publishing.

Carr-Saunders, A. (1948). The teaching of geography in colonial colleges: address to the Geographical Association. *Geography* 33: 1–7.

Child, E. and Barnes, T.J. (2019). Area studies and American imperial expansion without geography. *Journal of Historical Geography* 66: 43–54. https://doi.org/10.1016/j.jhg.2018.08.001.

Coppock, E., Derbyshire, A., Gilbert, H.S.A. et al. (1973). Study group reports. *Area* 5: 74–78.

Dixon, C. (1981). The Developing Areas Study Group. *Area* 13: 88.

Doornkamp, J.C. (1986). Professor Paul Harrison temple, 1936–1985. *Transactions of the Institute of British Geographers* 11 (2): 243–246.

Drayton, R. (2016). *The long 1970s: the endgame of decolonisation and the reconstitution of Imperialism Imperial and World History Seminar 15 February*. Institute of Historical Research, London.

Driver, F. (2001). *Geography Militant*. Oxford: Blackwell.

Fage, J. (1964). The centre for West African Studies, University of Birmingham. *Journal of Modern African Studies* 2: 577.

Farmer, B.H. (1973). Geography, area studies and the study of area. *Transactions of the Institute of British Geographers* 60: 1–15.

Gilbert, A. (1974). Developing Areas Study Group. *Area* 6: 74–75.

Graham, B. (1989). Bryan Wooleston Langlands. *Irish Geography* 22 (1): 61–63. https://doi.org/10.1080/00750778909478790.

Griffiths, M. and Baker, K. (2019). Decolonising the spaces of geographical knowledge production: the RGS-IBG at Kensington Gore. *Area* 52: 455–458. https://doi.org/10.1111/area.12586.

Healey, P. and Upton, R. (eds.) (2010). *Crossing Borders: International Exchange and Planning Practices*. London: Routledge.

Hickman, G.M. (1973). *The New Africa*. London: University of London Press.

Hickman, G.M. (1976). *The New Africa second edition*. London: Hodder and Stoughton.

Hickman, G.M. (1980). *The New Africa third edition*. London: Hodder and Stoughton.

Hickman, G.M. assisted by Hickman, R. and Hickman, L. (1990). *The New Africa fourth edition*. London: Hodder and Stoughton.

Hickman, G.M. and Dickens, W. (1960). *The Lands and Peoples of East Africa*. London: Longmans.

Hodge, J.M. (2007). *Triumph of the Expert: Agrarian Doctrines of Development and the Legacies of British Colonialism*. Ohio: Ohio University Press.

Horrell, M.A. (1961). *A Survey of Race Relations, 1958–1959*. Johannesburg: South African Institute of Race Relations.

Hountondji, P. (1990). Dependence in Africa today. *Research in African Literatures* 21 (3): 5–15.

Hoyle, B.S. (1975). Reviewed work(s): The Lands and Peoples of East Africa by G. M. Hickman. *Geography* 60 (3): 246.

Hoyle, B.S. (1989). Bryan Wooleston Langlands, 1928–1989. *Geography* 74 (2): 187.

Jackson, S. (1994). Monica Mary Cole 1922–1994. *The Geographical Journal* 160 (2): 244–245.

Kerr, D. and Mapanje, J. (2002). Academic freedom and the University of Malawi. *African Studies Review* 45 (2): 73–92.

Kolinsky, M. (1983). The demise of the Inter-University Council for Higher Education Overseas: a chapter in the history of the idea of the university. *Minerva* 21 (1): 37–80.

Kothari, U. (2005). Authority and expertise: the professionalisation of international development and the ordering of dissent. *Antipode* 37: 425–446. https://doi.org/10.1111/j.0066-4812.2005.00505.x.

Kwesiga, J.C. (2002). *Women's Access to Higher Education in Africa: Uganda's Experience*. Kampala: Fountain Publishers.

Luhanga, M.L., Mkude, D., Mbwette, D.J. et al. (2000). *Strategic Planning and Higher Education Management in Africa: The University of Dar Es Salaam Experience*. Dar es Salaam: University of Dar es Salaam Press.

Mabogunje, A.L. (1998). The history of the department. In: *50 Years of Geography in Nigeria: The Ibadan Story* (ed. O. Areola and S.I. Okafor), 1–18. Ibadan: University of Ibadan Press.

Mabogunje, A.L. (2011) *A Measure of Grace: The autobiography of Akinlawon Lapido Mabogunje*. Ibadan: Bookbuilders

MacMaster, D.N. (1989). Bryan Wooleston Langlands, 1928–1989, O.B.E., B.A. *Transactions of the Institute of British Geographers* 4 (3): 364–368.

Maddrell, A. (2009). *Complex Locations: Women's Geographical Work in the UK 1850–1970*. Oxford: Royal Geographical Society-Institute of British Geographers/ Wiley-Blackwell.

Mamdani, M. (2007). *Scholars in the Marketplace. The Dilemmas of Neo-Liberal Reform at Makerere University, 1989–2005*. Dakar: African Books Collective.

Mamdani, M. (2011). The importance of research in a university. Pambazuka News. Retrieved from: https://www.pambazuka.org/print/70932 (accessed 20 January 2021).

Mkandawire, T. (1995). Three generations of African academics: a note. *Transformation* 28: 75–83.

Morgan, B. and Moss, R.B. (1965). Geography and ecology: the concept of the community and its relationship to environment. *Annals of the Association of American Geographers* 55: 339–350.

Morgan, W.B. and Pugh, J.C. (1969). *West Africa*. London: Methuen and Co.

Morgan, W.T.W. (1973). *East Africa*. London: Longman.

Moss, R.P. and Morgan, W.B. (1967). The concept of the community: some applications in geographical research. *Transactions of the Institute of British Geographers* 41: 21–32. https://doi.org/10.2307/621324.

Moss, R.B. (1992). W. B. Morgan: an appreciation. In: *Agricultural Change, Environment and Economy: Essays in Honour of W.B. Morgan* (ed. K. Hoggard). London: Mansell Publishing.

Musisi, N.B. and Nansozi, K.M. (2003). *Makerere University in Transition: 1993–2000*. Oxford: James Currey.

Myers, G. (2001). Introductory human geography textbook representations of Africa. *Professional Geographer* 53: 522–532.

Naidoo, B. (1985). Censoring reality: an examination of books on South Africa. *Contemporary Issues in Geography and Education* 2 (1): 22–25.

Norcup, J. (2015). Geography education, grey literature and the geographical canon. *Journal of Historical Geography* 49: 61–74. https://doi.org/10.1016/j.jhg.2015.04.019.

O'Connor, A. (1989). Notes and news: Bryan W. Langlands, OBE. *African Affairs* 89 (354): 107–109.

Pietsch, T. (2013). *Empire of Scholars Universities, Networks and the British Academic World, 1850–1939*. Manchester: Manchester University Press.

Potter, R. and Catt, P. (1994). Monica Mary Cole, 1922–1994. *Transactions of the Institute of British Geographers* 19 (3): 373–377.

Potts, D. (1985). The Geography of Apartheid: the relationship between space and ideology in South Africa. *Contemporary Issues in Geography and Education* 2 (1): 2–7.

Prothero, M.R. (ed.) (1969). *A Geography of Africa: Regional Essays on Fundamental Characteristics, Issues and Problems*. London: Routledge and Kegan Paul.

Serunkuma, Y. (2018). The conditioning of a native informer: politics, economies and the academia as a marketplace. *Proceedings of the African Futures Conference* 2: 196–196. https://doi.org/10.1002/j.2573-508X.2018.tb00100.x.

Sicherman, C. (2005). *Becoming an african University: Makerere 1922–2000*. Trenton: Africa World Press.

Sidaway, J.D. (2013). Geography, globalization, and the problematic of area studies. *Annals of the Association of American Geographers* 103: 984–1002. https://doi.org/10.1080/00045608.2012.660397.

Steel, R.W. and Prothero, M.R. (ed.) (1964). *Geographers and the Tropics: Liverpool Essays*. London: Longmans.

Steel, R.W. and Watson, J.W. (1972). Geography in the United Kingdom 1968–1972: report to the 22nd International Geographical Congress at Montreal, Canada in August 1972. *The Geographical Journal* 138 (2): 139–153.

Stoler, A.L. and Cooper, F. (eds.) (1997). *Tensions of Empire*. Berkeley: University of California Press.

Udo, R.K. (2004). The past, present and future of geography in Africa: the case of Nigeria (speech given to the Africa specialty group of the Association of American Geographers, Philadelphia, USA). *African Geographical Review* 23 (1): 85–10. https://doi.org/10.1080/19376812.2004.9756180.

Wight, D., Ahikire, J., and Kwesiga, J.C. (2014). Consultancy research as a barrier to strengthening social science research capacity in Uganda. *Social Science and Medicine* 116: 32–40. https://doi.org/10.1016/j.socscimed.2014.06.002.

Williams, D. (2001). *A History of the University College of Fort Hare, South Africa, the 1950s: The Waiting Years*, vol. 4. New York: Edwin Mellen Press.

Zeilig, L. (2009). The student-intelligentsia in sub-Saharan Africa: structural adjustment, activism and transformation. *Review of African Political Economy* 36 (119): 63–78. https://doi.org/10.1080/03056240902885705.

# Chapter Eight
# Decolonising Geography Past and Present?

## Introduction

Ngũgĩ wa Thiong'o (1993, p. 65) notes that the end of empire in Africa provided only 'independence with a question mark'. Our book has explored the space around this question-mark, the terrain of uncertainty and incompleteness, but also of creativity, possibility, and optimism. This conclusion does two things. First, it summarises our key contributions to historical understandings of decolonisation by peopling the historiography of decolonisation and decolonising geography's histories. Second, and more substantively, we return to this question mark to consider the extent to which geography was decolonised in Africa in 1948–1998, and how we might learn from this earlier period to support contemporary decolonisation struggles.

## Decolonisation, its Histories and Geographies

### *Peopling the Historiography of Decolonisation*

The book has made key contributions to the interdisciplinary study of decolonisation by bringing people, their experiences, and agency, to the forefront. Recent high-profile interventions in the decolonisation debate in Africa have argued forcefully that much of the scholarship under this headline risks reifying colonialism as an all-powerful and all-encompassing category, flattening out histor-

*Decolonising Geography? Disciplinary Histories and the End of the British Empire in Africa, 1948–1998*, First Edition.
Ruth Craggs and Hannah Neate.

ical specificities, reinforcing binaries between coloniser and colonised, and in doing so failing to take African agency seriously (Táíwò 2022). Thinking through professional lives – individually and collectively – begins to dismantle some of the binaries and certainties that Táíwò (2022) is wary of, whilst also highlighting ongoing structures of coloniality which constrain African agency. Our account – through its use of collective biographies – foregrounds the diversity of articulations, priorities, and practices of this agency amongst African geographers.

Biographical approaches have the ability to re-insert the experiences, creativity, and constraints of individual lives into a field – the history of decolonisation – which has traditionally focused on nation states and political elites. Nationalist accounts of independence in Africa have reified state-level analyses whilst British histories of this period have sometimes failed to account for the domestic impact of decolonisation. Focusing on working lives in a period of unprecedented mobility (for some) in Africa provides a useful corrective in producing transnational, connected histories of the imperial metropoles and their former colonies (Stoler and Cooper 1997). These peopled accounts also powerfully illustrate the fallacy of theorisations that imply that nothing changed or the project of decolonisation failed (Olukoshi et al. 2020), by highlighting the affective experiences of decolonisation and the changes, opportunities and creativity of this period.

Focusing on working lives also provides additional opportunities for the historiography of decolonisation, by opening up new sources and spaces for research: cultural and educational institutions, professional associations, disciplinary networks. And alongside new sources and spaces, professional lives open up new timescales for research into decolonisation: careers lasted decades spanning constitutional transfers and the decades after. This reminds us of the need to explore legacies of empire and decolonisation beyond the immediate years of independence, and careers themselves provide a valuable lens through which to do so. Moving beyond the individual biographical accounts of recent studies (e.g. Guyot-Réchard 2022; Milford and McCann 2022), our account also highlights the value of looking across a cohort of professional lives to understand the texture of individual *and* collective experiences and contributions.

## *Decolonising Geography's Histories*

Our second contribution to accounts of decolonisation is to address questions of historiographical exclusions in histories of geography. The critical contextualisation and historicisation of the discipline from the perspective of geographical traditions from beyond the West is an important aspect of decolonising work (Daya 2020). Whilst the Nigerian geographer Akin Mabogunje – perhaps the most successful and well-known geographer on the continent – can hardly be understood as excluded by many measures, he and his colleagues are not a central part, or even a footnote, in many histories of geography which claim universality whilst consistently prioritising work in the Global North (e.g. Livingstone 1992). This book has begun to address these issues by focusing on the experience and contributions of African geographers primarily in Ghana, Nigeria, South Africa, Tanzania, and Uganda. Exploring diverse intellectual contributions across a wider range of spaces and experiences demonstrates the ways that individuals, responding to their own intellectual and regional contexts and priorities, shaped new theoretical and methodological innovations. African geographers developed crucial international research agendas, whilst also applying geographical insights in the specific context of post-colonial state-building. Whilst they absorbed insights from abroad, they were active agents in these intellectual developments, rather than passive receivers of new ways of practicing geography. Nevertheless, they were working within colonial structures – intellectual and institutional – which defined what were deemed successful career trajectories and important ideas. Viewing the discipline from the perspective of scholars from Africa and other parts of the world can help in 'Moving the centre' from which we theorise and understand the world geographically (Ngũgĩ 1993, see also Ferretti and Viotto Pedrosa 2018).

Histories of post-World War II geography have tended to highlight the Quantitative Revolution and subsequent disciplinary developments, focusing on the North American and British experience and focusing on 'great men' (e.g. Livingstone 1992; Barnes 2001; 2002; Garrison 2002). As Marcus Power and James Sidaway (2004, pp. 587, 595) note, there is value in pursuing the absences in these stories, to provide 'disruptive' alternatives to the ways that 'the discipline is conventionally narrated.' Nearly two decades ago Trevor Barnes identified geographical absences in the histories of the Quantitative Revolution, asking: 'why are places in Africa not ... there, or Asia, or Australia?' (2002, p. 508). In the book we have demonstrated that African places and people were there all along, both as key places and protagonists in the Quantitative Revolution (e.g. Mabogunje 1968), and as recorders and evaluators of this disciplinary history (e.g. Areola and Okafor 1998). Ghana and Tanzania were key empirical sites for the trial and exhibition of new methods and models, whilst in Nigeria, Mabogunje lead early innovative quantitative research and curriculum reform. South African geographers took part in a massive geographical reorganisation based on quantitative spatial planning models in the name of apartheid. Scholars in Tanzania and Nigeria were also central to the growing critique of the modernisation approaches that often accompanied quantitative work in Africa, contributing to new critical accounts of development drawing on new theories and methods. Histories of geography that do not acknowledge these African connections and contributions are necessarily partial and incomplete.

Diversifying histories of geography sheds light on significant relationships between geography and geopolitics that have yet to be acknowledged, and which intersected with struggles for decolonisation. Geography in the global south was often institutionalised in late colonial contexts, and universities were key sites for development funding tied to Cold War concerns (Livsey 2017). As countries became independent, geographers were engaged in regional planning for new post-colonial states. Whilst geography's role in empire-building has been widely acknowledged (Bell et al. 1995; Driver 2001), this book has provided one of the first detailed engagements with its relationship to decolonisation and post-colonial state-building. Moreover, whilst the impact of the Cold War context has been explored in detail with regard to Western, and particularly US geography (Barnes and Farish 2006), *Decolonising Geography?* has provided some of the first reflections on its impacts in geography in Africa. Opening up this relationship allows for further consideration of the impacts of this conflict on the shaping of the discipline of geography, its research foci, methods, and practitioners globally. In the US context, Cold War concerns drove quantification and area studies in the search for 'useful' strategic knowledge of foreign countries, and deprioritised radical and emancipatory scholarship. In Africa, it often had similar effects, contributing, alongside other, domestic, contexts, to the shaping of a discipline that only rarely challenged the underlying political economy of the continent.

## Learning from the Past

Our third contribution to decolonisation debates is to those focused primarily on the contemporary moment. Here our contribution is to bring the past and present into conversation, to ask what might be learned today, from these earlier debates. In this section we begin by asking to what extent projects of decolonising geography and the university were successful in the mid-late twentieth century, before moving on to consider what these answers mean for our theorising and practice in the twenty-first century.

### *To What Extent was Geography Decolonised in the period 1948–1998?*

The answer to the question to what extent was geography decolonised in the second half of the twentieth century depends very much on where is being discussed, and what is meant by decolonisation. In British geography, in this period, decolonisation was barely acknowledged, let alone engaged with. Despite empire shaping geography in the UK in important ways, there were few changes to courses or to the institutions of geography made to reflect decolonisation. Decolonisation was not seen as relevant as a topic for study, nor was it understood as a political process which could and should have ramifications on the discipline. It took until the 1980s for the colonial history of geography to begin to be addressed, and even this work remained focused on the past, not on the need for change in the present. It is only in the last decade or so that decolonisation debates have made any impact in British geography.

In Africa, in contrast, decolonisation brought about real change to the discipline of geography. In Africa in the mid-twentieth century, '"to decolonize" was the same thing as "to Africanize"', that is, for politicians and academics alike, replacing colonial institutions, structures and people with African ones was at the core of decolonisation as a process (Mbembe 2016, p. 33). Following decolonisation as constitutional event, the replacement of political leaders, civil servants and other professionals with Africans was a top priority. Universities contributed to this process by fulfilling the graduate 'manpower' [*sic.*] demands of newly independent nations; geography provided teachers, surveyors, resources assessors, and planners, who were all seen as crucial to building new nations. But African universities were also subject to this process of Africanisation themselves. The replacement of foreign (and mainly British) staff at the colonial universities founded in Ghana, Nigeria, Tanzania and Uganda was a high profile and symbolically important change which provided opportunities for African scholars, as well as impacting the careers of those replaced. But it also made a material difference to student experiences at these universities and helped to drive other forms of institutional and curricula change. The speed of change varied, depending largely on the political demands of national governments, with Nkrumah's government in Ghana demanding rapid transformation. Elsewhere changes took longer to unfold. Nevertheless, decolonisation as a process of Africanisation of staff in universities was largely achieved.

Whilst Africanisation of staff had a substantial impact for individuals, a focus on the replacement of people, rather than institutions, norms, and ideologies could cement colonial visions for the state as well as uneven power relations, this time between post-colonial elites and the rest of the population (Nkrumah 2009 [1964]; Fanon 2019 [1961]). It could also lead to what Fanon (2019 [1961], pp. 103; 105) described as a 'chauvinism' which became 'increasingly tinged with racism'. This occurred as Africanisation came to mean not the replacement of colonial positions with Africans from around the continent, reflecting pan-African and post-colonial federal ideals, but instead was associated with narrower visions which favoured only those of a particular nationality (e.g. Nigerianisation; Ugandanisation), sub-national region, or ethnic identity.

In the mid-twentieth century, decolonisation also meant institutional changes which aimed at breaking the links with Britain and creating specifically *African* universities. This required changes to the constitution of higher education institutions so they were no longer colleges of the University of London but independent universities. This break also allowed for other changes that followed: introducing more opportunities to study for honours degrees, and dramatically widening access through increases in student numbers and educational outreach. In loosening the ties to a hybrid London and Oxbridge model of higher education, space could be made for more students on campus in hostels rather than residential halls, and greater experimentation in institutional structures could take place. In South Africa, the structures of higher education reflected both British colonialism and white Afrikaner nationalism. The end of apartheid saw structural changes to universities, as well as to the institutions of geography with its two white dominated learned societies replaced in 1994 by one inclusive one after a long campaign. Decolonisation as institutional change did bring about substantial real changes, driven by university staff, including geographers. Simeon Ominde campaigned hard for Africanisation at Makerere College in the 1960s, whilst Lindisizwe Magi, Dhiru Soni, and Harold Wesso were integral to the changes in South Africa. However, from Ghana to South Africa this can be seen as an unfinished project with legacies of colonialism and apartheid still present within university structures, and, even more intractably, within the global university system.

In the mid-twentieth century, decolonisation also meant shifting disciplinary knowledges through changes to research and teaching in order to create an African geography. Our interviewees educated in the 1950s and 1960s generally reflected with wry amusement on the inadequacy and absurdity of the British focused geography they were taught, though this could have real impacts on career development, especially when it meant PhD studies which were of little relevance in African contexts. In the mid-twentieth century, for British and African geographers alike, rectifying this primarily meant localisation of content and case studies, for example through the introduction of new regional courses focused on Africa, the setting up of meteorological stations to collect local weather and climate data, and surveys of local areas to produce maps and case studies for teaching. It also meant the establishment of African geographical journals as important sites of knowledge production and dissemination. Decolonisation provided opportunities to define new areas of research, for example producing new historical geographies of Africa which destabilised colonial narratives of development and urbanisation by highlighting precolonial African urbanism (Mabogunje 1968). This period around and after constitutional decolonisation in African geography departments was one marked by interdisciplinary innovation, which, along with the wider political climate across the continent, combined to draw scholars to Africa. At Ibadan, when new theoretical and quantitative techniques put the department at the cutting edge of global research, and at Dar es Salaam where radical scholars from across the globe debated the political economy (and geography) of independence and dependency, new research directions drew scholars to Africa to be a part of it. New theories and ideas developed from and for African contexts, helping to move the centre to Africa, and provincialising Europe in the process (Ngũgĩ 1993; Chakrabarty 2009; Ndlovu-Gatsheni 2018; Sharp 2019). This mid-twentieth century period can be seen as an important moment in questioning and beginning to contest the hegemony of Anglo-American geography, a key aim of contemporary decolonisation practices, even if some of the openings of this period have subsequently been reversed.

However, unlike in other disciplines, such as history or English literature (Falola 2004; Ngũgĩ 1986), there was never a nationalist school of geography complete with innovative methods and sources to which scholars associated themselves at the time. Rather, there existed a vision for an African geography aligned closely with national development and state-building (Ominde 1971; Ojo 1978). To read Akin Mabogunje's (2011) autobiography, in which more than half is focused on his external development work, is to be left in little doubt that this is where he felt his contribution lay. As Mbembe (2016, p. 33) reminds us, in the mid-twentieth century, 'to decolonize was part of a nation-building project', and both British and African academics working in African universities often saw their roles within this context (even if the nations they imagined could look radically different). Nation building involved producing new national identities but also the development of new modern states. Geographers, deploying innovative quantitative methods, mapping, planning, and providing data about both populations and resources, often saw their roles as contributing to this development, and so to decolonisation. It is worth noting, though, that nation and state building could also be racist and violent, and geographers' work in South Africa contributed to the development of apartheid South Africa and its supposedly independent Bantustans. Moreover, much of this work continued with little change from the colonial era. In many places the 'geographers' toolkit' continued to facilitate 'extraction, territorial dispossession and the reconstruction of non-Europeans life-worlds.' (Radcliffe 2022, p. 28). This point was made forcefully by scholars working in Africa in the 1960s, 1970s and 1980s who contributed to the development of a new critical geography of development (Rodney 1972; Santos 1974; Slater 1977a; Ake 1982; Namasasu 1988; Nkrumah 2009 [1964]). Indeed, Milton Santos, the Brazilian geographer who worked at Dar es Salaam in the 1970s was highly critical of the role of planning, and with it, geography, in underdevelopment. 'There are many instances of the contributions of regional science, geography, and regional planning to the diffusion of capital' he argued, 'this is the case of, for example, the theories of central places, growth poles …' (1977, p. 89). In this way geographers often contributed both to the nation-building project of decolonisation, and to the further embedding of colonial-modernity in Africa, especially in relation to the environment, natural resources, and urban planning. Here, the extent to which geography was decolonised – or contributed to decolonisation – depends very much on your definition of this process.

South Africa in the 1980s also saw an upsurge of opposition to apartheid from within the geography community. In this period, the term decolonisation began to be used in South African scholarship to mean a questioning of the importation of Western theories to the South in ways akin to some contemporary debates that focus on epistemic justice, as well as to question the lack of focus on Black South Africans experiences and lives in research agendas (Mabin and Tomaselli 1973; Crush et al. 1982; Wesso 1992). Geographers also used their geographical skills as part of anti-apartheid campaigning in relation to particular struggles over housing, or direct action against the apartheid regime. They lead community education campaigns envisaged as part of wider movements for black consciousness. Whilst rarely using the term decolonisation at the time, these activist geographies probably look more like the contemporary calls for work with communities in direct pursuit of social justice in the name of decolonisation, than the practices outlined above.

There were many weaknesses in the decolonisation enacted in African universities in the mid-twentieth century. Gender disparities meant that geography departments were dominated by men (Mama 2005). Racism, sexism, and xenophobia remained commonplace in African universities in this period (as they did in Britain and other parts of the 'Global North'), even at the same time as disciplinary and institutional transformation was taking place. Universities remained primarily elite spaces, despite the changes made. Cold war geopolitics and continuing neo-colonial economic relationships often held back more revolutionary change. Focus fell on postcolonial state-building and Africanising institutions. For leaders such as Nkrumah, the primary impetus was for political change, as well as securing economic independence, not for any form of epistemological decolonisation, which was something which could come later (Ndlovu-Gatsheni 2018). This state-building focus of decolonisation work was quite different to the demands of contemporary university decolonisation movements.

As Sabelo Ndlovu-Gatsheni (2018) has argued, epistemological decolonisation was required in order to imagine and enact new independent political and economic futures. It wasn't possible to reimagine relationships between people and land beyond natural resources for capitalist exploitation without first questioning the disciplinary knowledges that prioritised land utilisation. It wasn't possible to create new *African* universities when they remained dominated by the demand for 'manpower' [*sic.*] for development and nation-building based on Western models. In many ways, then, the decolonisation of mid-twentieth century universities in Africa fell short, both of the demands and desires of those working so hard for it at the time, and of the aims of the movement today.

## *What Does This Mean for Decolonisation Struggles Today?*

Táíwò (2022) argues that it is confusing and unhelpful to utilise the same term to discuss the political transfer of power in the middle of the twentieth century and the more diverse project of epistemic and social justice which dominates debates today. Yet there have always been more continuities across time periods and decolonisation 'projects' than Táíwò (2022) allows (Marshall 2022). We argue that there are many lessons that can be learnt from this earlier period for today, even if this earlier period left decolonisation an unfinished business. In places like the UK, where decolonisation debates are much more recent, we can and should look to the experience and expertise of others in this area. Indeed, if, as Noxolo (2017a) argues, 'decolonization begins from the scholarship of black and indigenous peoples, and should be *led* by that scholarship' (p. 318, emphasis in original), then asking what Black African scholars were writing and doing in the name of decolonisation in earlier periods can make an important contribution. What can be learnt from their ideas and experiences?

*Change is possible, but wider structures continue to impede change*

Looking historically allows us to be optimistic that change is possible. In Ghana, Tanzania, and South Africa, universities changed dramatically after decolonisation and the end of apartheid, in ways that empowered African scholars and enabled different decolonising projects. Yet history also demonstrates some of the challenges that contemporary campaigns face. Whilst in the mid-twentieth century, many African governments enabled rapid transformation of universities, current ambitions to decolonise the university do not always have government support. Moreover, international contexts also provided limits on the changes possible. Whilst the 1960s saw post-colonial African universities held in high esteem domestically and internationally, this esteem was based, often, on their continued equivalence with Western counterparts (Livsey 2017). African universities in this period were also reliant to a large extent on foreign funding, both late colonial development money and, during and after decolonisation, on development aid often tied to Cold War diplomacy. International funding was tied to particular projects which reflected northern donors' (from both the West and the Eastern bloc) priorities for higher education in Africa. These broader structures, and the ongoing coloniality of the economic, political, and cultural contexts in which African universities operated, continued to structure African scholars' agency to pursue decolonisation of the disciplines and the university. Whilst the Cold War context is less important now, and African higher education no longer attracts the development aid it once did, external funding continues to impact African academics' professional lives, especially in relation to research and consultancy. Today's political and economic contexts are no less present in academic lives, and it is important to frame contemporary decolonising projects within these wider possibilities and constraints.

Since the 1990s, neo-liberalism has been the most important context for higher education in many parts of the world, including Africa and the UK. As Mahmood Mamdani has detailed in relation to Makerere University (2007), structural adjustment in the 1980s lead to debilitating austerity in many African universities and was followed by new development orthodoxies from the World Bank and other international donors which framed university education as an individual privilege rather than a collective project of national development. Though major impacts have also been felt in other parts of the world, including the UK, US, and Australia (Radice 2013), the neoliberalism of the academy has perhaps taken strongest root in African institutions with the most devastating of consequences for individual scholars and students (Daley and Kamata 2016).

The impacts of neoliberalism are not distinct from the struggle for decolonisation. Ndlovu-Gatsheni (2018, p. 186; see also deSouza 2019) has gone as far as to argue that from the late 1980s, in African universities, 'Markets became the major agents of coloniality'. Many have highlighted the ways in which neoliberal contexts also intersect with, disfigure, and undercut decolonising initiatives in Western universities (Holmwood 2018; Last 2018). Thus whilst we can learn from mid-twentieth century strategies for decolonisation in the African university and their successes and failures, we must also be mindful of the important differences in the global political and economic contexts which structure and constrain liberatory policies and projects. As Mbembe (2016, pp. 32; 37) has remarked, 'Is today's Beast the same as yesterday's or are we confronting an entirely different apparatus, an entirely different rationality – both of which require us to produce radically new concepts? ... my fear is that we might be fighting battles of the present and the future with outdated tools.'

*Clarity of meaning and impact is required*

It is important to be clear about what is meant by decolonisation when the term is used in research and to shape academic practice. As Fanon (2019 [1961]) highlighted, decolonising agendas could be distorted to become racist and exclusionary, and this remains a live issue in the twenty-first century (Mbembe 2016; Daya 2020). Co-option of decolonisation narratives for other, less progressive goals has always been a danger, and therefore we should not be surprised that it remains so today (Doharty et al. 2021). We also see in the history of these practices the watering down of radical changes in order to make them more palatable for university leaders and governments. These historical experiences should alert us of the need to be wary of how, exactly, decolonisation is put into practice.

*Decolonisation work is diverse*

Looking historically demonstrates that decolonising work (and it was often hard work) was very diverse. It involved campaigning for change within the university and beyond it, leading departments and facilitating institutional change, researching new topics, and redesigning curricula, consultancy work, and committees. It involved travel for study and supporting students, often against a backdrop of complex and sometimes unstable domestic and international political contexts. It involved editorial work for journals (some of which was done by male geographers' wives). The implication of this is that we need to look in a wide range of places to capture and acknowledge the hard work of decolonising disciplines and universities today.

Looking closely at geographers' decolonising work also demonstrates that there can be a disjuncture between research and theory on the one hand, and other parts of academic labour on the other. It is possible to publish important decolonial work whilst treating colleagues and students in ways which reinforce colonial hierarchies. On the other hand, many scholars working in Africa in the mid twentieth century (both African and British) published little which would be recognised as decolonial in relation to topic, methods or theory, but through their teaching, or university administration, they actively forwarded decolonisation processes. Whilst administrators often get a bad press as agents of bureaucratic marketisation in the twenty-first century university (see for example Mbembe 2016; Bhambra et al. 2020), in the mid-twentieth century they were sometimes crucial in pushing through radical policies (as well as hindering change). In taking seriously the full gamut of academic labour for decolonisation, we nevertheless agree with de Leeuw and Hunt (2018, p. 5) on the need to 'recognise decolonization [as] requiring grounded material practices, activism or a lived reality' rather than just theoretical contributions. Indeed, as Daley and Murrey (2022, p. 167) drawing on Lewis Gordon highlight, it is in the everyday practices 'where political change or stability reside.'

*Solidarity beyond borders and the university are important…*

Looking carefully at individual practices and ways of relating which helped to support decolonisation in the past, or which more broadly offered solidarity across borders despite ongoing coloniality, can provide lessons for the future. In our research, we came across many instances of friendship and solidarity which supported ongoing scholarship and decolonising work within and beyond the university. Geographers worked in solidarity at a transnational scale, supporting one another and leveraging the resources of Western universities to fulfil the requirements of colleagues working in poorly resourced departments in Africa. The creativity of African scholars and their international colleagues in producing funding applications and otherwise procuring valuable resources enabled geography departments to develop new programmes but also to continue to attract students and in doing so secure jobs and salaries. Whilst not touching the global inequalities of ongoing coloniality in education, much of this activity worked to support precarious colleagues and made a real difference to individual working lives.

*… but so are relationships within the university*

In addition to attending to global injustices, our research has highlighted the value of everyday practices and ways of relating, which could support both individual careers and wider projects of decolonisation, through countering racism, making opportunities, and dismantling structures which reinforce colonial versions of success. Whilst seemingly small, looking historically reveals these acts to be both radical and powerful. This is a reminder of the agency that we still have in the contemporary academy, wherever we are located, and the value (as well as challenges) of practices such as mentoring in challenging the persistent whiteness of geography (Faria et al. 2019). Shari Daya (2022, pp. 11–12) has argued 'we should be paying attention to, and fostering, the practices of care, collegiality, trust, and reciprocity that already exist ... Love and care, in the academy, are radical acts.' She continues

> we in geography departments need to turn our gaze inward, not in a return to parochialism but in self-critique, to see the role we have to play in creating more inclusive and just institutional cultures. In terms of our working practice, critical geographers should be paying attention to the everyday ways in which our actions, choices, and omissions perpetuate oppression and damaging hierarchies. (Daya 2020, p. 11)

For those located in the West, and benefitting from being white, as Daley and Kamata (2016, p. 98) suggest, there are particular responsibilities:

> Embedding anti-racist, decolonial scholarship within the academy requires white scholars, instead of treating their racialised identities as merely incidental, to interrogate how they might collude unintentionally in the perpetuation of white supremacy and thus the limitation of knowledge production in a globalized world.

As our work has shown, it is not just love and care that can be powerful, but also complaint: 'To make a complaint is often to fight for something. To refuse what has come to be' (Ahmed 2021, p. 26). However, as Sara Ahmed reminds us, complaining is difficult, repetitive, administrative, caring, and crucially embodied work: 'to complain is to become a container of negative affect, a leaky container, speaking out as spilling over.' (2021, p. 17). As our work has shown, decolonising work, within and beyond the university, is rarely career-enhancing, indicating the tensions in academic lives between political commitment and professional ambition.

Daley and Murrey (2022, p. 159) argue for what they call 'defiant scholarship': 'active scholarly practice that seeks to work against and outside colonial grammars and prevailing registers' in order to address coloniality in the university. Whilst Daley and Murrey's account suggests loud and unruly scholarship, something also invoked in the 'epistemic disobedience' described by Mignolo (2011) (cited by Daley and Murrey 2022), our research has shown that decolonising work can also be quiet and low key, and sometimes dull. This is a point Christine Noe (2022, p. 179) makes too: a 'lack of defiance in some cases does not mean absence of resistance to oppressive colonial legacies and structures .... defiant scholarship should endeavor to provide options and methodologies to capture silent resistance'. Defiance can also be dangerous, for individuals' careers, and sometimes even their lives (Noe 2022; Ouma 2022). This needs to be acknowledged. Learning from the past can illuminate the sorts of academic work that might contribute to transformation, as well as the barriers to that kind of work.

*Everyday working experiences matter*

Whilst we often know from our own experiences that material conditions – pay, infrastructure, security, funding for travel, teaching, and research – are hugely important in shaping geography and the university, little research engages in detail in the production of these conditions, and personal experiences of these inequalities. This lacuna allows for the persistence of attitudes which view weaknesses in scholarship and institutions from the global south as the result of individual failings rather than structural problems (Daley and Kamata 2016). In examining some of these structural problems in detail, and highlighting their impacts on individual careers, our research supports the calls of many in Africa and beyond for recognition of these structural problems, and for action to overcome them (Olukoshi 2006; Daley and Kamata 2016).

*What about the discipline of geography?*

In a recent review article, Clayton (2021, p. 12) asks 'What clues about the hopes and potential pitfalls of current decolonising projects – and not least, perhaps, regarding how and why geography today needs to be "decolonised" – might be gleaned from the study of historical geographies of "decolonisations", plural'? Our research has some important lessons for decolonising the discipline specifically.

By highlighting historical contingencies shaping the discipline, it is possible to unsettle the certainty of continued Anglo-American dominance that pervades today, undercutting many attempts at decolonisation. Whilst it is crucial to acknowledge this hegemony, we risk presentism if we fail to explore the genealogies of our contemporary situation. Although the United Kingdom and United States remain important in disciplinary histories told from the South, these relations have not been static. From Africa, looking across several decades, ebbs and flows of inclusion and exclusion are visible. In the 1950s and 1960s, geography at several African geography departments was understood as far more central to the 'international' discipline than other comparable institutions such as the University of Malaya in Singapore (now the National University of Singapore). Opportunities to participate in the fora which have (problematically) become associated with the 'best' scholarship – western journals, conferences and lecturing positions for example – have in some periods been more open to scholars from the global south than they are today (Mills and Branford 2022). And teaching positions in African universities, as well as journals published on the continent, were historically more internationally 'prestigious'. Understanding the shifting status of different national geography schools over time helps to understand the specific (and regionally differentiated) contexts through which present-day 'international' geography was produced.

This is important because as Tariq Jazeel (2019, p. 204) has highlighted,

> Geography's unequal geography lends to it a distinct teleology, which is to stress that our disciplinary formation has a trajectory, momentum and intellectual direction that together push back against our proclivities to routinely be accountable to the places about which we write ... the rationalities, normalizations and career ambitions that disciplinary geographers are trained to harbour in terms of publishing, conferencing, esteem and peer review ... constitute governmentalities that can effectively pull us away from the places, peoples and communities on which we work.

Müller (2021) has described a process of 'worlding geography' in order to open it up to more linguistic diversity, and as he points out we can also think in similar terms about gender, race, class, and geographical location, as other dimensions of 'epistemic struggle' (p. 1440). Understanding the shifting spaces, openings, and opportunities of the past, as well as the ways in which they have been foreclosed might help us to imagine and work for alternative futures in the discipline today.

This might involve reflecting on the space that exists (or doesn't) within the container of geography itself. As Daley and Murrey (2022, p. 172) have noted, 'Recent moves to decolonize geography or embrace pluriversal epistemes within geographical scholarship, nonetheless, have sometimes been structured in ways that further entrench something distinctive called "geographical thought"'. Reflecting on the historical formation of geography as a discipline, and the disciplining and exclusionary nature of this history (Hawthorne 2019) – but also the agency that individuals have claimed to shape and change it in the past – can be an important part of imagining a different future. Returning to Walter Rodney can help us to consider how we as individuals and communities navigate the disciplinary spaces of possibility. As he reflected, you need to fulfil some intellectual 'standards' in order to be able to challenge them, but the key is to know when to stop:

> There is a certain distance which one has to go in trying to meet the so-called standards. But beyond that it becomes self-defeating and ridiculous ... to see those standards as encompassing one's existence for all times is ... self-defeating as you simply become another bourgeois academic. The format in which you operate, the journals in which you publish, the language in which you express yourself, the people to whom you feel responsible for justifying this or that position, the books you are called upon to review and so on, makes you part of this bourgeois academic community. (Rodney 1990 [1975], p. 25)

Finally, histories of geography can provide additional examples of geographers using their disciplinary knowledges in pursuit of liberation. In South Africa in the 1980s, many geographers worked outside the academy with local community groups campaigning for housing, freedom of mobility, education and against state violence. They used their geographical training to support these projects and sometimes envisaged their work explicitly as part of a liberatory anti-racist campaign against the apartheid state. Decolonising work today demands that we look beyond the boundaries of the university (Parker et al. 2018; Cahuas 2021; Schildermans 2021), and looking historically underlines the value of such work but also the demands it places on those scholars involved: taking time away from research and publishing and in doing so damaging prospects of promotion, as well as in some contexts presenting real risks of violence and state detention.

## Conclusion

Ngũgĩ (1993, p. 28) reminds us 'that social and intellectual processes, even academic disciplines, act and react on each other not against a spatial and temporal ground of stillness but of constant struggle.' *Decolonizing Geography?* has highlighted the struggles for decolonisation that shaped ways of knowing and disciplinary directions, and struggles to decolonise disciplinary knowledges and the institutions of higher education in Africa.

Decolonisation is often homogenised, simplified down to two sides: coloniser and colonised. However, to reduce individuals to these categories can be deeply problematic. It does a violence to their lives, reducing them to representative 'stick figures' on either side, to use Cooper's (2019 p. 51) phrase, and thereby reducing the complexity of their experiences, emotions, and contributions. It also risks essentialising identities in ways which can be damaging, including for the decolonising work of today. Our biographical approach, grounded in post-colonial approaches which emphasise ambivalence and the imbrication of colonial and colonised societies (Daya 2020), attempted to capture the texture of decolonisation as experienced, practiced, campaigned for, and resisted, in all its complexities. It is from such a place that we argue geographers will be better situated to fight for decolonisation of the discipline and the university today. Walter Rodney argued that '[The guerrilla intellectual's] task is to operate within the aegis of the institution and the structure and to take from it and to transform it over time.' (1990 [1975], pp. 111–113). We hope that the insights into what decolonisation has meant in the past, and how that maps on to or challenges how the term is used in both theory and practice today, can provide materials through which to teach present and future geographers about the structures and epistemologies that they have inherited, and how to transform them.

## References

Ahmed, S. (2021). *Complaint!* Durham: Duke University Press.

Ake, C. (1982). *Social Science as Imperialism: The Theory of Political Development*, 2e. Ibadan: Ibadan University Press.

Areola, O. and Okafor, S.I. (eds.) (1998). *50 Years of Geography in Nigeria: The Ibadan Story*. Ibadan: Ibadan University Press.

Barnes, T.J. (2001). Retheorizing economic geography: from the quantitative revolution to the "cultural turn". *Annals of the Association of American Geographers* 91 (3): 546–565. https://doi.org/10.1111/0004-5608.00258.

Barnes, T.J. (2002). Performing economic geography: two men, two books and a cast of thousands. *Environment and Planning A* 34: 487–512. https://doi.org/10.1068/a3440.

Barnes, T.J. and Farish, M. (2006). Between regions: science, militarism, and American geography from World War to Cold War. *Annals of the Association of American Geographers* 96 (4): 807–826. https://doi.org/10.1111/j.1467-8306.2006.00516.x.

Bell, M., Butlin, R., and Heffernan, M. (eds.) (1995). *Geography and Imperialism, 1820–1940*. Manchester: Manchester University Press.

Bhambra, G.K., Nisancioglu, K., and Gebrial, D. (2020). Decolonising the university in 2020. *Identities* 27 (4): 509–516. https://doi.org/10.1080/1070289X.2020.1753415.

Cahuas, M.C. (2021). Reaching for El Mundo Zurdo: Imagining-creating-living Latinx decolonial feminist geographies in Toronto. *Gender, Place & Culture* 28 (9): 1213–1233. https://doi.org/10.1080/0966369X.2020.1786015.

Chakrabarty, D. (2009). *Provincializing Europe*. Princeton: Princeton University Press.

Clayton, D. (2021). Historical geography I: doom, danger, disregard – towards political historical geographies. *Progress in Human Geography* 45 (6): 1692–1708. https://doi.org/10.1177/03091325211011664.

Cooper, F. (2019). *Africa since 1940: The past of the Present*. Cambridge: Cambridge University Press.

Crush, J., Reitsma, H., and Rogerson, C. (1982). Decolonizing the human geography of Southern Africa. *Tijdschrift voor Economische En Sociale Geografie* 73 (4): 197–198. https://doi.org/10.1111/j.1467-9663.1982.tb00955.x.

Daley, P. and Murrey, A. (2022). Dismantling coloniality in contemporary African geographies. *Singapore Journal of Tropical Geography*. https://doi.org/10.1111/sjtg.12422.

Daley, P. and Kamata, N.W., in Powell, R.C., Klinke, I., Jazeel, T., Daley, P., Kamata, N.W., Heffernan, M., Swain, A., McConnell, F., Barry, A. and Phillips, R. (2016). Interventions in the political geographies of 'area'. *Political Geography* 57: 94–104. https://doi.org/10.1016/j.polgeo.2016.03.004.

Daya, S. (2020). On shaky ground: a response to Long et al. (2019). *South African Journal of Science* 116 (3–4): 1–3. http://doi.org/10.17159/sajs.2020/7932.

Daya, S. (2022). Moving from crisis to critical praxis: geography in South Africa. *Transactions of the Institute of British Geographers* 47 (1): 9–15. https://doi.org/10.1111/tran.12459.

de Leeuw, S. and Hunt, S. (2018). Unsettling decolonizing geographies. *Geography Compass* 12: e12376. 1–14. https://doi.org/10.1111/gec3.12376.

deSouza, P.R. (2019). The recolonization of the Indian mind. In: *Knowledges Born in the Struggle: Constructing the Epistemologies of the Global South* (ed. B. de Sousa Santos and M.P. Meneses). New York: Routledge.

Doharty, N., Madriaga, M., and Joseph-Salisbury, R. (2021). The university went to 'decolonise' and all they brought back was lousy diversity double-speak! critical race counter-stories from faculty of colour in 'decolonial' times. *Educational Philosophy and Theory* 53 (3): 233–244. https://doi.org/10.1080/00131857.2020.1769601.

Driver, F. (2001). *Geography Militant*. Oxford: Blackwell.

Falola, T. (2004). *Nationalism and African Intellectuals*. Rochester: University Rochester Press.

Fanon, F. (2019 [1961]). *The Wretched of the Earth*. Milton Keynes: Diana Publishing.

Faria, C., Falola, B., Henderson, J., and Torres, R.M. (2019). A long way to go: collective paths to racial justice in geography. *The Professional Geographer* 71 (2): 364–376. www.doi.org//:10.1080/00330124.2018.1547977.

Ferretti, F. and Viotto Pedrosa, B. (2018). Inventing critical development: a Brazilian geographer and his Northern networks. *Transactions of the Institute of British Geographers* 43: 703–717. https://doi.org/10.1111/tran.12241.

Garrison, W.L. (2002). Lessons from the design of a life. In: *Geographical Voices: Fourteen Autobiographical Essays* (ed. P. Gould and F.R. Pitts), 99–123. New York: Syracuse University Press.

Guyot-Réchard, B. (2022). Stirring Africa towards India: Apa Pant and the making of post-colonial diplomacy, 1948–54. *The International History Review* 44 (4): 892–913. https://doi.org/10.1080/07075332.2022.2093941.

Hawthorne, C. (2019). Black matters are spatial matters: black geographies for the twenty-first century. *Geography Compass* 13: e12468. https://doi.org/10.1111/gec3.12468.

Holmwood, J. (2018). Race and the Neoliberal University: lessons from the public university. In: *Decolonising the University* (ed. G. Bhambra, D. Gebrial, and K. Nişancıoğlu), 37–52. London: Pluto Press.

Jazeel, T. (2019). *Postcolonialism*. London: Routledge.

Last, A. (2018). Internationalisation and Interdisciplinarity: sharing across boundaries? In: *Decolonising the University* (ed. G. Bhambra, D. Gebrial, and K. Nişancıoğlu), 208–230. London: Pluto Press.

Livingstone, D. (1992). *The Geographical Tradition*. Oxford: John Wiley and Sons.

Livsey, T. (2017). *Nigeria's University Age: Reframing Decolonisation and Development*. Basingstoke: Palgrave Macmillan.

Mabin, A. and Tomaselli, K. (1973). Perspectives of South African towns and cities. *Fifth South African Students Geographical Conference*, Stellenbosch.

Mabogunje, A.L. (1968). *Urbanization in Nigeria*. London: University of London Press.

Mabogunje, A.L. (2011). *A Measure of Grace: The Autobiography of Akinlawon Ladipo Mabogunje*. Ibadan: Book Builders.

Mama, A. (2005). Gender studies for Africa's transformation. In: *African Intellectuals: Rethinking Politics, Language, Gender and Development* (ed. T. Mkandawire). London: Zed.

Mamdani, M. (2007). *Scholars in the Marketplace. The Dilemmas of Neo-Liberal Reform at Makerere University, 1989–2005*. Dakar: African Books Collective.

Marshall, J. (2022). Review: decolonization and its discontents. *The World Today*. Retrieved from: https://www.chathamhouse.org/publications/the-world-today/2022-08/review-decolonization-and-its-discontents.

Mbembe, J.A. (2016). Decolonizing the university: new directions. *Arts and Humanities in Higher Education* 15 (1): 29–45. https://doi.org/10.1177/1474022215618513.

Mignolo, W.D. (2011). Geopolitics of sensing and knowing: on (de) coloniality, border thinking and epistemic disobedience. *Postcolonial Studies* 14 (3): 273–283. https://doi.org/10.1080/13688790.2011.613105.

Milford, I. and McCann, G. (2022). African Internationalisms and the Erstwhile Trajectories of Kenyan community development: Joseph Murumbi's 1950s. *Journal of Contemporary History* 57 (1): 111–135. https://doi.org/10.1177/00220094211011536.

Mills, D. and Branford, A. (2022). Getting by in a bibliometric economy: scholarly publishing and academic credibility in the Nigerian academy. *Africa* 92: 839–859. https://doi.org/10.1017/S0001972022000481.

Müller, M. (2021). Worlding geography: from linguistic privilege to decolonial anywheres. *Progress in Human Geography* 45 (6): 1440–1466. https://doi.org/10.1177/0309132520979356.

Namasasu, O. (1988). Towards a decolonized radical Southern African geography. *Geographical Education Magazine* 11: 1–13.

Ndlovu-Gatsheni, S.J. (2018). *Epistemic Freedom in Africa: Deprovincialization and Decolonization*. London: Routledge.

Ngũgĩ wa Thiong'o (1986). *Decolonising the Mind: The Politics of Language in African Literature*. Portsmouth: Heinemann.

Ngũgĩ wa Thiong'o (1993). *Moving the Centre: The Struggle for Cultural Freedoms*. Portsmouth: Heinemann.

Nkrumah, K. (2009 [1964]). *Consciencism: Philosophy and Ideology for De-Colonization and Development with Particular Reference to the African Revolution*. New York: Monthly Review Press.

Noe, C. (2022). Intellectuals at the hill: scattered pieces of defiant African scholarship. A commentary on Patricia Daley and Amber Murrey's 'Defiant scholarship: dismantling coloniality in contemporary African geographies'. *Singapore Journal of Tropical Geography* 43: 177–179. https://doi:10.1111/sjtg.12425.

Noxolo, P. (2017a). Introduction: decolonising geographical knowledge in a colonised and re-colonising postcolonial world. *Area* 49: 317–319. https://doi.org/10.1111/area.12370.

Ojo, G.J. (1978). Thirty years of geographic thought in Nigeria. *The Nigerian Geographical Journal* 21 (1): 3–23.

Olukoshi, A. (2006). African scholars and African studies. *Development in Practice* 16 (6): 533–544.

Olukoshi, O., Hormeku-Ajei, T., Balaji, A., and Nayar, A. (2020). Reclaiming Africa's early post-independence history. *Africa Is A Country*, 2020, https://africasacountry.com/2020/07/reclaiming-africas-early-post-independence-history (accessed 20 December 2022).

Ominde, S.H. (1971). Geography and African development. In: *Studies in East African Geography and Development* (ed. S.H. Ominde), 1–8. London: Heinemann Educational Books.

Ouma, S. (2022). Navigating the landscape of defiant scholarship in and beyond Africa: on archives, bridges and dangers. A commentary on Patricia Daley and Amber Murrey's 'Defiant scholarship: dismantling coloniality in contemporary African geographies'. *Singapore Journal of Tropical Geography* 43: 180–185. https://doi.org/10.1111/sjtg.12423.

Parker, P., Holland, D., Dennison, J. et al. (2018). Decolonizing the academy: lessons from the graduate certificate in participatory research at the University of North Carolina at Chapel Hill. *Qualitative Inquiry* 24 (7): 464–477. https://doi.org/10.1177/1077800417729846.

Power, M. and Sidaway, J.D. (2004). The degeneration of tropical geography. *Annals of the Association of American Geographers* 94: 585–601. https://doi.org/10.1111/j.1467-8306.2004.00415.x.

Radcliffe, S. (2022). *Decolonizing Geography: An Introduction*. Cambridge: Polity Press.

Radice, H. (2013). How we got here: UK higher education under Neoliberalism. *ACME: An International Journal for Critical Geographies* 12 (2): 407–418. Retrieved from. https://acme-journal.org/index.php/acme/article/view/969.

Rodney, W. (1972). *How Europe Underdeveloped Africa*. London: Bogle-L'Ouverture Publications.
Rodney, W. (1990 [1975]). *Walter Rodney Speaks: The Making of an African Intellectual*. Trenton: Africa World Press.
Santos, M. (1974). Geography, Marxism and underdevelopment. *Antipode* 6 (3): 1–9.
Schildermans, H. (2021). *Experiments in Decolonizing the University: Towards an Ecology of Study*. London: Bloomsbury Publishing.
Sharp, J. (2019). Practising subalternity: postcolonial Tanzania, the Dar School and pan-African geopolitical imaginations. In: *Subaltern Geographies* (ed. T. Jazeel and S. Legg), 74–93. Athens: University of Georgia Press.
Slater, D. (1977a). Geography and underdevelopment—part II. *Antipode* 9 (3): 1–31. https://doi.org/10.1111/j.1467-8330.1977.tb00089.x.
Stoler, A.L. and Cooper, F. (eds.) (1997). *Tensions of Empire*. Berkeley: University of California Press.
Táíwò, O. (2022). *Against Decolonisation: Taking African Agency Seriously*. London: Hurst Publishers.
Wesso, H. (1992). *The Colonisation of the Geographical Mind: A Critical Contextual Analysis of the Institutionalisation and Establishment of Geography as an Academic Discipline in South Africa*. Unpublished PhD Thesis. University of the Western Cape.

# Index

Note: Page number followed by "*f*" denotes figures and page number followed by "*t*" denotes tables respectively




*Decolonising Geography? Disciplinary Histories and the End of the British Empire in Africa, 1948–1998*, First Edition.
Ruth Craggs and Hannah Neate.